D1029159

# CHRONOLOGY OF ECLIPSES AND COMETS

## AD 1 – 1000

# CHRONOLOGY OF ECLIPSES AND COMETS

# AD 1 – 1000

D. JUSTIN SCHOVE

in collaboration with

Alan Fletcher

THE BOYDELL PRESS

First published 1984
The Boydell Press
an imprint of Boydell and Brewer Ltd
PO Box 9, Woodbridge, Suffolk IP12 3DF and
51 Washington Street, Dover, New Hampshire 03820, USA

British Library Cataloguing in Publication Data

Schove, D. Justin (in collaboration with Alan Fletcher)
Chronology of eclipses and comets AD1 to 1000.
1. Eclipses, Solar – History 2. Comets – History
I. Title
523.2 QB451

ISBN 0-85115-406-9



THE SPECTRUM OF TIME

The Spectrum of Time project was initiated in 1948. International collaboration soon enabled me to check dates of Astronomical and Meteorological phenomena and of Famines and Epidemics. Some of the earlier papers reproduced in 'Sunspot Cycles' (1983, denoted here as SC) express acknowledgement to many who helped. I thus sent all eclipse reports to Dr Fletcher who asked me to incorporate his studies of European eclipses in this book. With new results from Ice-cores, Varves and Tree-rings we have a chronological framework for the history of the environment illustrated by maps and charts. The collections, which include xeroxes of sources, are at present at my home.

D. J. Schove

Printed in Great Britain by St Edmundsbury Press
Bury St Edmunds, Suffolk

C O N T E N T S

# ILLUSTRATIONS

## ABBREVIATIONS USED

| | |
|---|---|
| AA | *Auctores Antiquissimi* in the MGH series |
| AC | Annals of Clonmacnoise |
| AI | Annals of Inisfallen |
| AJ | *Astronomical Journal* |
| AM | Anno Mundi (Year of the World) |
| AN | *Astronomische Nachrichten* |
| ASC | The Anglo-Saxon Chronicle in the MGH series<br>Old English version in Rolls series, 2 vols., 1861<br>New ed., by D. Dumville & S. Keynes, in progress (Boydell & Brewer) |
| AT | Annals of Tigernach (Irish Annals) |
| AU | Annals of Ulster |
| BEHE | Bibliothèque de l'Ecole des Hautes Etudes, Paris |
| BTC | The Bonn volume containing Theophanes Continuatus ... Symeon ... George ... ed. I. Bekker, Bonn 1863 |
| Calv. | Calvisius 17th C |
| CHFMA | Les Classiques de l'Histoire de France au Moyen Age. Société d'Edition 'Les Belles Lettres', Paris |
| Chr.Min. | *Chronica Minora* I, 1892, II, 1894 = MGH (AA) *9* |
| CMHF | F. Guizot, Collection des Mémoires relatifs à l'histoire de France. Paris 1823-1835 |
| CS | Chronicon Scotorum |
| CSCO | *Corpus Scriptorum Christianorum Orientalis*. Scriptores Syri |
| CSEL | *Corpus Scriptorum Ecclesiasticorum Latinorum* |
| DJS | D.J. Schove (Author) |
| EHR | *English Historical Review* |
| Ginz. | Ginzel 1899 |
| JBAA | *Journal of the British Astronomical Association* |
| MGH | *Monumenta Germaniae Historica* (See AA, SS, SrG, SrL, SrM) |
| MN | *Monthly Notices of the Royal Astronomical Society*, London |
| MT | Mean Time |

ABBREVIATIONS USED

| | |
|---|---|
| Mur. | L.A. Muratori<br>Rerum Italicarum Scriptores, Milan, originally 1723-1751 |
| Newc. | Newcomb 1878 |
| Newt. | R.R. Newton |
| Obs. | Observatory (Periodical published by the Royal Greenwich Observatory) |
| Opp./<br>Oppolz. | Oppolzer 1886 or 1962 (reprint) |
| Petav. | Petavius 1627 |
| PG | J.P. Migne. Patrologia Graeca |
| PL | J.P. Migne. Patrologia Latina |
| PO | Patrologia Orientalis (ed. Graffin and F. Nau) |
| RE | Pauly-Wissowa-Kroll. Real-Encyclopaedia der Classischen Altertumswissenschaft (1894+). Stuttgart |
| RFA | Royal Frankish Annals (See RFA & Scholz in bibliography) |
| Scal. | Scaliger 16th C |
| Schr./<br>Schroe. | Schroeter 1924 |
| SE | Seleucid Era |
| SpE | Spanish Era |
| SrG | Scriptores rerum Germanicarum in usum scholarum (MGH) 1839-1969<br>Often a more handy version than SS |
| SrL | Scriptores rerum Langobardicarum et Italicarum (MGH) 1878 |
| SrM | Scriptores rerum Merovingicarum (series of MGH) 1884-1920 |
| SS | Scriptores in the MGH series (Main series 1826+) founded by C.H. Pertz |
| Str. | Struyck 1740 |
| T-C | Tycho-Curtius 1666 |
| UT | Universal Time (Midnight to Midnight on the Greenwich Meridian) |

PREFACE

Dr Alan Fletcher (1903-1981), whose scholarly knowledge of Astrononomy, History and Languages made this book possible, died before it had been accepted for publication. At Cambridge he had been a Wrangler with Distinction and had gained the Tyson medal in Astronomy, and he later became Reader in Applied Mathematics at Liverpool University. For short periods he worked at Brown University and had been a visiting Professor at Yale. He was well-known for the help he gave historians on problems of chronology, and in 1970 at a Historical Association symposium on Historical Eclipses I invited him to explain what could be done by collaboration between scientists and historians in this field.

At that time I was attempting to date early Auroral and Meteorological phenomena as part of the so-called Spectrum of Time project, dendrochronologists needing more accurate dates than were then available. For this it was necessary to have the correct dates of comets and eclipses. Dr P-Y Ho, then working at Cambridge with Dr Needham, helped us with the comets, but for eclipses Dr Fletcher was the obvious expert to consult.

The 1970 symposium included papers by Stephenson, some of whose diagrams are included in this book, and contributions from astronomers working on Maya chronology, and a solution to this problem is given here. The first result of the conference was to inspire us with the need to collaborate with one another and with R.R. Newton (who sent us preprints of his work), to produce a reliable list of eclipses and comets of the first millennium AD.

The supposed dates and quotations of West European, Byzantine and Middle Eastern eclipses I sent to Dr Fletcher with short typescripts. Dr Fletcher combined the three into a single chronology, having first worked his way through the great MGH Scriptores volumes and checked the original Greek and Latin quotations. We were thus able to solve chronological problems that had puzzled historians. Furthermore, he traced the earliest historical identifications for many of the eclipses. His work was recognized by grants from the Leverhulme Trust, for which we wish to express out thanks. After his official retirement, the University of Liverpool granted him an honorary Fellowship, his departmental heads (Professors L. Rosenhead and J.G. Oldroyd) offering facilities that enabled him to continue his work in the Department. Since the death of Professor Oldroyd, Professor C. Michael has enabled me to work with Dr P.J. Message and with Mrs. J. Seed in respectively preparing and typing the material for publication. This has been organized especially by Mrs. M. Dandridge and by Mrs. J. Murray, who was responsible for most of the camera-ready typing and for correcting a number of our inconsistencies.

I have had to omit some of the eclipses that were once considered relevant - many of these would never have been noticed west of the Baltic or north of the Sahara. I have also omitted some discussions of past controversies, but the interested reader can find Dr Fletcher's version of the original MS in the Department or in the Spectrum of Time collections.

For precise tracks, computer programs and print-outs and much pre-publication material I should like to thank especially Dr R. Kudlek (Hamburg), Professor O. Gingerich (Harvard), Dr J. Meeus (Louvain), Professor R.R. Newton (Johns Hopkins), and Dr R.S. Stephenson (Durham). Encouraging and helpful comments in reading parts of the MS have been given by historians, notably Mr R. Latham (Public Record Office) and Dr A.P. Smyth (Kent), and our thanks are due to all - Academics and Librarians - who have helped in the project, and to the British Academy for generous grants. Dr Fletcher saw and approved my Appendixes A and B but he did not live to check the first typescript; however, in updating this, the continued interest and scientific help of Dr P.J. Message and the expert proof reading and historical advice of Kenneth Harrison, have proved invaluable. For help with the computer indexing I am very grateful to Mr G. Smye-Rumsby, BA, and Mr K. Daws, BSc. Many collaborators in the project have been named in articles or in the text.

Dr Fletcher wished to dedicate his efforts to his wife, Marjorie, for her continued help. I should like to thank my wife, Vera, and my daughters, Ann, Mary and Hilary (respectively graduates in History, Geography and Physics), but they wish me to dedicate my own efforts to the memory of a great scholar, Alan Fletcher.

INTRODUCTION

## 1. ECLIPSES AND COMETS

Eclipses and comets were recorded in early chronicles because they were believed to account for events that were still in the future. Today we find these records help us to date events in the past. They provide a chronological framework for History, dating other phenomena in the same context. Famines, plagues and wars were frequently ascribed to events in the heavens, and the corrected dates of the eclipses and comets enabled us to date the meteorological and political events that came before and after. Our list of early recorded eclipses begins in 2095 BC and our calculated dates of Halley's comet seem to go back to c.1403 BC.

The earliest eclipses and comets have slight documentation and there is little to add to the published discussions by F.R. Stephenson, R.R. Newton and others cited in our bibliography. The later eclipses and comets are usually accurately dated by historians. Our concern in this book is therefore with the period AD 1 - 1000: in that millennium records are available from various sources but the dates assigned by historians are often incorrect.

Our object has indeed been, not to analyse each eclipse and comet scientifically, but to identify and list them in chronological order for purposes of scientific chronology. This is useful because distinguished historians have performed feebly, and even perpetrated elementary 'howlers' in trying to make their own identifications (e.g. AD 447, 878, 904). Those scientists who have prepared chronologies of the weather and other natural phenomena have made further errors through accepting dates printed incorrectly; sometimes even in the manuscripts the dates are incorrect. However, dates of historical solar eclipses can usually be calculated and in our check-lists (Appendix A) we have indicated the alternatives where uncertainties persist. The comet dates in our Appendix B often differ from those conventionally accepted by European historians but they are reliably based on a comparison of records from both West and East.

## 2. ECLIPSE IDENTIFICATION

The first attempt to use eclipses as a chronological framework for History was made about the time of Julius Caesar in the first century BC. An eclipse mentioned in Plutarch's Romulus XII (see the Loeb ed. of Plutarch I, 21) was supposed to have preceded the foundation of Rome and was dated (incorrectly) as 754 BC, so making it an auspicious event for the founding of the city in 753 BC.

The AD/BC system of dating years was not invented until the sixth century AD but this, together with the Old Style Julian calendar in which four consecutive years always contained the same number (4 x 365¼ = 1461) of days, made it easy for Renaissance scholars to calculate the dates of historical eclipses, using known astronomical cycles. With computers we are now more readily able to calculate accurate tracks of even solar eclipses (Meeus and Mucke 1985, Stephenson and Houlden 1985). The lack of reliable dates in many of the chronicles prior to the ninth century AD still prevents us from solving every chronological problem, and probabilities have been indicated quantitatively in our tables (Appendix A).

Astronomers know the precise days of solar eclipses and the precise nights of lunar eclipses. It is not open to the historian to argue that an eclipse could have occurred at some other date in the same season, or at the same time in an adjacent year. Indeed, whenever there is an eclipse in the corresponding season of the following year it is never on the same calendar date but about 11 days earlier or about 19 days later. This behaviour is in fact a useful first rule in determining the correct year of a 'dated' eclipse.

However, frequently the date of an eclipse is not given and even the year is known only approximately (cf. Schove 1948, 1950, 1960, 1961, 1968). Then historians turn to the standard eclipse lists of Oppolzer 1962, a source we discuss below. Many of the earlier eclipses have thus been 'dated' by what R.R. Newton 1969, 287 calls the identification game (see also Newton 1972, 1979 and Schove 1971). Newton, whose ideas are followed below, asserts "Most of what the technical literature has to say about ancient solar eclipses is wrong". False primary reports, inaccurate secondary quotations and wrong scientific deductions account for the errors. Small eclipses are not noticed. Those provided by astronomers for historians (e.g. V. Grumel, La Chronologie, 1958, Paris) are mostly unobservable in practice.

2.1 FALSE PRIMARY REPORTS

Many authorities accept as 'observations' reports that were not correct even at the time they were written. Four types of false eclipses are thus Assimilated, Literary, Magical and Calculated.

2.11 Assimilated eclipses arise because a chronicler associates the time of one event by relating it to another. If a chronicler makes an eclipse contemporary or nearly contemporary with another event about a year or less away, and if he makes no great drama of the coincidence (cf. AD 17, 46, 610-2, 622, etc.), we can reasonably assume unconscious assimilation of the record and accept at least the reality of the eclipse.

In Norway a great battle, in which king Olaf was killed, took place on 29 July 1030; a month later on 31 August there was an eclipse of the Sun at Stiklestad in the same region. The account in the Heimskringla links the two events "But when the battle began a redness came over the heavens and likewise over the sun, and before it ended it grew as murk as night ...". This report was written by Snorre in the thirteenth century but today it is regarded as the only extant 'primary' source. The eclipse of AD 29, as we explain below, was perhaps assimilated in Asia Minor with the (unknown) Crucifixion date. The view that the Crucifixion was associated with a lunar eclipse and dated 3 April AD 33 is considered (but rejected) in our Addenda.

2.12 Literary eclipses are consciously introduced into early history, and the Roman 'eclipse' of 754 BC was just such an invention: we now know that there was no solar eclipse visible in Italy between 777 and 745 BC. In Spectrum of Time studies we read sources backwards, stopping when take-off point is reached and treating earlier material as borrowed or legendary. In Rome, take-off point in this sense is the 5th century BC (Livy). The take-off point for Ireland is c.555 (and c.655), Japan c.610, England c.664, Korea c.1000, Annam c.1100, and so on.

Early Chinese History, like the Roman, was "fabricated backwards as time itself moved forward" (Needham). The story of the beheading of drunken astronomers who failed to predict an eclipse is pure invention. Nevertheless, some astronomers have asserted that its date as 22 October 2136 BC is established "beyond all reasonable doubt". (See Needham 1959, 189 for this forgery of the 4th century.)

2.13 Magical eclipses are especially those with religious or political overtones. Solar eclipses, as Newton drily remarked, have a remarkable tendency to happen during battles, at the deaths of great personages or at the beginning of great enterprises. On the other hand, if we accept that king Herod was ruling at the birth of Christ we can date the lunar eclipse of Josephus and so infer that Christ was born before 4 BC.

The eclipse of AD 29 is a magical eclipse insofar as it adds religious significance to the Crucifixion. The Ghost Eclipse of Felix (c.303-4) is a typical eclipse of this kind. Chinese sources give eclipses political significance but record the correct dates of even those that are invisible (e.g. AD 780 in Schafer 1977, 169ff).

2.14 Calculated eclipses are in this millennium inserted into the sources especially in China (cf. Needham 1959, 420ff): in Europe some similar predictions were so inserted in the early ninth century (cf. AD 810 July 5).

In the first millennium AD the assimilated eclipses are the most common of the four types of false reports.

2.2 INACCURATE SECONDARY QUOTATIONS

Eclipses were borrowed as part of the 'international trade in marvels' by one chronicler from another and in the process they were often misdated. Such eclipses must be termed 'ghosts'. On the other hand, Bede's eclipses of 538 and 540, as we explain below, must have been genuinely recorded in the Mediterranean as the eclipses were striking but not visible in England. The 'Irish' eclipse of '445' is similarly shown to be a borrowed record from a SW European observation in 447.

2.3 FALSE SCIENTIFIC DEDUCTIONS

Historians often misdated an eclipse because the year was only vaguely recorded. The eclipse of Ithaca, near Troy, was dated by scientists as 1178 BC April 12, with the Trojan War as 1198/88 BC. These dates were still accepted as reasonable in the 1970 edition of the Cambridge Ancient History but we now know that the Odyssey, which knit together local traditions, was not written down until the eighth century, a century in which striking solar eclipses could have been seen at Ithaca in 765 and 711 BC.

In the first millennium AD many of the false dates arise because these are indeed eclipses listed in the Oppolzer catalogue at the years in question. The eclipses were often far too small to be noticed by ordinary people. Similar errors occur in the BC period: the eclipse of the poet Ennius usually ascribed to 400 BC probably refers to the total eclipse of 402 BC. When the "sun gave an omen" to the Hittites it was probably the total eclipse of 1340 BC rather than the partial one of 1335 BC, although neither date fits the alternative chronologies now current (cf. J. Lehmann, The Hittites, 1977, London, Collins, 205-206). The Chinese Book of Odes eclipse usually ascribed to 776 BC is probably dated 754 BC as this was much more conspicuous, although the Confucian eclipse of 709 BC is genuinely 'total' and was evidently observed.

2.31 Maya chronology corrected.

The Maya of Central America made a special study of lunar eclipses, as their Eclipse Table (cf. Aveni 1980, Table 17, 176) explained in Appendix C will reveal. Archaeologists have usually assumed that there was no calendar change between the Maya of the first millennium AD and the Aztecs of the sixteenth century. This assumption is not satisfactory astronomically, although a change of two days is a little better. Even this, however, makes the key date of their eclipse table a date when an eclipse was centred over the Indian Ocean. A new correlation, 86.4 years or 31,541 days on from that given in popular books, makes the base AD 842 March 30 when a lunar eclipse was visible on both sides of the Atlantic. The 842 date marks the beginning of an eclipse cycle of 11,960 days or 32¾ years, a cycle used also by the Chinese for their predictions.

There are many astronomical arguments for adopting the new chronology for the period of the Classic Maya and these are explained elsewhere (especially in Schove 1984b and 1984a). Apart from the Eclipse Table itself there are other dates in the same Dresden Codex (a medieval Maya Codex now preserved at Dresden) that are separated by eclipse cycles of either 11,960 days (solid lines in fig.1) or 9,360 days; these cycles are known as the Thix and the Fox and are respectively 36 and 46 times the Maya Sacred Year of 260 days. In the Venus table additions are indicated of

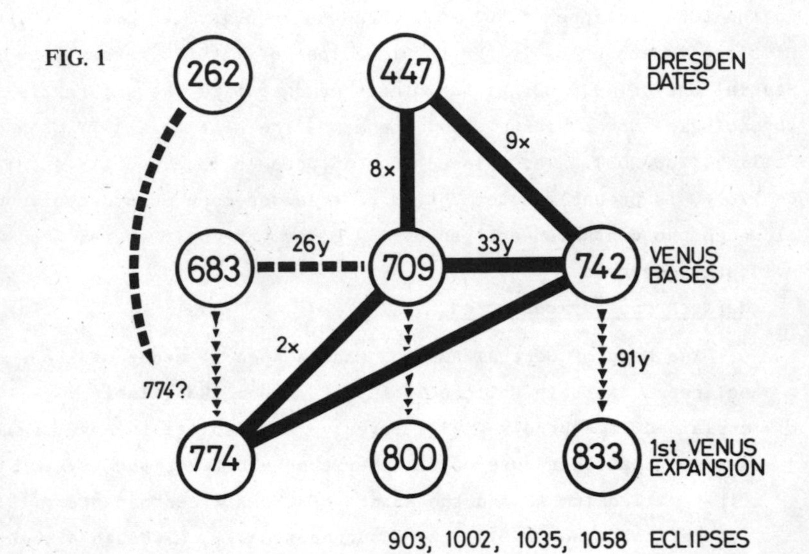

## ECLIPSE CYCLES IN THE DRESDEN CODEX

AD years derived from Maya dates (as in Appendix C) are circled.
Eclipse cycles of near 26, 32½ and 91 years (and their multiples) are shown.

D J Schove 1984

33,280 days or near 91 years: this also is an Eclipse cycle. There are also, in the still largely undeciphered text, glyphs which are known to represent eclipses. Dates of obvious eclipse significance are those associated with the Venus table as explained in Appendix C, and the AD years only are circled in fig.1. Some other dates separated by approximately similar intervals are likewise represented in the same figure. However, these eclipse cycles do not work exactly over long periods, and the Maya seem to have attempted to adjust the dates. The first date in AD 292 differs from the AD 774 date by 20 cycles of 9,360 days less one month, the second date in AD 447 is 21 days later than a total solar eclipse in Yucatan and is only approximately 8 cycles of 11,960 days earlier than the main Venus base of AD 709. Such coincidences could be accidental. On the other hand, the difference between the dates in AD 904 and 1297 is an exact multiple of 11,960 days. Both the 904 and 1297 dates are known to be calculated in eclipse half-years from the 842/874 dates of the Eclipse Table, as explained by J.E.S. Thompson in 'Commentary on the Dresden Codex' (Amer. Philos. Soc. Memoir 93, Philadelphia) 1972, p.74. The dotted line represents 123 eclipse half-years, but that is not an eclipse cycle. Empirical adjustments appear to have been made and most of the next few dates in the Codex (in 903, 1002, 1010 and 1058), although not exact multiples of 11,960 days, do correspond within a few days (exactly in the case of 1002 March 1) to visible eclipses.

The failure of the cycles and of the Maya predictions over long periods appears to have led to the calendar change. A slight change in the Short Count dates would suffice to make the lunar, Venus and eclipse tables viable for forecasting. This could however mean a change in the Long Count dates of 86.4 years, such a change being almost an exact multiple of the lunar month, the eclipse half-year and the Venus year. The Aztec records about AD 1500 prove that even small solar eclipses were watched for and recorded.

We consider specific Maya dates relevant to eclipses in the final sections of the fifth, seventh, eighth and ninth centuries. Our figure shows that the AD years of the Maya dates in 447, ..... 709, 742, 774 and another set in 842, 874, 903/4, ..... 1002, 1035, ..... 1297 approximate to Arithmetical progressions. The dates, except in 1297, fall close to eclipse dates in the new correlation. A chronological list is given in Appendix C with the

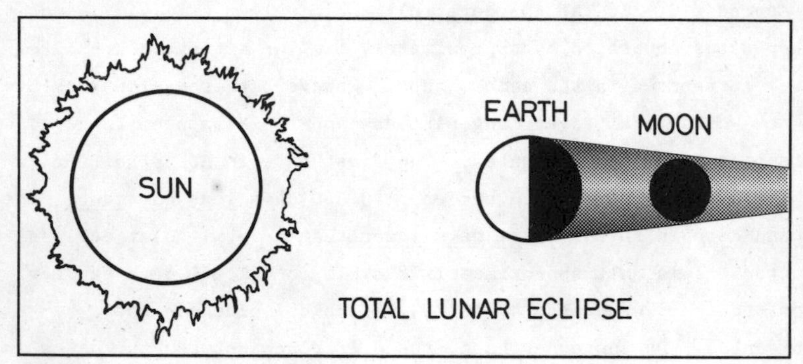


FIG. 2

FIG. 3

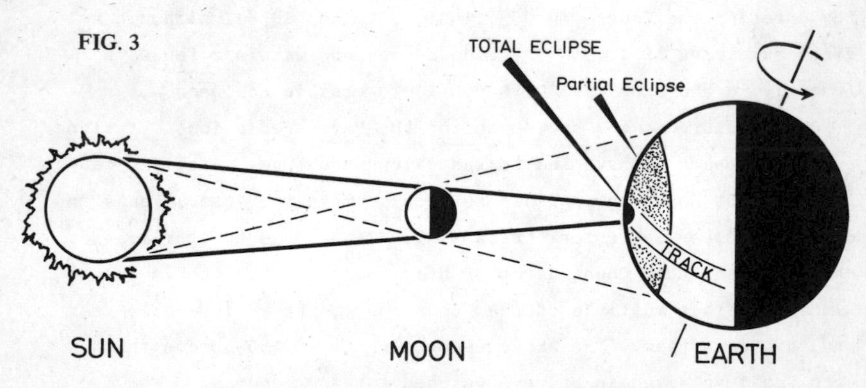


The Sun-Moon-Earth in Syzygy and regions (exaggerated) of Total (solid) and Partial (dotted) Solar Eclipse.

G SMYE-RUMSBY 1984

Maya Long Count equivalent as given by Thompson (op.cit.).

## 3. THE GEOMETRY OF ECLIPSES

The essential geometry of eclipses is illustrated in figs. 2 and 3. The Earth moves round the Sun and the Moon moves round the Earth. Every night the Sun is eclipsed by the Earth, but for an eclipse in the usual sense the three bodies must line up in what is termed syzygy. If the Earth is in the middle, as can happen at Full Moon, we have a lunar eclipse (fig.2); if the Moon is in the middle, as can happen at New Moon, we have a solar eclipse (fig.3). Our diagram is not to scale, but shows the path of totality and the region in which the eclipse is partial.

The Earth moves round the Sun once a year in what is known as the plane of the ecliptic, the other planets moving nearly in the same plane but in different periods. The Moon moves round the Earth once a month, and it might therefore be supposed that there would be two syzygies each lunation, one at Full and the other at New Moon.

However, the plane of the Moon's orbit is not that of the ecliptic, but is inclined to it at a small angle (5°) and it is only when the two planes intersect that an eclipse is possible. This happens at intervals of 148/178 days. This intersection occurs along what is known as the line of the nodes, the node cycle or eclipse half-year averaging 173½ days.

### 3.1 LUNAR ECLIPSES

A Lunar Eclipse occurs when part or all of the Moon is in the shade; such an eclipse is obvious to anyone seeing the Moon at all. This means that the inhabitants of that half of the Earth facing the Moon at the time can see the eclipse provided local weather permits. For visibility at a particular place, therefore, we need to decide first whether the Moon was visible in the region in question. At a lunar eclipse the Sun and Moon are opposite one another, or at antipodal points of the celestial sphere. If the Sun is well above the horizon, the Moon is well below and invisible. A famous historian, Plummer, plodding manfully but without astronomical knowledge through the eclipses of the Anglo-Saxon Chronicle, came to a lunar eclipse of about 904, but refrained from confirming that year, because he failed to realize that all the alternative eclipses (e.g. 905) in neighbouring years were either small in magnitude or invisible in England. At 10.10 a.m. on 21 May in any year, we hardly need an almanac to tell us that the Sun is well up above the horizon in England, so that Plummer could have inferred that the eclipsed Moon would be well below the horizon, and invisible. Simple tables of sunrise and sunset are very similar for all years in Gregorian New Style dating and can be estimated by any historian who possesses a current diary. Nevertheless, by such simple procedures we have in numerous cases been able to correct the historian's dates for medieval lunar eclipses.

The historian can find all too easily the details about lunar eclipses in the second part of Oppolzer (1887 and 1962, 325-375). This gives calculated dates, times (reckoned from midnight), magnitudes in digits (number of twelfths of diameter eclipses, less than 12 meaning partial, 12 or more total) and the sub-lunar point, that is, where the Moon at mid-eclipse is in the zenith. Schroeter's list of 'European' lunar eclipses (1923, 169-306) on the other hand omits certain eclipses (after AD 1000) which were genuinely observed in Europe.

Descriptions of lunar eclipses vary widely and astronomers have been puzzled as to the reasons. On some occasions the Moon has various shades and may look bloody or coppery, on others it does completely 'disappear'. The 'horrid black shield' which covered part of the Moon in AD 743 is a typical description of a partial lunar eclipse.

## 3.2 SOLAR ECLIPSES

Solar eclipses are either Total, Annular or Overall Partial, indicated by t, r (for Ring) or p in the final column of Oppolzer's tables. We find that the overall partial p-type eclipses of Oppolzer were never noticed and even the annular r-types were often missed. Most of the early records relate to eclipses that were total, either at the place of observation or within a few hundred miles of the track of totality.

3.21 Total Solar Eclipses (t-type). Totality occurs where the dark cone of the Moon's shadow reaches the ground. As this cone is only just long enough, the diameter of the terrestrial region affected is never greater than about 270 km at any one moment. Moreover, the Earth rotates from west to east and the shadow races eastwards, varying slightly to north or south because of the Moon's own movements; at the equator the speed of this dark shadow is about 365 m/sec. Among primitive peoples this is a path of panic. The eclipse record by Livy in Rome in 188 BC was seen several hours later by the Chinese, and as records of the comet of 190 BC are found in the same sources we have a striking confirmation of chronology. Stephenson's revised tracks of some significant Chinese eclipses are given in figs. 5, 6 and 7 (cf. also Stephenson and Houlden for others).

The total phase seldom lasts more than five and never more than eight minutes, but the time seems like hours to any witness who does not understand what has happened. An entry in the Irish Annals (AI) referring to a 'dark morning' in 594 is the earliest genuine dated eclipse from the British Isles, and illustrates the impression made by totality.

Long before totality commences, Venus is usually visible, but during totality, planets and a few stars may be seen. At the total eclipse of 136 BC (fig.9) Babylonian astronomers saw Venus, Mercury, Jupiter and Saturn as well as the brighter stars (cf. Stephenson and Clark 1978, 31).

Total eclipses are rare; at any one place the average is three times in a millennium. The next total eclipse in the British Isles for instance will be in 1999, and then only the extreme south-west will be affected. Sawyer 1972 considers that when 'The Sun stopped in the middle of the sky' (Joshua 10, 12-14) Joshua was referring to the eclipse of 1131 BC (see fig.4), rather than the alternative eclipse which was not total at Gibeon. The century is uncertain and Sawyer's interpretation has been criticized, but it is supported by a Russian account (in the Lavrentievsky Chronicle 1846, Vol.I, in the St. Petersburg Collection) of an eclipse in AD 1230 when "people thought that the sun had begun to move backwards". A similar claim in the Anglo-Irish chronicle of Henry of Marlborough for the year 1407 (sent to me by Professor Cashman) presumably refers to the partial eclipse of 1406 June 16 (or possibly to an aurora).

Darkening of the Sun can also follow volcanic eruptions and we discuss the dark Suns of the 530s, 620s and 797 from this viewpoint in the light of the new dates for eruptions determined from the ice-layers of the Greenland ice-cap.

3.22 Annular Eclipses (r-type). An Annular Eclipse (cf. 873, 891) occurs when the Moon is a little further away from the Earth than usual; this happens because the Moon's orbit round the Earth is

FIG. 4

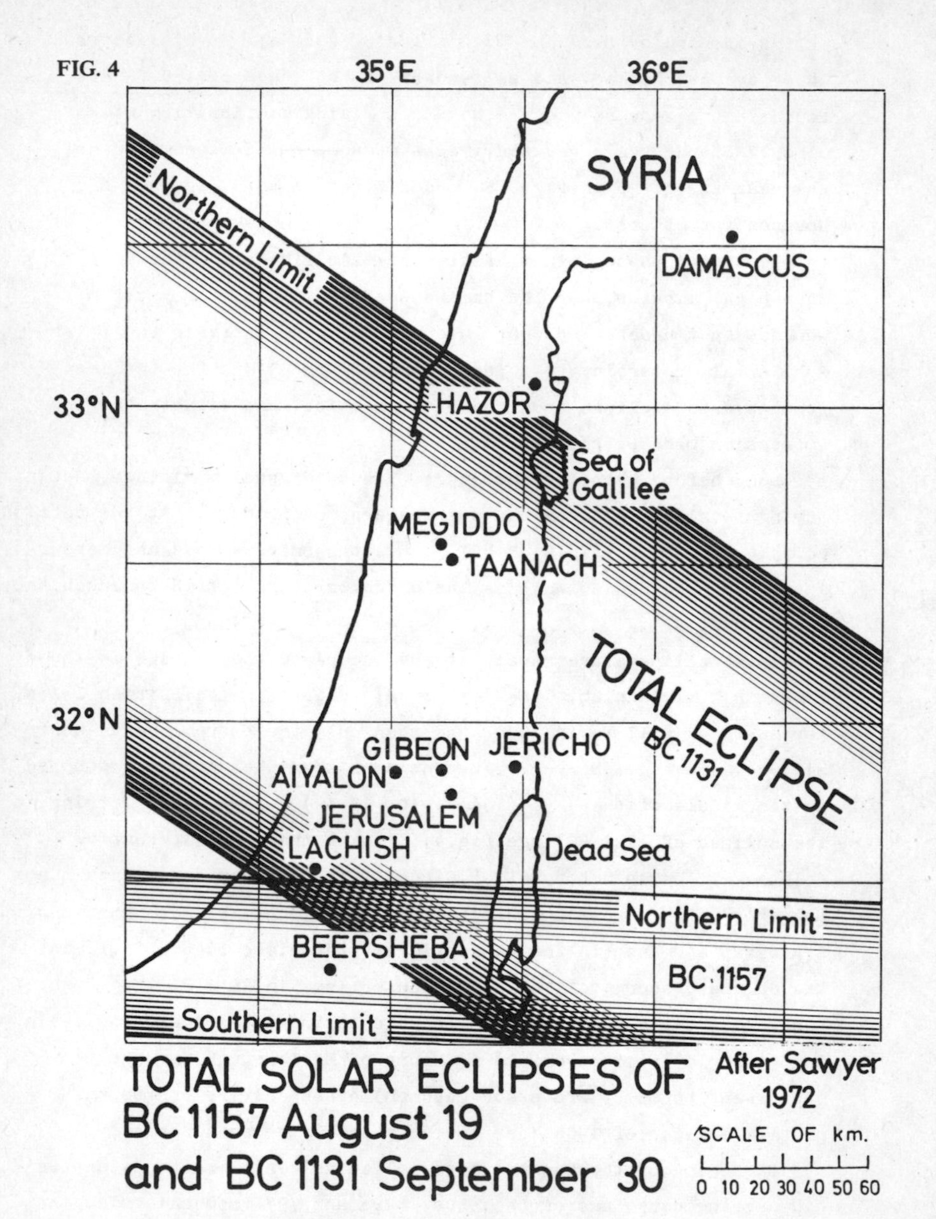


The eclipse of 1131 Sept. 31 may explain why the 'sun stood still' at Gibeon in the time of Joshua.

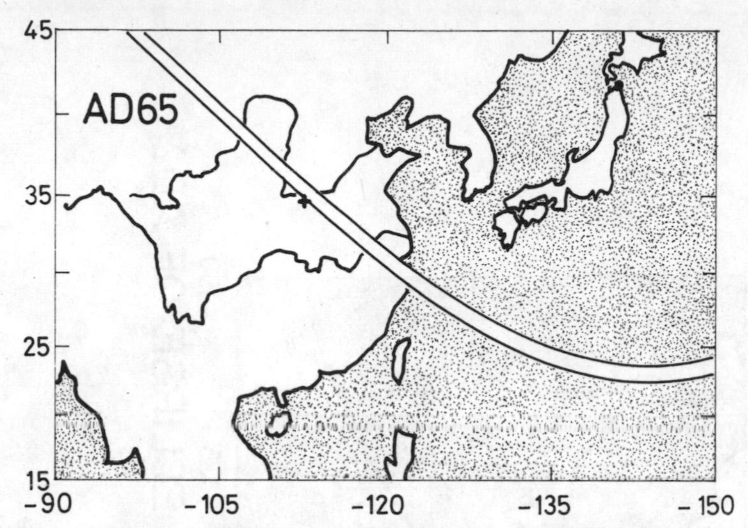


FIG. 5 THE TOTAL ECLIPSE OF AD65
After Stephenson & Houlden (in progress)
Assuming $\Delta T$ = 2 hours 30·3 mins (see fig. 11)
+ marks the city of Loyang in figs. 5 & 6

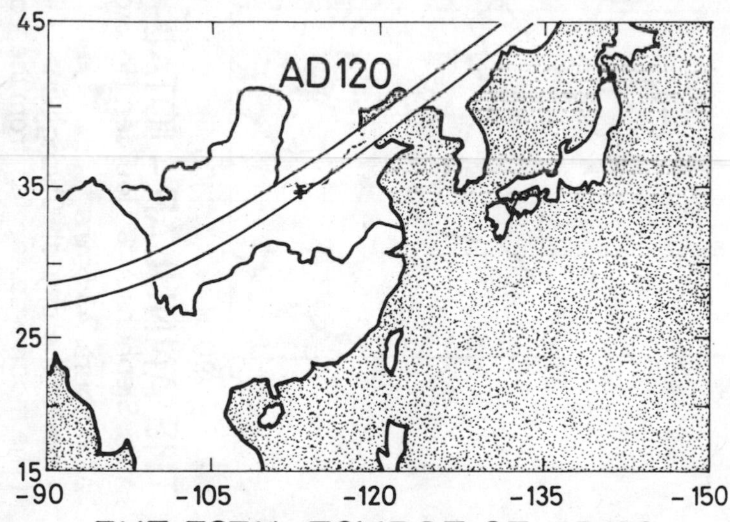


FIG. 6 THE TOTAL ECLIPSE OF AD120
After Stephenson & Houlden (in progress)
Assuming $\Delta T$ = 2 hours 19·5 mins

FIG. 7

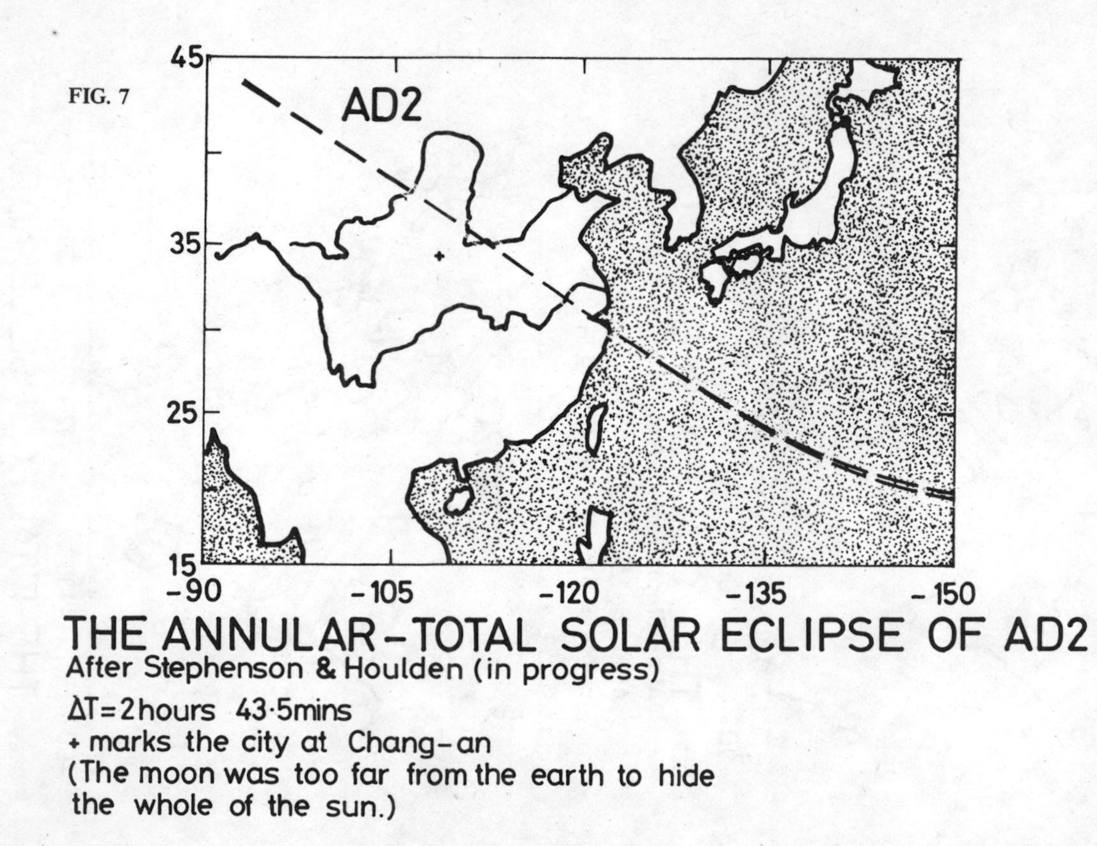


THE ANNULAR-TOTAL SOLAR ECLIPSE OF AD2

After Stephenson & Houlden (in progress)

ΔT=2hours 43·5mins

+ marks the city at Chang-an

(The moon was too far from the earth to hide the whole of the sun.)

slightly elliptical. The name comes from the Latin word annulus, a ring, for at place in the central belt instead of the blackness of a total eclipse a brilliant ring of light surrounds the darkened disc of the Moon. The intensity of daylight may not be greatly reduced so that an annular eclipse may not even be noticed. We have found hardly any clear descriptions of annular eclipses before AD 1600. However, a good description comes from Nishapur in Iran in 873 (see Ninth-century section). In 1263 an Icelandic account tells us that when king Hakon lay in Widehall Bay, off NE Scotland, "great darkness came upon the sun" (from Frisbok's Hakon's son's Saga, cc.326/328 in Codex Frisianus, 573-576, translated by A.O. Anderson 'Early Sources of Scottish History', Vol.II, 608 and 615-616. 1922. Edinburgh. Also Rolls Series 88, 347).

In recent atlases of eclipse tracks, the annular-total belt is tentatively indicated by two dashed lines for the southern and northern limits as in part of the track for AD 2 (fig.7). Annular-total eclipses are classified with fully total eclipses as Central. Strictly, this terminology so applied ignores a slight complication which we shall disregard.

3.23 Partial Solar Eclipses (t- or p-type). To the north and south of the belt of totality (or annular totality) there is a wider belt in which the t- or r-type solar eclipse is Partial; this is shown in an exaggerated form in our figure. The observer in this zone is not exactly in direct line with the Moon or Sun, so that a solar crescent is still visible. Such eclipses are more frequent than is usually supposed, for they occur about once every 2½ years at any given location. However, the loss of light is smaller than heavy clouds would produce and partial eclipses usually passed unnoticed by the astronomically-unsophisticated chronicler. Many erroneous identifications have been made by historians who have imagined that small partial eclipses (e.g. that of 400 BC, the so-called eclipse of Ennius noted above) would have been observed by ordinary people. Astronomers, and those who have been forewarned, may notice an eclipse of magnitude 0.70 (cf. AD 808) if they see it in a reflection, at sunset or through thin cloud or haze, and then the moon-shaped black crescent suggests that a large bite is taken out of the Sun. The average person notices a thin solar crescent of a solar eclipse only when the magnitude reaches 0.99.

3.24 The Magnitude. The magnitude of even a solar eclipse used to be estimated in twelfths, but the fraction of the solar diameter obscured by the Moon is now expressed on a scale from 0 to 1. The subjective impression made is not in proportion, a solar eclipse of 0.80 often being difficult to perceive. Medieval solar eclipses from Europe recorded in the chronicles were analysed by calculated magnitude, and over 55% belonged to the class 0.90/1.00 (cf. Newton 1974, Table 3). In China the Confucian Spring and Autumn Annals show a large number also in the range 0.60/0.90, which suggests that at that time they too were indeed usually observed: the Sun was probably watched at the New Moon through its reflection in water, as was known to be the case in later centuries.

## 4. CATALOGUES AND MAPS

The 'Eclipse Canon' of Oppolzer (1887, 1962), already mentioned, lists the dates and details of 8,000 solar eclipses and 5,200 lunar eclipses occurring between 1207 BC November 10 (Julian or Old Style) and AD 2161 November 17 (Gregorian or New Style). Some information often adequate for a historian lies in the third part (pp.377 ff.), for this provides circumpolar maps of the central solar eclipses with their dates, total eclipses being indicated by a single solid line and annular by a single dotted line. Possible records might be found from places which on these maps are within 2 cm on either side and up to 1 cm at either end.

The tracks in Oppolzer are, however, not sufficiently precise for the location of total eclipses. The attempt was made to indicate the sunrise, the midday and the sunset points of each eclipse mapped, but even these points are often 100 km in error in the Ancient World, and the curved lines interpolated between them are for several reasons only first approximations.

The tables in Oppolzer can be examined first - the dates are in order, the latitudes and longitudes help in spotting a position on the maps, and the letters t, r and p indicate whether the eclipse was total, annular or overall partial. The overall partial eclipses are effectively invisible - they are those in which the Moon's shadow cone misses the Earth, passing north or south, according as to whether Oppolzer's tabulated quantity gamma ($\gamma$) is greater than +1 or less than -1. The occurrence in an early record of an overall partial eclipse implies a calculation or prediction, not an observation.

Extensions backward in the Ancient World were attempted for the

Middle East by P.V. Neugebauer 1931, in his Canon of Solar Eclipses, 4200 BC - 900 BC (and Lunar back to 3450 BC). G. van den Bergh did not know of this work (personal communication to DJS) but analysed the periodicities in Oppolzer in his 1955 work on the periodicity of eclipses (van den Bergh 1955) and produced an independent catalogue for the earlier period. His maps are essentially extrapolations from Oppolzer, based on such cycles as that of 11 years or 135 lunations, after which eclipses can repeat themselves in the same places. One of the cycles noted was that of 1,841 years and 1 month: eclipses of the next few years will partly repeat the pattern of the mid-second century AD, e.g. AFRICA AD 146 Feb 28/1987 Mar 29, CHINA AD 146 Aug 25/1987 Sep 23, PACIFIC AD 147 Feb 17/1988 Mar 18. Neugebauer and van den Bergh were essentially in agreement, not only with one another but also with the new and more reliable atlas of solar eclipses published by Kudlek and Mickler (1971) which goes back to 3000 BC for the Middle East.

All these atlases were published without any confirmatory empirical evidence from before the 8th century BC (cf. however fig.4). On account of the uncertain variability of the Earth's rotation it is difficult to specify solar eclipse paths reliably in the distant past. However, back to 1375 BC the tracks are essentially in conformity with the subsequent calculations and identifications of Muller and Stephenson 1976, and certainly more accurate than the tracks on Oppolzer's maps. Totality in Kudlek and Mickler's atlas is again indicated by solid lines, but this time both the northern and southern limits of totality are shown; dashed lines indicate annular-total eclipses in the same way, and times are indicated as well. The dates are given on their maps only as Julian Day Numbers (JDN cf. Appendix C), but reference to their tables supplies the dates in negative (Astronomical) form. Their earliest calculated eclipse for Thebes is -2972 equivalent to 2973 BC; the latest eclipse mapped is that of AD 59 (JDN 1742 on p.199).

For the Mediterranean World from 900 BC to AD 600 the atlas of Ginzel, 1899, has stood the test of time (cf. Boll 1919), and when it overlaps in the Middle East with the modern work of Kudlek and Mickler, 1971, it does so consistently. Total eclipses are shown by Ginzel in green, annular eclipses in buff and annular-total in brown. The more precise calculations for some initial classical eclipses are essentially in agreement, although calculated tracks now make it possible (Muller and Stephenson 1975, 522-3) to use e.g. the eclipse of 310 BC seen by

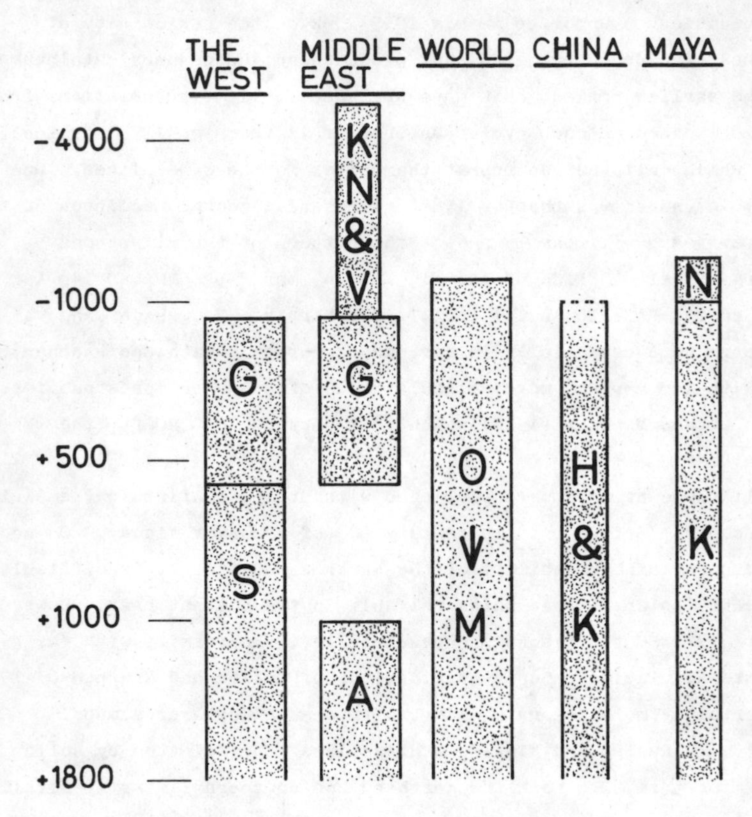


A = Gray 1968 (Solar)
G = Ginzel 1899
H = Stephenson & Houlden
K = Kudlek etc: (Solar) 1971
M = Meeus etc: 1966/85
N = Neugebauer 1931
O = Oppolzer 1887, 1962
S = Schroeter 1923
V = Van den Bergh 1955
W = Newton (Lunar) 1977

FIG. 8

D J Schove 1984

the tyrant Agathocles, on a voyage from Syracuse to Carthage, to indicate that he probably sailed by the route through the Straits of Messina. Ginzel's maps include S. Europe, N. Africa and S.W. Asia.

From AD 600 to 1800 the standard work is that of Schroeter 1923, who gives maps of the Western solar eclipses, this time including N. Europe in his purview. This work was not used by Newton but his identifications are essentially consistent with the maps as drawn. Schroeter omits overall-partial eclipses but as far as solar eclipses in temperate Europe are concerned this does not matter. His maps do not include some late Medieval Russian solar eclipses that were observed. (In his lunar eclipse lists he omits partial eclipses, so that Oppolzer's work is still the more useful for identifications.)

Additional maps (cf. Gray 1965 for Africa cf. Schove 1968) and modern calculations (cf. O'Connor 1952 for the British Isles and cf. Schove 1955a) have proved useful for certain regions.

In Spectrum of Time studies (Schove 1983a. See bibliography) we have been fortunate in receiving and also using Dr. Kudlek's unpublished maps and tables, notably for China and Central America. His results are generally consistent with Stephenson's independent calculations (cf. Stephenson and Houlden in progress). For the dating of Shang lunar eclipses in China, R.R. Newton has published 'A Canon of Lunar Eclipses for the years -1500 to -1000'. 1977. (Johns Hopkins).

Modern eclipses are fully and accurately treated in the work of Meeus 1966 and collaborators, which is more than a mere continuation of Oppolzer and which covers the period AD 1898 to 2510. Historical lunar eclipses are treated in Meeus 1969, and solar eclipses in Meeus and Mucke 1985; these works give more accurate tracks than are available in Oppolzer. Our chart (fig.8) summarizes this section.

## 5. PREDICTION

### 5.1 LUNAR ECLIPSES

Lunar eclipses are more common and are more easily predicted than solar eclipses. In the late third millennium BC, in Sargon's Akkad, lunar eclipses were already being predicted; they occurred only at Full Moon in the middle of the Mesopotamian month and then only when the path of the Moon cut the ecliptic, a region of the sky with which the Mesopotamians were especially familiar. Prediction became simpler still in the first millennium BC when cuneiform records of dated observations were kept in Nineveh and Babylon. Nevertheless, their predictions were often wrong. The predicted lunar eclipses for 249 BC on April 18 and October 13 did not take place (see Newton 1976, 131-2, Table IV 3) and a further false prediction was made for 104 BC Jan 8.

In China we have evidence of the false prediction of a lunar eclipse in 172 September 3, but the periodicities evident in the Annals (see below) enabled them to make many successful predictions.

The art of predicting lunar eclipses culminated among the Maya, whose eclipse tables, as preserved in the Dresden Codex, provide reliable predictions for almost every eclipse through any thirty-three year period. Their list gives the intervals after a first eclipse when another might occur somewhere in the world. The tables predict even solar eclipses as well, but they must have been based empirically on long series of observations of lunar eclipses (Appendix C and Schove 1984a).

## 5.2 SOLAR ECLIPSES

Eclipses of the Sun are much more difficult to predict, since their visibility from any one point on the Earth depends not only on the distance of the Earth from the Sun and from the Moon, but also on the relative dimensions of these three bodies, and this was not known in early times. The possibility of a solar eclipse at New Moon was nevertheless considered by the Babylonians (and probably the Maya) whenever the Moon was near the plane of the ecliptic; both they and the Chinese had lunar calendars and a solar eclipse was no doubt envisaged on the first day of every month. The Maya recorded the day of the Moon, but at first they numbered the days from first visibility, except for special Katun-ending dates (cf. Schove 1984b, 21).

The eclipses in Far Eastern sources are often based not on observations but on calculations, which were inserted in the Dynastic Histories at a later date. The eclipse of 720 BC may have been added in this way: it was of magnitude only 0.50 at Chü-fu. It could just have been seen at sunrise, but cycles of 317 lunations (25½ years) which separated the observed eclipses of 695 and 669 BC may have been used later to extrapolate the cycle backwards to calculate the date of the preceding eclipse in the sequence.

In the work of Ssuma-chien there are some genuine observations of Early Han eclipses before 100 BC. In the Dynastic History of the Early Han, written in the 1st century AD, there are numerous additional (mostly calculated) eclipses. In his translation Professor H.H. Dubs made particular studies of these eclipses but he did not appreciate that many of them were calculations added to the primary sources. Indeed, it is often difficult, in sources from both China and Japan, to distinguish between the observed and the calculated, unless additional details are provided.

## 6. THE SCIENTIFIC PROBLEM

The value for scientists of the historical accounts of eclipses lies in what is known as the problem of the accelerations, a problem which was tackled seriously first by Fotheringham (cf. Fotheringham 1920). In the past twenty years improved calculations have been made, especially at Newcastle and Liverpool in England by R. Stephenson (now at Durham) and at Baltimore in the United States by R.R. Newton (now at Johns Hopkins) in the various papers cited in our bibliography.

A total solar eclipse at a particular place provides very precise information about both the motion of the Moon and the rotation of the Earth. The deceleration or negative acceleration of the Moon was first inferred from eclipse records by Halley in 1695, but a later Astronomer Royal, Spencer-Jones, considered eclipses since Halley and attempted to determine also the negative acceleration of the Earth's rotation. Recent small changes have now been measured very precisely (Stephenson and Morrison 1982). Tidal friction is gradually slowing down the Earth's rotation about its axis; it is also gradually decreasing the angular velocity of the Moon in its orbit about the Earth. These two decelerations are reflected in a very gradual increase in the length of the day and the (lunar) month respectively. Other non-tidal effects such as the varying atmospheric circulation and sea-level are also involved (cf. Newton 1972, Stephenson 1975, Stephenson 1982, 173).

The change in the length of the day (relative to the fixed year) is indicated by the number of daily ridges found on fossil corals - this was 400 in the Early Cambrian but 390 about three hundred million years ago and is now only 365 - the lunar month was near 31 days in the Early Cambrian and is now only 29½ (cf. Newton 1969, 825. See also Addenda).

Stephenson's two maps for the eclipses of 136 BC and AD 484 show the overall effect: ancient eclipses occurred well to the east of

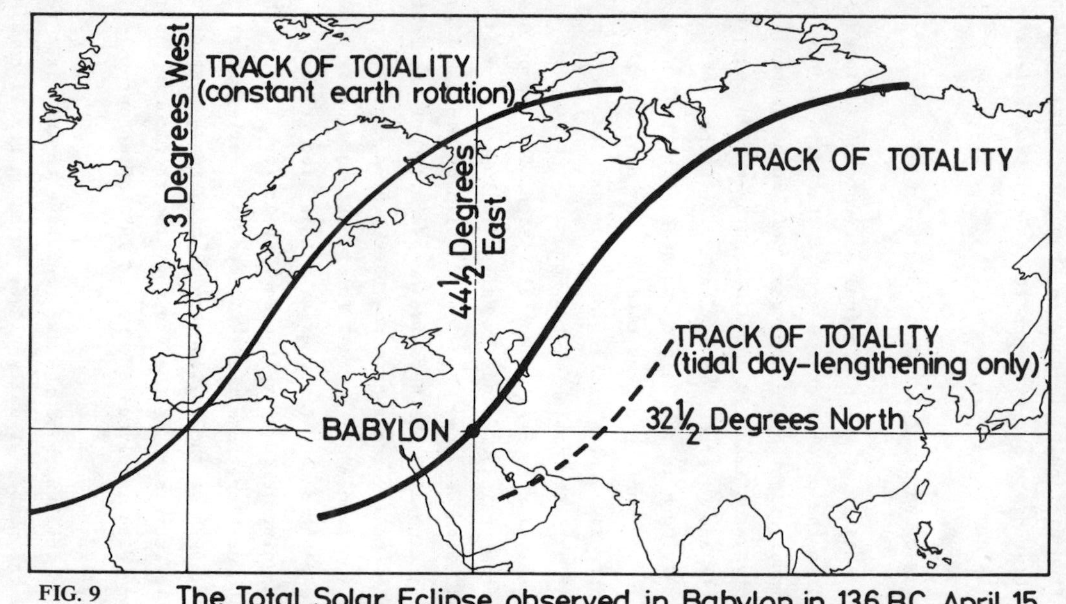


FIG. 9 The Total Solar Eclipse observed in Babylon in 136 BC April 15 would have been seen in Europe if the rate of the earth's rotation were the same as it is now.

(After Stephenson 1982, 180)

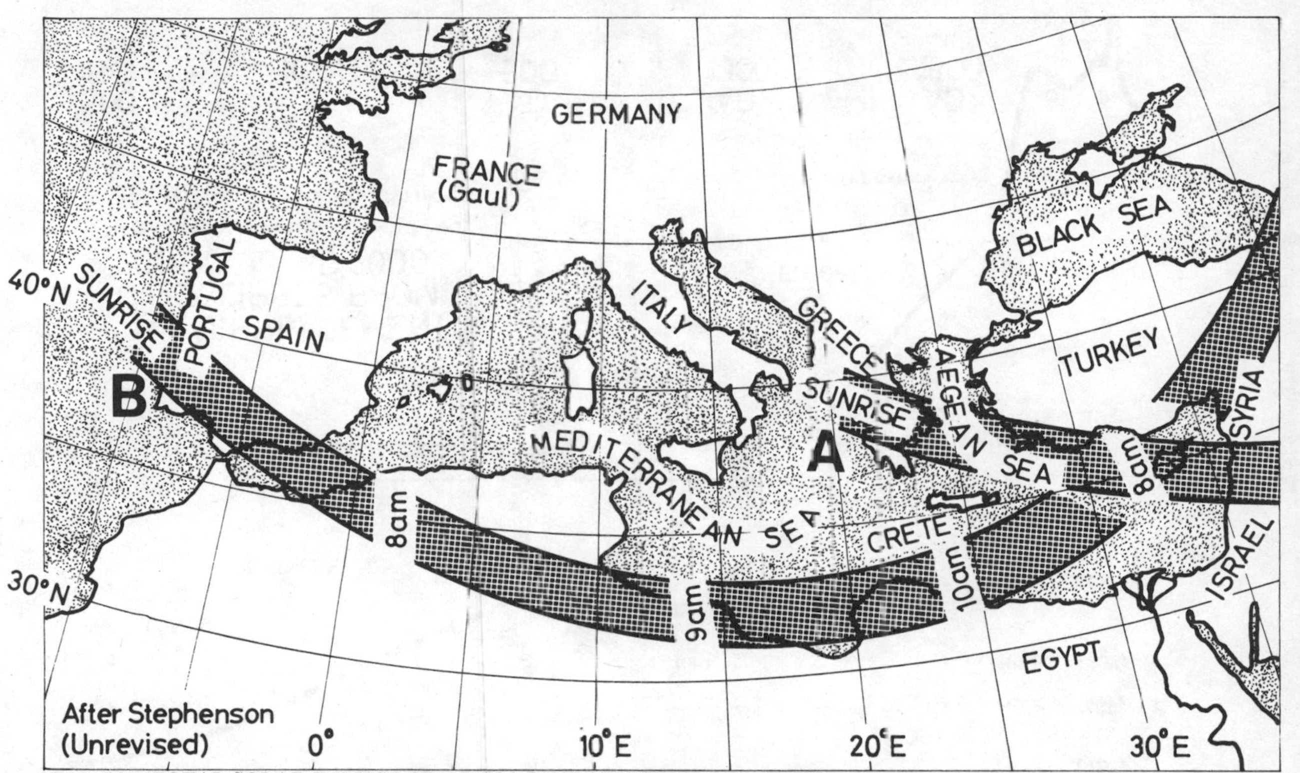


FIG. 10 TOTAL SOLAR ECLIPSE OF AD 484 JANUARY 14 Observed in Athens

The track through Athens must be correct (see text AD 484). The track would have begun further West if the earth's rotation had been constant. However, Dr Stephenson now finds (see fig. 11) $\Delta T$ near AD 500 only 1 hour not 2 hours so that the displacement is near 15° not the 30° implied by the track beginning in Portugal.

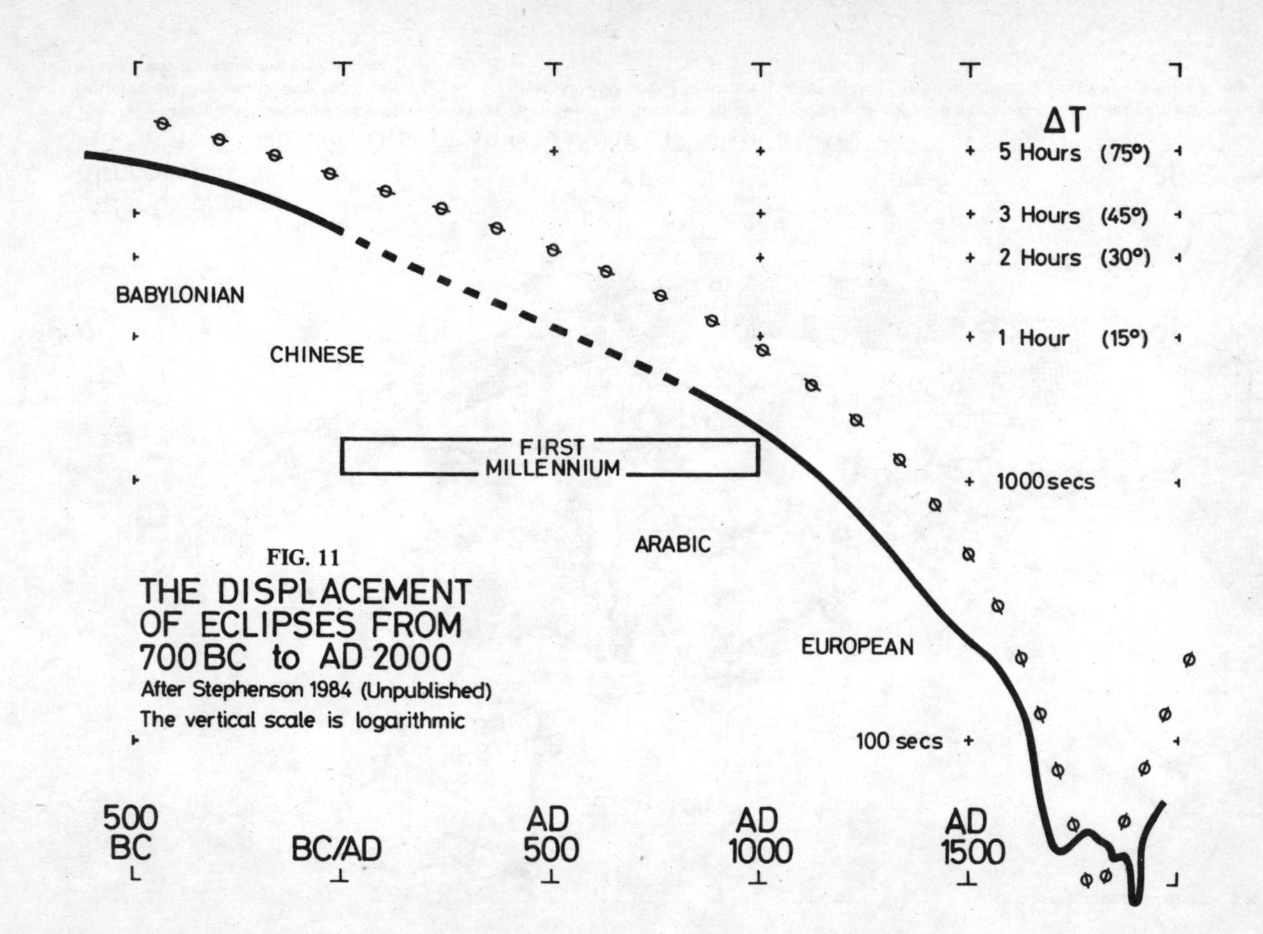


FIG. 11

THE DISPLACEMENT OF ECLIPSES FROM 700 BC to AD 2000

After Stephenson 1984 (Unpublished)

The vertical scale is logarithmic

the positions expected if the length of the day had been constant. In 136 BC the discrepancy was near 47½° of longitude - that is $\Delta$T in fig.9 is over 3 hours. By AD 484 (fig.10) it would be near 30° ($\Delta$T near 2 hours) according to the tidal effect used for his map, but Stephenson points out that the latest results (in fig.11) make the displacement near 15° ($\Delta$T 1 hour). (Personal communication April 1984.) However, there are not many eclipses in the first millennium that enable to specify the $\Delta$T value and the curve is therefore indicated by dashes for most of this period. The values adopted for the Chinese eclipse-maps (figs.5, 6 and 7) are 2 hrs 30.3, 2 hrs 19.5 and 2 hrs 43.5 mins respectively; these fit the assumption that the observations were usually made at the dynastic capitals. In the 9th and 10th centuries Islamic eclipses provide accurate dates of $\Delta$T and the curve (simplified from a detailed unpublished chart by Stephenson) is reliable from then on. The scale is non-linear and is drawn to illustrate the small but precisely known irregularities of recent decades. The details for different periods will be discussed by Stephenson and Morrison in Phil. Trans. Roy. Soc. 1984/5. However, we can read approximate values from fig.11.

The dots show the values expected from the tidal effect by itself

$\Delta$T = 10,000 seconds at 30 BC
1,000 seconds at AD 1150
100 seconds at c.AD 1640

The uncertainty in the first millennium adds significance to the places of observation of solar eclipses in China and Europe in our period. A list of usual total and near-total eclipses to 1567 was published by Stephenson in J. Brit. Astron. Assoc. 89(3), 1979, 246-247 and the table below gives a slightly enlarged version for the period up to AD 1000. There is no disagreement about the total change and the eclipse catalogues have empirically accepted realistic values. The problems have been to determine whether the rate of acceleration has changed from century to century and to separate the terrestrial and lunar accelerations.

In the United States, Newton critically rejected, as we have noted, the various eclipses for which the evidence appeared unsound and made a statistical study of the remaining eclipses - Classical, European and Islamic. He concluded that the variations in the rate from century to century were occasionally significant.

## USABLE PRE-TELESCOPIC OBSERVATIONS OF TOTAL AND NEAR-TOTAL SOLAR ECLIPSES

| *Julian date* | *Place* | *Lat. N* ° ′ | *Long. E/W** ° ′ | *Observations* |
|---|---|---|---|---|
| BC | | | | |
| 1375 May 3 | Ugarit | 35 47 | 35 47 | Sun put to shame; went down in daytime |
| 709 July 17 | Chu-fu | 35 32 | 117 01 | Total |
| 601 Sept. 12 | Ying (?) | 30 20 | 112 15 | Total |
| 549 June 12 | Chu-fu | 35 32 | 117 01 | Total |
| 198 Aug. 7 | Ch'ang-an | 34 21 | 108 53 | Annular |
| 188 July 17 | Ch'ang-an | 34 21 | 108 53 | Almost complete |
| 181 Mar. 4 | Ch'ang-an | 34 21 | 108 53 | Total; dark in daytime |
| 147 Nov. 10 | Ch'ang-an | 34 21 | 108 53 | Almost complete |
| 136 Apr. 15 | Babylon | 32 33 | 44 17 | Total; 4 planets seen |
| 89 Sept. 29 | Ch'ang-an | 34 21 | 108 53 | Not complete, like a hook |
| 80 Sept. 20 | Ch'ang-an | 34 21 | 108 53 | Almost complete |
| 35 Nov. 1 | Ch'ang-an | 34 21 | 108 53 | Not complete, like a hook |
| 28 June 19 | Ch'ang-an | 34 21 | 108 53 | Not complete, like a hook |
| 2 Feb. 5 | Ch'ang-an | 34 21 | 108 53 | Not complete, like a hook |
| AD | | | | |
| 2 Nov. 23 | Ch'ang-an (?) | 34 21 | 108 53 | Total |
| 65 Dec. 16 | Kuang-ling (?) | 32 26 | 119 27 | Total |
| 120 Jan. 18 | Lo-yang | 34 47 | 112 41 | Almost complete; on Earth it was like evening |
| 360 Aug. 28 | Chien-k'ang | 32 02 | 118 47 | Almost total |
| 516 Apr. 18 | Chien-k'ang (?) | 32 02 | 118 47 | Annular |
| 522 June 10 | Chien-k'ang (?) | 32 02 | 118 47 | Total |
| 840 May 5 | Bergamo (?) | 45 42 | 9 40 | Sun hidden from world, then shone again |
| 912 June 17 | Cordoba (?) | 37 53 | 4 46* | Total; darkness covered the Earth |
| 968 Dec. 22 | Constantinople | 41 01 | 28 59 | Sun deprived of light (very clear account of corona) |
| 968 Dec. 22 | Farfa | 42 13 | 12 42 | Sun in darkness |
| 975 Aug. 10 | Kyoto | 35 02 | 135 45 | Total; Sun colour of ink |

## USABLE ECLIPSES

This is part of a table given by Dr F. R. Stephenson in *Jnl. Brit. astr. Assn.*, 89, 1979, 246.

The above list is only part of Stephenson's table, which includes later eclipses up to AD 1567.

Other eclipses discussed in this book might now be considered e.g. AD 179 China, AD 59, 71, 164, 447, 484, 512, 534, 590, 592, 594, 601, 733, 807, 813, 840, 878, 939, 966.

In England, Stephenson and his collaborators select those eclipses (cf. Stephenson and Clark 1978, 43) which appeared to be on the one hand reliable as to time and place and, on the other hand, critical inasmuch as there was a large north-south component in the track. They conclude that the acceleration of the rate of change in the motions has been practically constant, but point out that some possibly periodic variability is not inconsistent with the eclipse-evidence. This variability from century to century is, nevertheless, in the second millennium AD shown to be less than 100 seconds of time!

The American and the British investigators differ in their choice of significant eclipses. We need better evidence as to the place in which critical eclipse observations were made than given by the conventional name of the chronicle. This we have endeavoured to provide in our text and in our tabular summaries. Meanwhile, although the scientists disagree about the irregularities of certain centuries, we know that since classical times the retreat of the Moon from the Earth has been at the average rate of 6 cm per year or 120 metres in total. Space scientists expect to determine soon the correct rate with even more precision.

## THE EARLIEST DATED ECLIPSES

Eclipse-dating has now been extended back by Huber (1982) to 2095 BC July 25, the lunar eclipse of the last year of Sulgi, the second ruler of the Third Dynasty at Ur.

Other early lunar eclipses have been dated 2053 BC April 13/14 and 1659 BC February 9. The first dated solar eclipse followed a fortnight later on February 23 but there is no record of the important solar eclipse of 1650 BC. These dates fit with the long chronology for the Venus Tables (1702 BC) and for Hammurabi (1848 BC), the end of the Babylonian dynasty being 1651 BC. This long chronology, suggested originally on astronomical grounds by Sidersky (1944), a chemist, has been confirmed by both Weir (1982) and Huber (1982).

## 8. PATTERN OF CHAPTERS

The pattern adopted for each century is similar and consists of:

1. An <u>Introduction</u> describing briefly the cultural geography of the sources of the period.
2. The <u>Chronology of Western Eclipses</u>. Western here includes North Africa, Arabia and S.W. Asia. Each eclipse is treated separately and an attempt is made to locate the primary sources.
3. <u>Chinese eclipses</u> deemed to be of scientific interest. These are very few as we can seldom separate the predictions from the observations and there is no way of detecting which observations were made at the Capital of the period. Reference is made, however, to forthcoming works by Stephenson and Boulton (1985) and by Cohen and Newton (1985).
4. <u>Japan or India</u>. An occasional eclipse of special interest is noted from the seventh century onwards.
5. <u>Mayan eclipses</u> that were predicted and probably observed by the Maya in Mesoamerica are noted from the fifth century onwards. (See also Appendix C.)

## 9. THE SPECTRUM OF TIME

The method adopted is explained in Schove 1983a and fig.14 which is reproduced from p.106 of 'Sunspot Cycles' by kind permission of Van Nostrand.

### 9.1 COMET CHRONOLOGY

Comets (Appendix B) can be better dating criteria (Schove 1975) than lunar eclipses as they are much less frequent. However, historians have had the same difficulties with identification. In place of the calculated eclipse date-list of Oppolzer, historians use the standard comet-catalogues of Pingré (1783/84) and his successors and these all contain numerous errors. I have found the catalogue of my collaborator Dr. P-Y Ho (1972, plus a few corrections in Hasegawa 1979, Table 3) very reliable, as the Far Eastern dates nearly always agree with the corrected Western dates. However, only from AD 634 can Chinese comets be corroborated by Japanese investigations and I have rejected as fabrications all Korean comets prior to AD 1000.

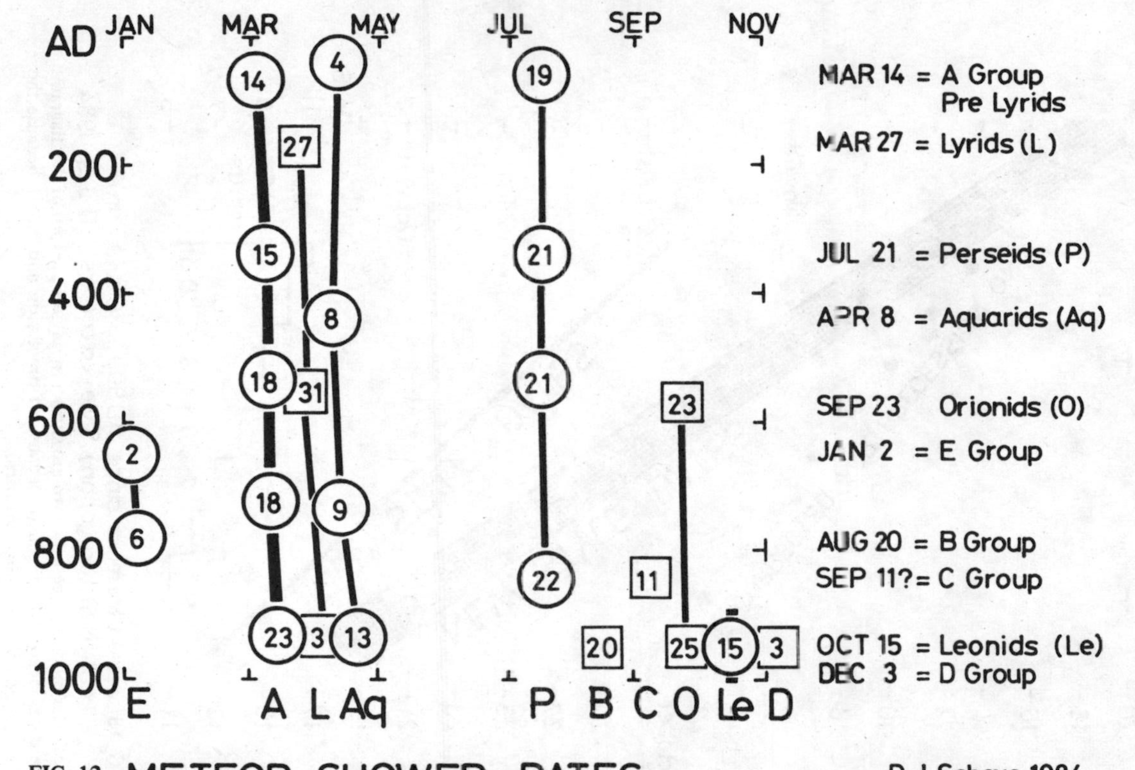


FIG. 12 METEOR SHOWER DATES in the first millenium AD

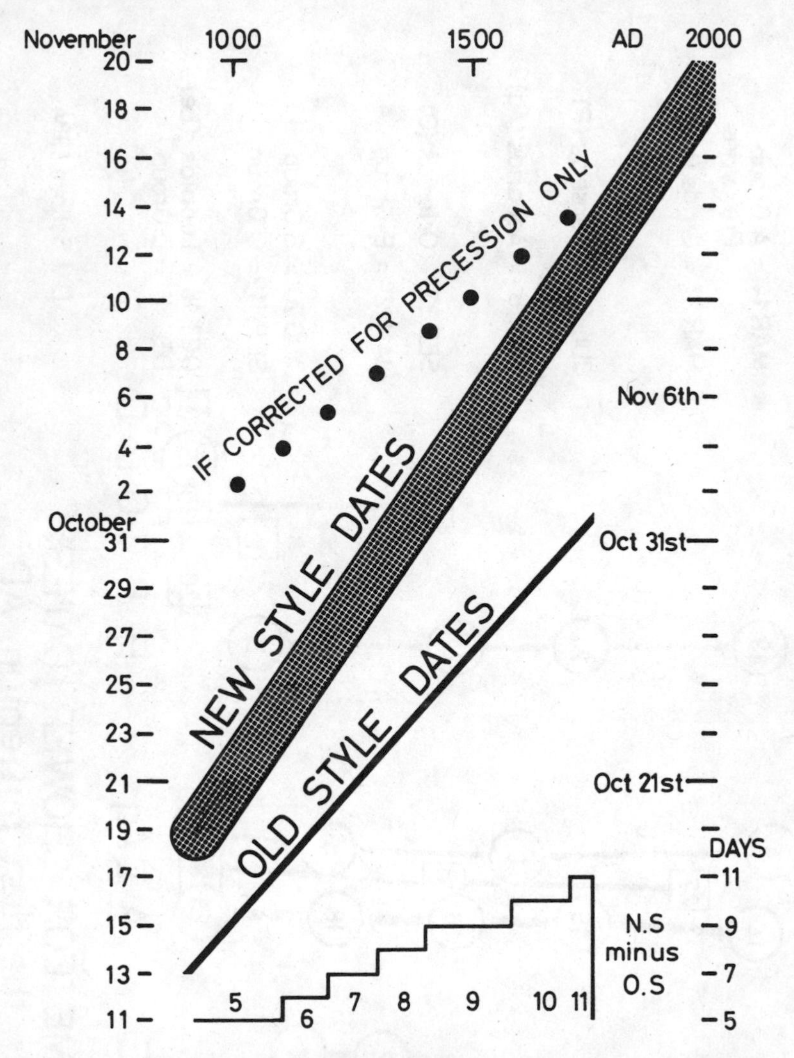


FIG. 13 Leonid Meteors since AD 900 in Old (Julian) and New (Gregorian) Style calendars. D J S 1984

**The differences between the two are shown in days at the foot of the diagram. The dates become later partly because of precession and partly because the orbit changes through the centuries.**

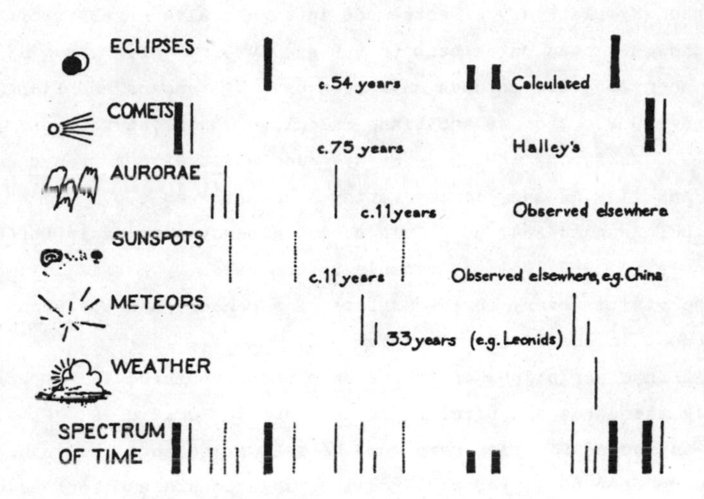


**FIG. 14 THE SPECTRUM OF TIME**
**From D. J. Schove 'Sunspot Cycles' 1983, p.106 (Van Nostrand Reinhold, New York and Wokingham, UK)**

The early dates of Halley's comet have been estimated by various methods with inconsistent results. Schove 1955, 288 noted that there were 8 appearances in a millennium, so that extrapolating backwards from AD 2066±, 1066 and 66 he reached -934 or 935±5 BC and queried whether 965 BC would be a suitable date for the comet of David that pointed to Ornan's threshing-floor and led to the building of the Temple at Jerusalem. A less crude method was used by Kamienski who explained why planetary effects made intervals alternately short and long, and estimated dates back to 929 and 1005 BC (cf. Schove 1956a). Modern sophisticated methods have been used by Yeomans and Kiang 1981 and Brady 1982. Yeomans and Kiang calculated dates back beyond 986 BC ... as far as 1404 BC. Brady found a fit with supposed comets in 620 and 1140 BC and, incorporating such dates as c.993 BC and c.1444 BC, reached 2647 BC. This comet, already seen again through telescopes in 1983, will be visible to the naked eye again in 1986. It is so visible every three-quarters of a century and was very conspicuous in the first millennium AD.

Western descriptions of comets have been assembled in typescript but only the dates and brief references are included here. Up to AD 634 an 'identification game' on the assumption that the Chinese dates were correct worked well; slight adjustments had to be made to the Western dates in Barrett (1978) and Hasegawa (1979). From AD 634 the Chinese and Japanese records usually corroborated one another, and when they did not the Western accounts provided a 'casting vote'. To eliminate aurorae and meteors from comet-catalogues the Latin terminology of portents given by Dall 'Olmo (1980) proved very helpful.

Many dating errors occur in European comets (e.g. in Chambers 1889). Bede's comets of '678' and '729' are here dated 676 and 730; Bede seems to have placed the 730 comet one year early so that it preceded the death of two men - assimilation once again (cf. Schove 1975, 407 n.7 and cf. Schafer 1977, 110-111). No independent Western source, such as Leo the Isaurian, has been found that can confirm the exact year. The comet associated with the Archbishop of Canterbury hitherto dated 995 (after the Anglo-Saxon Chronicle) seems to be tactfully misdated from 989 to fit Sigeric's death (994/5) instead of his accession. A clue is provided by a date 'August 10' preserved in Florence of Worcester from an earlier version of the Anglo-Saxon Chronicle, as this is close to the Chinese date of the 989 comet. Likewise the 'ghost' comets of

904 (_recte_ 905) and 959 (_recte_ 961/962) were, I suggest, invented to fit the birth and death of the Byzantine Emperor Constantinus VII (cf. Rodgers 1952). Halley's is the only periodic comet in this millennium that has so far been distinguished: dates in AD 66, 141, 218, 295, 374, 451, 530, 607, 684, 760, 837, 912 and 989 (cf. Mucke 1976) have been confirmed by computer calculations (Yeomans and Kiang 1981, Brady 1982). Novae and Supernovae as listed by Stephenson and Clark 1978, 64 have been included in our tables; differential diagnosis is not always certain, but radio and X-ray waves are still reaching the Earth from parts of the sky specified as locations of Chinese 'guest stars': among these 185, 386 and 393 are thus selected as possible supernovae by Stephenson and Clark 1978, 74.

The reasons for the corrections to the usual dates of Western comets have been fully explained in my unpublished typescript comet-chronology. I had hoped to find a collaborator of the calibre of Dr. Fletcher who would prepare a full version for publication. A short chronology is nevertheless in preparation. I can meanwhile assure historians that 'dates' of the many 'European' comets not mentioned in our tables are excluded with good reason.

The scale of marks adopted is given in Appendix B. In general European chroniclers from about AD 300 until the late sixteenth century noted comets with scores of 3 or more. Chinese observers noted (albeit briefly) comets with scores of 2 and often 1 as well - except in the periods 452-500, the 520s, 542-559 and 712-729. Japanese astronomers begin to record from before 634 and frequently record comets of score 2, except c.726-744. Korean records in the _Samguk Sagi_ do not often coincide with European comets, and the dates in the first millennium AD were found to be magical and fictitious. From the early eleventh century, however, Korean records are good. Byzantine records usually include comets with scores of 3 or more from AD 336 to 594/5 and from '904', _recte_ 905 to 1106. West European sources are complete in this respect from AD 1097; previously, however, they were good only in the ninth century. It is possible that in the following periods, when we are dependent almost entirely on evidence from one part of the world, some comets worthy of 3 points have been omitted:

(a) Pre-336 and c.403-413. Chinese evidence appears to be missing c.306-28 and Roman comets of c.319 and c.408-13 are not confirmed.

(b) Later fifth century. The Chinese reports are very brief (460, 461, 464/5, 467, 483, 498). Whereas the Chinese phenomenon of 467 "might have been an auroral display" (Ho 1962), the Byzantine evidence makes it clear that there was a very bright comet.

(c) Early seventh century. All sources except China are weak in the first quarter of this century.

(d) Late eighth century. Chinese and Japanese sources record no comets from between 773 and 814 (cf. e.g. Schafer 1977, 113). This period, like the period since 1910, was indeed without any bright comets as West European evidence is sufficient to rule out the possibility of comets scoring 4 or more points.

(e) 913-940. At present this is a period of confusion. If further Islamic records can be found the chronology could be clarified.

## 9.2 METEORIC SHOWERS

Meteor showers when 'the stars fall down' are sometimes useful for dating but as they often recur at the same date in several successive years they provide better evidence for the calendar date than the year. The usual dates of such showers in the Old Style calendar of the first millennium AD are shown in fig.12. It is for instance not clear whether the Leonid meteor showers were observed first in AD 900 in northern Europe or in AD 902 when displays were seen in Mediterranean countries (Schove 1972, Dall 'Olmo 1978, cf. Zhuang 1977 and Schafer 1977, 94ff. for the Far East). We include in fig.13 the dates of the Leonid meteor showers since AD 900, and show these in both the Old Style (Julian) and New Style (Gregorian) calendars, indicating at the foot of the diagram the difference in days between the two systems.

## 9.3 AURORAE

Aurorae which follow cycles of 11 years and of 4 weeks and which are sometimes seen in Europe and Asia on the same night can be useful for dating, and the chronology of visions and portents has made it possible to prepare tables of the years of sunspot maxima back to c.200 BC (Schove 1983, 14 and 208). International displays seen in both West and East occurred in e.g. 937 Feb 14, 1014 Apr 29 and 1138 Oct 6-7.

Dates of sunspot maxima are tabulated in Schove 1983a, Appendix B; dates of minima were incorrectly given in Appendix A in the first printing and are given here for the first millennium AD. Medieval visions which occurred within three years of these dates are not likely to have been auroral, but many such visions occurring 3 or 4 years afterwards are explained by the aurora borealis.

## 9.4 OTHER PHENOMENA

The chronologies of these phenomena have been taken into account in dating the eclipses of the period AD 1-1000. References to weather, tree-rings and earthquakes, and even local gossip incorporated in the sources, have helped us to indicate the probable location of the observers. Often this location is far from the country in which the extant chronicles were written, as will be exemplified in eclipses taken from Icelandic (878), Scandinavian (691), Irish (447, 512) and English (538, 540) sources.

The chronology of eclipses and comets as elaborated here has already been useful in dating other events such as epidemics (Schove 1972/4) and famines. Recent advances in dendrochronology (reviewed e.g. in Schove 1983c and 1983d and in ed. Schove and Fairbridge 1985) will soon be used by historians to check how far meteorological factors can explain historical changes. In the meantime the various dates collected for the Spectrum of Time provide a chronological framework for the sources of Western History.

YEARS OF SUNSPOT MINIMA ESTIMATED FROM AURORAL AND SUNSPOT OBSERVATIONS, AD 4 - 1501
(Dates in parentheses are uncertain.)

| | AD 4 to 1398 | 17 to 1411 | 28 to 1422 | 38 to 1434 | 49 to 1445 | 60 to 1455 | to 1468 | to 1480 | to c.1489 | to 1501 | MEAN REMAINDER |
|---|---|---|---|---|---|---|---|---|---|---|---|
| 0+ | AD 4 | 17 | 28 | 38 | 49 | 60 | ( ) | ( ) | ( ) | ( ) | c.5? |
| 100+ | ( ) | ( ) | 22 | (34) | 46 | 57 | 68 | (80) | (92) | (105) | c.1? |
| 200+ | (5) | (15) | (26) | (38) | (51) | (63) | (74) | (87) | 98 | 107 | c.6? |
| 300+ | 7 | 16 | 26 | 37 | 48 | 57 | 67 | 79 | 91 | 103 | c.4 |
| 400+ | 3 | 15 | 26 | 37 | 46 | 57 | 70 | 82 | 95 | 107 | c.3 |
| 500+ | 7 | 17 | 28 | 40 | 52 | 62 | (73) | (82) | (93) | 104 | c.6 |
| 600+ | 4 | 14 | (25) | (38) | 50 | (60) | 70 | 79 | 91 | 102 | c.4 |
| 700+ | 2 | 13 | 26? | 38 | 49 | 59 | 71 | 82 | 92 | 102 | 4 |
| 800+ | 2 | 13 | 24 | 35 | 46 | 56 | 68 | c.80 | c.91 | 100 | 2 |
| 900+ | 0 | (10) | 22 | 33 | 46? | 58 | 70 | 81 | 90 | 100 | 1 |
| 1000+ | 0 | 11 | 24 | 36 | 48 | 61 | 72 | 83 | 93 | 105 | 3 |
| 1100+ | +5 | 15 | 24 | 34 | 44 | 55 | 68 | 80 | 89 | 98 | 1½ |
| 1200+ | -2 | 9? | 22? | 35 | 46 | 56 | 69 | 81 | 92 | 102 | 1 |
| 1300+ | +2 | 12? | 22? | 34 | 47 | 58 | 68 | 78 | 87 | 98 | c.1 |
| 1400+ | -2 | 11 | 22 | 34 | 45 | 55 | 68 | 80 | 89? | 101 | c.1 |

Omitted in error from D.J. Schove, 'Sunspot Cycles', 1983, Appendix A, p.369.

The Remainder is obtained by dividing the figures on the left by 11.

# C H R O N O L O G Y

# THE FIRST CENTURY

## INTRODUCTION

## OCCASIONAL ITALIAN RECORDS

The first three centuries of the Christian era, when the Roman Empire was at its height, are a Dark Age as far as natural phenomena are concerned. In the Republican period Livy and Obsequens had reported portents from the Pontifical Annals; in the Imperial period the works of contemporary historians have seldom survived, and there are no Dynastic Histories of the kind that were compiled in China. Papyrus and numismatic evidence may, nevertheless, match some of the undocumented eclipses and comets in our catalogue.

Comet chronology until AD 600 depends mainly on the reliable Chinese records, and the Graeco-Roman dates in the standard catalogues (e.g. Barnett 1978) often need adjustment accordingly. Western dates were sometimes deliberately falsified even in later centuries.

Eclipse records of the first century are so few that elaborate attempts to adjust the usual chronology could still be discussed in the nineteenth century. The Babylonian and Alexandrian astronomers must have had records of such eclipses as 38, 35, 22 and 10 BC and have used their records to predict the small eclipses of AD 5 and 45. However, their records are lost, and our information comes mainly from Italy, the Western half of the Empire being more important than the Eastern in what is known as the Silver Age of Latin Literature.

Our knowledge of the eclipses of the period - as indeed of the emperors themselves - would be different if the eyewitness sources - such as the Universal Chronicle of Cornelius Bocchus or the Roman History of Velleius (c. AD 29) - were ever found e.g. at Herculaneum. At present we have to rely on such 'Annals' as those of Tacitus, which cover the period 51/64 but which were not written until about AD 100. The Dead Sea Scrolls found so far mention neither comets nor eclipses, and the Babylonian astronomical diaries did not continue into the AD period.

In the first four centuries of our era there are only three eclipses dated by month and day in the Julian calendar in the original sources, namely the solar eclipses of AD 45, 59 and 346; the reckoning of years in the usual AD system was not invented until about AD 525, so that the three eclipses mentioned, two of which lie in the first century, have an evident importance for chronology.

Julius Caesar's reformation of the calendar (in 45 BC on the usual scale) was misunderstood by the pontifices, leap years being at first inserted in what we call every third year, so that error gradually grew up; this error, when discovered, was gradually corrected. The whole process occupied about half a century. But it is believed that the Julian calendar in its definitive form, with a leap year every fourth year, was kept up with absolute regularity from at latest AD 8 until the Gregorian reform of AD 1582 (and of course with equal regularity until various later dates in other countries). An invaluable confirmation of this statement is provided at AD 45 and 59 by the solar eclipse records mentioned above.

A number of our eclipse identifications in the first century AD depend upon our acceptance of the year-chronology now almost universally adopted, which we believe to be reliable. We believe, for example, that the usual lists of consuls (Samuel 1972, p.256 ff.) enable consular years to be converted safely into years AD throughout the first few centuries of our era, and that the years of accession customarily quoted for individual emperors are correct.

It is true that in the past there have been unorthodox chronologies. Authors who wrote several centuries ago worked on the assumption that Augustus died, say, in AD 16 or 17, rather than in AD 14. But nowadays such variant dates are hardly credible, and we should not refer to the matter at all if it were not for the unorthodox chronologies of Seyffarth 1878 and Stockwell (below). Seyffarth's thesis was that on two occasions consules suffecti had been mistaken for ordinary consuls, so lengthening the roll of consular years by two years which did not really exist; he considered that the consuls customarily listed for the usual AD 46 and 78 were only suffects. For AD 79 and onwards, Seyffarth's year-numbers AD agree with the usual ones. But the usual AD 47 to 77 he labels AD 48 to 78, the usual AD 1 to 45 he labels AD 3 to 47, and similarly for BC dates; for example, he contends that the Julian reform of the calendar dates from Jan.1 of 43, not 45, BC, and that Caesar was killed in 42, not 44, BC.

A book on eclipses and chronology must mention the work of Seyffarth, who was trained in the tradition of literature, philology and history, and who made much play with eclipse identifications. To make a complete refutation of over 100 pages of very detailed work, supported by much learning, would be a time-consuming task which we have not attempted. But we were able to satisfy ourselves that it is unnecessary to pay extravagant attention to Seyffarth's chronology where it deviates from the normal.

For one thing, Seyffarth's mathematical treatment of eclipses is wrong; for the empirical correction which he applies to times of New Moon and Full Moon (among other things) virtually cancels out the usual 'acceleration' term in the Moon's elongation, known in Seyffarth's time and well established since. But most of all we reject Seyffarth's chronology because, as we explain in more detail under the solar eclipse of 59 April 30, we are satisfied that his alternative identification of Pliny's eclipse as that of 60 October 13 is wrong, disagreeing as it does with the month and day given in the original source. Thus we regard Seyffarth's chronology as disproved, and the usual chronology as confirmed, at a vital point.

A further variant chronology is that of J.N. Stockwell, who was an astronomer of some distinction rather than a historian. He contended that Roman events over several centuries should be placed one year earlier than is customary. See AJ 12, 1892, 121-5. This reference is given merely as an example; there is a whole controversy between Stockwell (revisionist) and W.T. Lynn (orthodox) in the AJ 10-17 (1891-6) and the Observatory 18-19 (1895-6). As far as the first century AD is concerned, Stockwell contended that, for example, Augustus died in AD 13 (not 14), and the year of the four emperors was AD 68 (not 69).

We now proceed to consider the eclipses in chronological order, as far as is possible in a century in which competing identification of the same eclipse are so frequent. We shall often refer to eclipses of the Sun and Moon in forms such as S.59 April 30 and M.71 March 4; until the year AD becomes greater than 31 we shall insert a comma after the year-number (e.g. S.5, March 28) in order to avoid confusion about which is the year-number and which the day-number.

### AD 5 S.5, March 28 PARTIAL SOLAR ECLIPSE IN S.W. MEDITERRANEAN

Solar eclipse (AD 5, March 28 on the usual chronology). Cassius Dio lv.22 (Loeb ed., Vol.VI, p.450-1) says that in the consulate of Cornelius (Cinna) and Valerius Messalla a partial eclipse of the Sun occurred. He gives no month or day.

The orthodox equivalent of the consular year is AD 5, and on that basis there is virtually no doubt that the eclipse is that of AD 5, March 28. This had a magnitude of about 0.40 at Rome around 5 p.m. which astronomers consider might be sufficient to be noticeable with a low Sun. The identification goes back at least to Calvisius 1620 (295) and Petavius 1627 (821).

The central line in the longitude of Italy was in the latitude of the Sahara.

Calvisius 1605 (464) identified the eclipse wrongly as S.5, Sept. 22, but Calvisius 1620 (295), the first posthumous edition, gave S.5, March 28.

Our date AD 5, March 28, refers to the proleptic Julian calendar (i.e. the Julian calendar continued backwards on the same principle which operated for well over fifteen centuries before AD 1582).

This eclipse is almost unique among eclipses recorded before AD 600 inasmuch as its date is definite yet its central path is too far south to be plotted on the maps of Ginzel (1899). It could just have been observed in southern Spain or North Africa, regions where mining and road-building were actively pursued in this half-century. Many first century writers, including Seneca, came from Spain, as did L. Cornelius Bocchus, author of a lost 'Universal' Chronicle of this period, a source that was perhaps known to Dio. Possibly this small eclipse had been predicted by astronomers and was noticed because it was expected.

<u>AD 14</u> M.14, Sept. 26-27 TOTAL ECLIPSE IN S.E. EUROPE

This is the usual, and evidently correct, identification of an eclipse of the Moon described in Tacitus, Annals, i, 28 (Loeb ed., tr. J. Jackson, vol.ii, 1931, p.290). The eclipse is mentioned also in later writers, especially Dio, lvii, 4 (7, 1924, p.122). The eclipse occurred fairly soon after the death of Augustus, as the Pannonian legions mutinied in the hope of extorting better pay and conditions of service from the new emperor, Tiberius, or, alternatively, deposing him. Drusus, son of Tiberius, was sent with praetorian cohorts to Pannonia, and managed to take advantage of the eclipse to quieten the troops.

Augustus died on August 19. Assuming that the year of his death was AD 14, the identification of the eclipse as that of AD 14, Sept. 27 is the only possible one. It goes back at least to Scaliger 1583 (241). The eclipse was a rather deep total one; Meeus and Mucke, like Oppolzer and Ginzel before them, find magnitude at least 1.6, with mid-eclipse occurring around 5.30 a.m. mean time at Naupontus (near modern Ljubljana in Yogoslavia), in the vicinity of which the revolt occurred. Umbral eclipse began about 3.40 a.m., totality about an hour later.

The view of totality would have been ended by dawn and moonset in any case, but Tacitus says it was ended by clouds. It may be noted that the date given in the Loeb edition (early morning of Sept. 26) is correct only if days are reckoned in the old (pre-1925) astronomical fashion, from noon to noon; in the usual civil reckoning, from midnight to midnight, the eclipse took place in the early morning of Sept. 27.

In a discussion of lunar eclipses in the Roman period P. Bicknell in 'The Witch Aglaonice and Dark Lunar Eclipses in the Second and First Centuries BC' (JBAA 93, 1983, 160-163) notes that the Moon did not normally disappear completely in the first century AD (in contrast to some earlier eclipses as in 168 and 63 BC) and he discusses whether this is related to the decreased sunspot activity of the period 20 BC/ AD 290 (cf. Schove 1983, Appendix C, p.373).

The question was raised very early whether the interval of about 39 days between the death of Augustus on Aug. 19 and the eclipse on Sept. 27 is sufficient to accommodate the events which have to occur: news of the death of Augustus travels from Nola and Rome to Pannonia; the revolt occurs; a messenger travels from Pannonia to Tiberius; Drusus and his cohorts make their way to Pannonia, presumably from Rome. Ginzel 1899 (197) considers the time to be sufficient.

17 S.17, Feb. 15 FALSE YEAR FOR ECLIPSE OF AUGUSTUS
SOLAR ECLIPSE AFTER HIS DEATH

In ancient times the Sun was often credited with an eclipse around the time of the death of a famous person. Such a statement usually amounts to no more than a stock literary compliment. Augustus seems to be no exception. The earliest such statement which we have encountered is far from contemporary; it occurs in Dio, lvi, 29 (7, p.66). Among portents before the death of Augustus he mentions that the Sun was totally eclipsed. If we assume that Augustus died on August 19 of the year AD 14 we have to go back to BC dates to find any earlier eclipse even remotely tallying. The first eclipse of the first century AD which was central anywhere near Italy occurred on Feb. 15 of AD 17. The track in Ginzel 1899 passes from Libya via Greece to the Danube delta, and the track thus crossed the Mediterranean. But this eclipse (whose consideration goes back at least to Paul of Middelburg who died in 1534 as bishop of Fossombrone) was certainly well after the death of Augustus.

Eusebius (ed. Schoene, II, 1866, 146) does not claim totality, or even that the eclipse occurred _before_ the death of Augustus. Eusebius-Jerome (ibid. 147, or ed. Fotheringham, 1923, 253) has "Defectio solis facta et Augustus ... moritur". But the relaxation of conditions scarcely helps; in the years AD 8 to 16 inclusive, one searches in vain for any solar eclipse of relevance to the Romans. But Struyck 1740 (98) mentions the (North American) solar eclipse of AD 13, April 28 (29 is a mistake) as an eclipse which may have been computed, though not observed.

We conclude that the _solar_ eclipse of AD 17 is the probable basis for the reports of Dio and Eusebius, but its identification score in our table is assessed as 6 points only, as there is possible confusion with the observed lunar eclipse of AD 14.

_29_ S.29, Nov. 24 'CRUCIFIXION' SOLAR ECLIPSE IN ASIA MINOR

It would be out of proportion to treat extensively here of a subject on which so much has been written without reaching any final conclusion. We limit ourselves to giving a brief outline and a few of the main references.

It is well understood that any daytime darkness at the Crucifixion itself (Matt. xxvii, 45; Mark xv, 33; Luke xxiii, 44-45) must have been meteorological (e.g. cloud, fog, dust-storm), since the Passover occurs near Full Moon, at which only a _lunar_ eclipse can occur.

The only historical mention of an actual eclipse roughly around the epoch of the Crucifixion dates from about a century after the event, in the 'Olympiads' of Phlegon, of Tralles in Caria, a freedman of Hadrian (117-138). This work is extant only in fragments, but the relevant passage is quoted through Eusebius (Schoene, II, 148; see Eusebius-Jerome, ed. Fotheringham, 256) in Syncellus (Bonn ed., 614), and appears rather similarly in the Chronicon Paschale (Bonn ed., I, 412). It speaks of a great solar eclipse at the sixth hour of the day, which almost all investigators have identified as that of AD 29, Nov. 24. This was total or nearly so in Bithynia about 11 a.m. Since Phlegon wrote primarily about Asia Minor, and no other great eclipse of the Sun occurred in the Near East around the time in question, the identification is practically certain, and the passage evidently preserves the memory of genuine observation. Phlegon is quoted as having put the eclipse in the fourth year of the 202nd Olympiad, and this would be a useful piece of information if it could be relied upon, but '202nd Olympiad' is more plausible than 'fourth year'. Reliance on Olympic reckoning is always risky if the exact year is needed, but on the most common calculation (variants exist) the 202nd Olympiad means AD 29-33, and its fourth year means AD 32-33, when there was no great solar eclipse in the Near East. On the eclipse of Phlegon, see, for example, Ginzel 1882 (684, 691), 1899 (198-200), Boll 1909 (2360), Fotheringham (MN 81, 1920-1, 112), P.V. Neugebauer 1930 (B27), Kudlek & Mickler 1971 and Newton 1979 (42, 208).

Fotheringham attributed the first identification to Kepler, Eclogae Chronicae, 1615, 126.

Unfortunately, although the record of the eclipse of Phlegon is not itself particularly weak, inasmuch as the identification is virtually certain, the eclipse is too loosely set among other events of the time to be of much use chronologically. With regard to the Crucifixion, the only connection is a vague 'at which time'. While this obviously agrees well enough with dates AD 29, March 25 and AD 30, April 7, the connection is so indefinite that the date of the eclipse (AD 29, Nov. 24) cannot really be said to conflict crucially even with AD 33, April 3. Most investigations into the date of the Crucifixion arrive at one or other of the three spring-dates mentioned, all of which were Fridays. However, the date needs also to be Nisan 14, or perhaps 15, in the Jewish calendar (so as to be at the beginning of Passover), and AD 29, March 25 does not satisfy this requirement; for example, see Table 140 in J. Finegan, Handbook of Biblical Chronology, Princeton, 1964. This leaves only AD 30, April 7 and AD 33, April 3, and of these two Finegan's discussion leads him to some preference for the first. This would partly agree with Phlegon, as both the eclipse and the Crucifixion would at least fall within the same year in a number of Eastern calendars which begin the year in summer or autumn (e.g. Greek, Byzantine, Syrian, as well as Jewish civil).

Discussions of the date of the Crucifixion abound. Valuable references may be found in Finegan (loc. cit., p.285); see also Ginzel 1899 (198-200). Attention may be drawn to a number of papers published in 1930-4 in the Astronomische Nachrichten, vols. 240-2, 252, 254. Further references may be found in the Astronomischer Jahresbericht, 16, 33 (1914, Emanuelli), 30, 17 (1928, Schoch). 49, 80 (1948-9, Boneff), and 51, 94 (1951, Hennig). For views on how a darkening of the Sun came to be mentioned, see J.F.A. Sawyer, J. Theol. Stud. (N.S.) 23, 1972, 124-8, and, in lighter vein, W.H. White, J. Brit. Astro. Assn., 56, 1946, 72-3. Struyck 1740 (102) mentions the possibility that the eclipse of Phlegon may have been S.71 March 20, the Olympiad being wrong by ten (212 instead of 202). He refers to controversy in 1732-4 between W. Whiston and A.A. Sykes. See Addenda for supposed Lunar Eclipse of AD 33 (Nature

45 S.45 Aug. 1 PREDICTED 'ECLIPSE' OF CLAUDIUS

Dio, lx, 26 (Loeb ed., tr. E. Cary, vol.vii, 1924, p.433), speaking of the emperor Claudius (AD 41-54 on the usual chronology), says:

"Since there was to be an eclipse of the sun on his birthday, he feared that there might be some disturbance in consequence, inasmuch as some other portents had already occurred; he therefore issued a proclamation in which he stated not only the fact that there was to be an eclipse, and when, and for how long, but also the reasons for which this was bound to happen."

Slightly earlier (lx. 25; Loeb p.429) Dio mentions the consuls as Marcus Vinicius (second time) and Statilius Corvinus; these are the consuls for AD 45 on the usual chronology. The narrative reads as though the eclipse itself occurred under these consuls, especially since, after mentioning the eclipse and discoursing on the causes of eclipses, Dio immediately says that when that year expired the consuls who succeeded were Valerius Asiaticus (second time) and Marcus Silanus; these were the consuls for AD 46 on the usual chronology. However, the link between eclipse and consuls could conceivably be questioned.

But considerably earlier (lx. 5; Loeb p.379) and in a different context, Dio happens to have given a much more compelling piece of information: Claudius was born on the first day of August. This vital clue may be found also in the earlier author Suetonius (Claudius, beginning of Chap. 2).

It seems beyond doubt that the birthday of Claudius did fall on August 1. It is true that his birth occurred in 10 BC (on the usual chronology), so that in this connection August 1 refers to the imperfectly operated Julian calendar of the time, and not to the mathematically regular proleptic Julian calendar now used by astronomers. But this is irrelevant; the birthday of Claudius would be celebrated on August 1, and by the time of his reign the regular Julian calendar was well established. Thus there is only one possible identification of the eclipse, namely 45 Aug. 1. The band of totality crossed Africa, with Oppolzer's approximate noon point falling in the Red Sea. The maximum phase at Rome occurred about (probably somewhat after) 9 a.m. local apparent time. The magnitude at Rome was not very great (about 0.3), but, as Ginzel 1899 (201) and Newton 1979 (209) pointed out, we do not know whether the eclipse was seen at Rome; we are dealing with a prediction rather than an observation. See also C.J. Westland, Nature 151, 1943, 111. Neugebauer 1975 (2, 666) discusses how far prediction was really possible.

Dio's appears to be the oldest extant reference to the eclipse.

Whether Tacitus mentioned it we do not know, because the time of the eclipse falls in one of the lost books of his Annals.

The identification given goes back at least to Scaliger 1583 (241); Petavius and other noted chronologists also accepted it.

Ginzel 1899 (201), with his brief "Seyffarth (457) 47 June 25", appears to suggest that Seyffarth favoured an unorthodox date, but this is not so. Seyffarth accepted the inevitable S.45 Aug. 1; his unorthodoxy consists in shifting certain consuls and emperors by a year or two relative to this fixed date. The eclipse of 47 June 25-26, which was lunar (not solar), is mentioned by Seyffarth in a different context.

46 M.46 Jul. 6 ECLIPSE OF CLAUDIUS AND VOLCANIC ERUPTION

Sextus Aurelius Victor (Caesars, Chap. 4, Claudius, 12), writing three centuries after the event, says "in the Aegean Sea a large island suddenly appeared, during a night on which an eclipse of the Moon had occurred". The passage refers to the sixth year of Claudius (AD 46 on the usual chronology) and the 800th year of Rome (usually AD 47). The seeming inconsistency has engaged chronologists at least since Scaliger 1598 (437-8), 1629 (466-7), but we shall rely below on contemporary writers for the best indication of the year.

We know of no other ancient statement connecting the eclipse with the island, but there are a number which mention islands only; these differ in detail, but do at least agree in mentioning island formation in the Thera group (modern Thira or Santorin, 25°.4E., 36°.4N.). All which specify a year indicate one in the range AD 44 to 49, except that some modern editors interpret a corrupt passage in Pliny (see below) as referring to AD 19 rather than AD 46. Useful maps of the Thera group are given in F. Hiller von Gaertringen, Thera, Vol. I, Berlin, 1899, which includes further references (see especially pages 63-4 and 39). The eruption of Thera 1650-1400 BC is now much discussed as linked with the decline of the Middle Minoan civilisation in Crete.

The only contemporary accounts known to us which mention a year are by Seneca and Pliny. Seneca (Nat. Quaest. ii, 6 and vi, 21) alludes to the appearance of an island near Thera and Therasia in the consulate of Valerius Asiaticus (AD 46 on the usual chronology).

Pliny (Nat. Hist. ii, 87 or 89, §202 and iv, 12, §70) apparently refers to the same event, but unfortunately is badly afflicted with the variant readings (see, for example, the Teubner edition). One has to decide whether the consuls given in the manuscripts are those of AD 19 or 46, listed in Bickerman 1968 (184) or Samuel 1972 (256ff.).
AD 19 (AUC 772) M. Junius Silanus Torquatus & L. Norbanus Balbus,
AD 46 (AUC 799) D. Valerius Asiaticus II & M. Junius Silanus.
To us the manuscript readings, as far as they are quoted in the Teubner edition, seem to adumbrate the consuls for AD 46 at least as clearly as those for AD 19, and we shall assume AD 46, disregarding the AD 19 mentioned in, for example, the Loeb edition (Vol. I, p.332).
Pliny gives the further vital piece of information, fortunately untroubled by any variant reading affecting sense: he gives the date of the birth of the island as viii idus Julias, or Julii (i.e. July 8).

Of later authors who give some indication of year in mentioning the island formation in the first century AD (there were other similar upheavals during historic time), we may refer to three, interpreting consuls on the usual chronology:

Dio (lx. 29, Xiphilinus epitome; Loeb ed., Vol. VIII, pp.2-5), consuls of AD 47, and year of Rome 800;

Eusebius-Jerome, Chron. Canones (ed. J.K. Fotheringham, 1923, p.262, or ed. A. Schoene, Vol. II, 1866, 1967, p.153):
(a) in Fotheringham, Olympic year 206, 1 (normally AD 45-46), and 5th year of Claudius (AD 45 or perhaps 46);
(b) in Schoene, Ol. 206, 4 (normally, though not in Schoene, AD 48-49), and 8th year of Claudius (AD 48 or perhaps 49).
In Schoene, the Armenian version (p.152) seems to support (a), and the Syrian epitome (p.211) is arranged to agree with (b).

Cassiodorus, Chron. (Chr. Min. 2 = AA 11, 1894, 137), consuls of AD 45.

There was no lunar eclipse in AD 45. The lunar eclipses of AD 46 to 48 given in Oppolzer are:

| | | Mid-eclipse (UT) | Magnitude | Half-duration Partial | Half-duration Total |
|---|---|---|---|---|---|
| (1) | 46 Jan. 11 | 7.28 a.m. | 0.69 | $87^m$ | - |
| (2) | 46 July 6 | 7.11 p.m. | 0.65 | 85 | - |
| (3) | 46 Dec. 31 | 9.5 p.m. | 1.82 | 112 | $52^m$ |
| (4) | 47 June 26 | 3.5 a.m. | 1.62 | 110 | 49 |
| (5) | 47 Dec. 21 | 4.18 a.m. | 0.45 | 73 | - |
| (6) | 48 June 14 | 5.38 p.m. | 0.29 | 60 | - |

The most favoured identification is (3). Since this total eclipse happened on the New Year's night of AD 46-47, it neatly justifies both AD 46 (especially Seneca) and AD 47 (Dio). The identification goes back at least to Calvisius 1620 (314) and Petavius 1627 (II, 307), and is adopted in Ginzel 1899 (201) and Boll 1909 (2360).

But we question the correctness of the identification; we think (2) is more probable. The partial eclipse in the evening of 46 July 6 would be distinctly noticeable, with nearly two-thirds of the Moon's diameter darkened about 8.53 p.m. Thera mean time, more than 1½ hours after sunset (7.15 p.m.). Pliny's date, July 8, refers to the island formation, not the eclipse (mentioned only by Victor), and it is known that in eruptions an island may take more than 24 hours to be formed.

Of the other four eclipses, (1) was invisible at Thera, and only slightly visible at Rome; it is not known where the eclipse was observed (if it was observed), though Rome is sometimes assumed. Seyffarth 1878 seems divided in his opinion; on p 417 he gives (6) for the Aurelius Victor passage, but AD 48 is too far from Seneca's date.

It may be added that the small lunar eclipse of AD 19 July 5 was invisible at Rome and Thera. Thus only the eclipse of 46 July 6 fits Pliny's mention of July.

49 S.49 May 20 NILE DELTA SOLAR ECLIPSE (UNRECORDED)

The approximate track of annularity illustrated in Oppolzer starts at sunrise in the Atlantic, crosses Africa to Libya, and passes over the Mediterranean to Central Asia Minor and a noon point near the Caucasus region, then via the Caspian into NE Asia. The accurate track of Ginzel 1899 (Map X), whose maps are consistent with modern calculations (cf. Meeus and Mucke 1983), shows the Near Eastern part of the band of annularity as running somewhat differently, over the Nile delta, Western Syria, Northern Euphrates, Caucasus and Caspian.

59 S.59 April 30 SOLAR ECLIPSE OF NERO IN ITALY AND ARMENIA

This eclipse in the reign of Nero is chronologically important as the second of the two eclipses in the first century AD dated with month and day in an original source, in this case Pliny. In Nat. Hist., ii, 70 or 72, §180 (Loeb ed., Vol. I, 1944, pp.312-3), he says: "An eclipse of the Sun a few years ago, in the consulate of Vipstanus and Fonteius, on the day before the calends of May, was observed in Campania between the 7th and 8th hours of the day, and by Corbulo campaigning in Armenia between the 10th and 11th hours of the day". Campania is the region around Naples (14°.3E., 40°.8N.). Ginzel 1899 (202) calculates the Armenian eclipse for Artaxata (44°.54E., 39°.90N.), in the east of Armenia, but Erzurum (41°.5E., 39°.9N.) should be kept in mind, see below. On the usual chronology, the consulate of Vipstanus and Fonteius means AD 59.

Tacitus, Annals, xiv, 12 (Loeb ed., tr. J. Jackson, Vol.IV, 1937, pp.126-7) mentions what is probably the Italian end of Pliny's eclipse, namely a sudden obscuration of the Sun connected with (apparently after) the assassination of Agrippina during the festival of Minerva (March 19-23).

Tacitus, Annals, xiii, 41 (Loeb ed., Vol.IV, pp.70-73) also refers to darkness at Artaxata, but since this is placed in the previous year (consuls Nero III and Valerius Messalla) and does not read at all like an eclipse, it could be considered as a misdated reference to the Armenian end of Pliny's eclipse. Indeed, Jackson says that April 30 would be too early in the season; the legions could not possibly be at Artaxata, and were probably still at Erzurum.

Dio (lxi, 16, Xiphilinus epitome; Loeb ed., Vol.VIII, 1925, pp.72-3) mentions the eclipse without month or day: "in the midst of the sacrifices that were offered in Agrippina's honour in pursuance of a decree, the Sun suffered a total eclipse and the stars could be seen". Agrippina was assassinated in the neighbourhood of Baiae (west of Naples), but the reference to totality and the appearance of stars can hardly be true of Rome or Campania, and is probably a case of artistic licence.

Martianus Capella, De Nupt., Book vi (ed. A. Dick, 1925, pp.294-5) says, probably abbreviating Pliny injudiciously, that the eclipse was seen at the 7th hour of the day in Campania and at the 11th hour of the day in Armenia. His date is wrong (for xi. Kal. Mai. read ii. Kal. Mai.).

The eclipse was only partial at both places of observation, which were both to the north of the track of totality; this runs from South America to a noon point off Morocco, and then to Tunisia and Syria and through Persia. See Ginzel 1899, Kudlek & Mickler 1971, and Newton 1979 (187, 210 and 469).

The eclipse was, however, moderately large in both Campania and Armenia; calculators generally find maximum magnitude about 0.84 in both regions. One might gather from Pliny's account that the difference in local times of mid-eclipse in the two regions would be about 3 hours (doubtless seasonal hours, each equal to about 1.15 equinoctial hours).

In fact, calculators find a difference of about 2 seasonal hours (2.3 equal hours). If Pliny means that mid-eclipse in Campania occurred about 1½ seasonal hours after noon, his estimate seems about half an hour too early; in Armenia, 4½ seasonal hours after noon seems about half an hour too late. However, Pliny is not giving fractions of hours, and there is no doubt about the identification of Pliny's eclipse.

One is bound to believe that Pliny really meant April 30, and in that case the eclipse as recorded in Pliny occurred on 59 April 30. It is only a short step to the conclusion that the assassination of Agrippina also fell in AD 59, and we see no reason to doubt it.

62 M.62 March 13-14 ECLIPSE OF HERO OF ALEXANDRIA

The lunar eclipse of Hero (AD 62 March 13-14). We have long known that Hero of Alexandria lived sometime between -200 and +300 (he mentions Archimedes and is mentioned by Pappus). In Chap.35 of his Dioptra (Opera, ed. H. Schöne, 3, 1903), Hero uses as a numerical example a lunar eclipse which occurred ten days before the spring equinox, at the fifth hour of the night at Alexandria. O. Neugebauer 1938, 1939, believing that Hero probably used an eclipse he had seen, investigated lunar eclipses between -200 and +300, and found only a single one, namely M.62 March 13-14, which tallied. Hence he tentatively suggested that Hero lived before Ptolemy (against the conjecture of certain earlier scholars). Neugebauer referred also to a paper of 1923 by A. Rome. The eclipse was further considered in 1950 by A.G. Drachmann, who agreed with Neugebauer in thinking that Hero lived before Ptolemy. The references are:

A. Rome, Ann. de la Soc. Sci. de Bruxelles, 42 (1923), Mém., 234-258.

O. Neugebauer, K. Danske Vidensk. Selskab, Hist.-filol. Medd., 26(2), 1938 and 26(7), 1939. cf. also Bull. Amer. Math. Soc. 54, 1948 (1033).

A.G. Drachmann, Centaurus, 1, 117-131, Copenhagen 1950.

The identification M.62 March 13-14 is repeated, and put in modern context, in Neugebauer 1975 (2, 846).

67 (not 64) S.67 May 31 FIRST SOLAR 'ECLIPSE' OF APOLLONIUS

Two solar occurrences (the other about AD 95) are mentioned in the life of Apollonius of Tyana by Philostratus, a work whose status is uncertain. It relates to real people, but some critics believe that not all its incidents are real, so that it has only the kind of truth possessed by a historical novel; Boll 1909 (2360) refers to it as "Philostrats Apollonius roman". Other critics appear undeterred by the work's suspicious leaning towards the supernatural. The question mark which hangs over the work makes it not worth while to consider the solar occurrences at length.

The first occurrence, which is explicitly described as an eclipse, is mentioned in Book iv, Chap.43 (in the Loeb ed., tr. Conybeare, Vol.I, 1912, pp.450-3). Shortly before the eclipse, Apollonius met the consul Telesinus, who held office in AD 66.

Seyffarth 1878 (458) identified the eclipse as S.67 May 31. Ginzel 1899 (202), though sceptical about genuineness, concurred. He found magnitude 0.77 at Rome. The track of annularity in both Ginzel and Oppolzer passes over Spain and ends in or near Egypt.

The Loeb edition indexes the eclipse as "? AD 64". This refers to S.64 August 1, whose track of totality runs too far north (over Central Scandinavia).

Clearly the eclipse of 67 fits better than the eclipse of 64 and we have allocated 4 points for identification.

For the second solar occurrence, see later, under AD 95 approx.

c.68-69

## M.c.68-69 DIO'S ECLIPSES OF VITELLIUS

Lunar eclipses about AD 68-69. Various eclipses, some difficult to identify satisfactorily, are mentioned in Cassius Dio lxiv/lxv. 8 and 11, now extant only in the epitomes of Xiphilinus (11th century) and Zonaras (11th-12th cent.); in the Loeb edition of Dio, Vol.VIII, 1925, pp.232-3 and 236-9. We shall denote the two passages by (A) and (B). Both relate to the brief 'reign' of Vitellius, mid-April to December in the 'year of the four emperors', AD 69.

(A), giving bad omens for Vitellius, says "The Moon, contrary to precedent, appeared to suffer two eclipses, being obscured on the fourth and on the seventh day (τεταρταία καί ἑβδομαία)".

(B), referring apparently to the confused skirmishing which led up to the second battle of Cremona, says "The great confusion ... in the camp of Vitellius was increased that night by an eclipse of the Moon. It was not so much its being obscured ... as the fact that it appeared both blood coloured and black and gave out still other terrifying colours".

(A) occurs, in different contexts, in Xiph. and Zon.; (B) in Xiph. only. The contribution of Xiphilinus is well seen in Boissevain's edition of Dio (Vol.III, 1955 reprint, pp.122, 125, 618, 619). This, like the Loeb edition, gives extracts from Zonaras, of whom complete editions also exist. For example, (A) is given, in Zonaras xi. 16, in the Bonn edition (Vol.II, by Pinder, 1844, p.489) and the Teubner edition (Vol.III, by Dindorf, 1870, p.48).

It is uncertain whether (A) and (B) together refer to two or three eclipses. In the narrative, (A) seems to occur within a month or two of July, and (B), just before the second battle of Cremona, is believed to refer to the second half of October. Thus it is unnecessary, and indeed barely permissible, that (B) should refer to the second of the two eclipses mentioned in (A).

The eclipses available, which are the only lunar eclipses of AD 67-70 visible in Italy, may be taken from Oppolzer, with their dates, mid-eclipse times (in UT, i.e. Greenwich mean time reckoned from midnight), maximum magnitudes, and half-durations:

| | | UT | Mag. | Half-duration | |
|---|---|---|---|---|---|
| | | | | Partial | Total |
| (1) | M.67 Nov. 9 (Mon.) | $20^{h}01^{m}$ | 0.21 | $51^{m}$ | - |
| (2) | M.68 May 5-6 (Th.-Fri.) | 23 32 | 1.42 | 109 | $44^{m}$ |
| (3) | M.68 Oct. 29 (Sat.) | 6 03 | 1.59 | 110 | 46 |
| (4) | M.69 April 25 (Tu.) | 4 39 | 0.87 | 94 | - |
| (5) | M.69 Oct. 18 (Wed.) | 21 04 | 0.92 | 96 | - |

Seyffarth 1878 (460) makes (A) and (B) together relate to (2) and (3), presumably meaning that (A) relates to (2) and (3), and (B) to (3). The mention of blood-red and black in (B) suggests a total eclipse, so (3) fits; but the two-eclipse solution is unattractive. The corresponding three-eclipse solution would be that (A) refers to (1) and (2), and (B) to (3). But either solution is scarcely available except to supporters of AD 68 for the year of the four emperors; among astronomers, J.N. Stockwell leans towards this (AJ 12, 1892, 122; Observatory, 18, 1895, 57). Seyffarth himself supported AD 70, and was able to place the eclipses in AD 68 only by assuming a most implausible confusion of consuls. Before leaving Seyffarth, note that he appears to rely on Dio as requiring one of the eclipses to have occurred "on Oct. 29", in other words, to be (3). We have not noticed this in modern editions of Dio, but, genuine requirement or not, it fits a suggested identification made below.

In Calvisius 1620 (327-8), (A) is identified with (4) and (5), and (B) with (5). This does not make (B) refer to a total eclipse; but there is no way of making it do so, if (B) is intended to refer, as it almost certainly is, to AD 69. We have no record of any calculation making (5) total. Dio's description is perhaps literary rather than objective; alternatively, the colour effects were of meteorological origin.

As a three-eclipse solution seems preferable, one wonders whether (A) may not refer to (3) and (4), and (B) to (5). In the usual chronology, (3) fell in the reign of Galba, but is not too far back to be linked with (4); both (3) and (4) would be cut off by sunrise.

It is difficult to explain why (A) says "on the fourth and on the seventh day". Days of the Moon, days of our 7-day week, and days of the Roman 8-day week, all fail. Yet the numerals, being spelled out, can hardly be written off as copying errors. One notes that (3) and (4) occurred respectively on 4. Kal. Nov. and 7. Kal. Mai., and the numerals 4 and 7 make one wonder whether the injudicious epitomizing may not be to blame. If this seems far-fetched, it is a measure of the difficulty involved in making complete sense of the epitomized Dio. Ginzel 1899 (202) is sceptical about the reliability of the record. He gives further references, as also does Boll 1909 (2360).

It may be mentioned that Struyck 1740 (99) had the idea of Roman dates badly translated into Greek, but his explanation of "fourth and seventh" involved one solar and one lunar eclipse, namely S.69 April 10

(4.id.Ap.), invisible at Rome, and M.69 April 25 (7. Kal. Mai.).

On (A), it is difficult to say more than that lunar eclipses did occur around the relevant times, but that Dio, his epitomists, and the scribes have somehow, between them, transmitted to us little useful information about them. But the eclipses are recorded as breaking the rules, and genuine eclipses break no rules.

On (B), the identification of the eclipse as (5) seems to have obtained at least a fair measure of acceptance in the literature.

It may be added that Tacitus (Hist. iii. 23; Loeb ed., Vol.I, 1925, p.368), in describing the night battle at Cremona, mentions no eclipse, but does say that the Moon, "rising in the middle of the night" behind the Flavians, favoured them, compared with the Vitellians. As the Moon, to rise at night, would be beyond the full, this fits reasonably well; the Moon was full, and eclipsed, on 69 Oct. 18 and the battle is sometimes dated 69 Oct. 24-25. There is no reason why Dio and Tacitus should not both be correct; eclipse before the main battle, late evening moonrise during it.

We conclude that Dio's references may probably refer to the eclipses of 68 Oct. 29, 69 Apr. 25 and Oct. 18.

71 M.71 March 4 and S.71 March 20 PLINY'S ECLIPSE PAIR

For a year (discussed below) which is unclear, but certainly during the reign of Vespasian, the elder Pliny (Nat. Hist., Book ii, §57; Loeb ed., Vol.I, p.206) says, speaking of eclipses of the Sun and Moon, "For it has happened even in our time that both stars were missing within 15 days". This must be taken to refer primarily to the Rome-Naples region.

There is very little doubt about the identification. All calculators find the partial lunar eclipse of 71 March 4 to have had a magnitude of not less than 0.4 in mid-evening at Rome, and the annular total solar eclipse of 71 March 20 (whose track of centrality, in Ginzel 1899, Libya, The Peloponnese, and the Danube delta) of not less than about 0.8 in mid-morning. Granted clear weather, both eclipses would be plainly visible.

The difficulty is to know what Pliny really wrote about the year; most versions make him refer in some way to Vespasian and his son Titus. For present purposes, it gives a sufficient idea of the corrupt passage, with the chief variants, to quote it as "imperatoribus Vespasianis patre III (or IV) filio (or filio II, or filio iterum) consulibus (or eos, or cos, or eius)". A lunar eclipse often occurs a fortnight before or after a solar eclipse.

But the consuls in the Stockwell, usual, and Seyffarth chronologies, respectively, belong to years AD as follows (the brackets enclosing the Stockwell numbers are explained later):

| | Stockwell | Usual | Seyffarth |
|---|---|---|---|
| Vespasian II & Titus | (69) | 70 | 71 |
| Vespasian III & Nerva | (70) | 71 | 72 |
| Vespasian IV & Titus II | (71) | 72 | 73 |

Thus Vespasian's colleague in his third consulate (usual AD 71) was not Titus, but Nerva.

Ginzel 1899 (205) responds to the difficulty that Nerva was not the son of Vespasian by supposing, with Zech, that Pliny referred merely to "patre III", and that mention of his son is a later interpolation.

Those who wish to try to form an opinion of their own may start by consulting, among texts, the Teubner edition (Vol.I, Mayhoff, 1906, p.145); and, among astronomers and chronologists, apart from Ginzel 1899 and others quoted above: Calvisius 1620 (330) or 1650 (457); Seyffarth 1878 (460); Stockwell, AJ 10, 1891, 187; Lynn, Obs. 14, 1891, 235 and 19, 1896, 204, 331. One does not need to accept Seyffarth's chronology to enjoy his spirited account of the textual difficulties. Stockwell, rightly or wrongly, uses regnal years rather than consular years at this point; he takes the third year of Vespasian to commence on 70 July 1, instead of the usual 71 July 1. But the consular years (commencing on January 1) which we have given in brackets under Stockwell in the table above can scarcely misrepresent him; his whole thesis is that the usual year-numbers need reducing by unity. Pliny's eclipse was discussed again by Newton 1979 (cf. 215-216) who considered it insufficiently certain in place or time to be used scientifically.

Here we have accepted the year was AD 71 and allocated 8 points for both eclipses.

It may be added that various astronomers refer to S.71 March 20 in its other capacity as one of the candidates for the chronologically vague 'eclipse of Plutarch'. So Ginzel 1882, 1884 (539); Stockwell, Obs. 19, 1896, 330; Lynn, ibid., 204, 331; Fotheringham, MN 69, 1908, 29 and 81, 1920, 110, P.V. Neugebauer 1930 (B27); Newton 1970 (114); Muller & Stephenson 1975 (475, 522); and Muller 1975 (8. 35f, A 3.3). A similar remark applies to S.75 Jan. 5 (the acceptable solar eclipse mentioned above as rejected in the Pliny context by Hofmann and Ginzel because paired with a rather ill-fitting lunar eclipse). For the 'eclipse of Plutarch', see below.

5 or
3

## POSSIBLE SOLAR ECLIPSE OF PLUTARCH

Eclipse of Plutarch (probably AD 75 or 83). This eclipse, much discussed over the centuries, still lacks definitive identification. Plutarch wrote in Greek, but his works are usually known by their Latinized titles. One of his essays, "De facie quae in orbe lunae apparet", on the Face in the Moon, (Loeb ed., Moralia, Vol.XII, 1957, pp.1-223, ed. H. Cherniss) contains (see pp.116, 120, also 8-12) a possibly 'literary' (fictional) dialogue (between real persons) in which a "recent" eclipse of the Sun "beginning just after noonday" is mentioned, apparently as total (many stars were seen). No place or time is given. Unfortunately Plutarch's dates are only roughly known (say about AD 46 to about AD 120). His native region was Chaeronea and Delphi (not far apart), but he was much travelled (he visited Alexandria) and he resided for a considerable time in Rome. He put the eclipse passage in the mouth of one Lucius, of Carthage.

Formerly it was usually supposed that the report referred to either Chaeronea or Delphi, but Rome was always a possibility, and was revived by F.H. Sandbach after detailed examination of what is known about the lives of the various individuals mentioned by Plutarch in connection with the symposium (Class. Quart. 23, 1929, 15). However, Newton 1970 (117) says "I do not take the passage from Plutarch to be a description of a specific eclipse. If it be one, it is unidentifiable both in time and place".

Yet mention of an eclipse may have arisen from some eclipse having been seen by someone, e.g. by Plutarch (preferably) or Lucius (possibly); indeed, the participants in the symposium are reminded of the eclipse, as though it had been widely noticed.

It is tempting to try to identify this underlying eclipse, but even in the post-Oppolzer period identifications have ranged from AD 59 to AD 118. However, the solar eclipses of

(a) 71 March 20, (b) 75 Jan. 5, (c) 83 Dec. 27

are now most favoured, and here we have allocated 3 points to each, noting that Newton 1979, 214, does not see any reason to believe that the observation was real.

We shall discuss these three eclipses briefly, but perhaps sufficiently for the general reader. We shall then, for the sake of anyone who wishes to investigate further, run through the various modern identifications in chronological order, and mention a few older writings.

First consider the tracks.

(a) S.71 March 20 was annular-total. The main advantage of this identification is that an unusually narrow band of centrality happens to go through, or very near, Chaeronea. The probability of this happening by chance is small. The argument has recently impressed Muller & Stephenson 1975 and Muller 1975, as it previously impressed Ginzel and others.

(b) S.75 Jan. 5 was total. Its main attraction is that quite a narrow band of totality went through, or near, Carthage, where (as Muller 1975 notices) it might well have been seen by Lucius. The track also went through Southern Italy, where some of the Romans might possibly have seen it. The eclipse was not total at either Rome or the Chaeronea-Delphi district; but it was the only first-century eclipse whose central track passed over Italy at all.

(c) S.83 Dec. 27. The only reason why this possibility still remains is that the eclipse was total at or near Alexandria, where it might perhaps have been seen by Plutarch. This identification is virtually useless if the place of observation was Carthage, Italy or Greece.

These considerations of track point primarily to (a), but time of day also needs consideration. In that respect, (a) is not very satisfactory; it did not "begin just after noonday". It was total (or nearly so) at Chaeronea around 11 a.m. (b) is more satisfactory, it was total in Southern Italy around 3 p.m., and earlier at Carthage. (c) is also moderately satisfactory; it was total near Alexandria around 3 p.m. Much evidently depends on how seriously one takes Plutarch's rough indication of time. There is no really compelling identification. But the modern tendency to accept either AD 71 or AD 75 means that astronomy does give some support, however modest, to the view that the "De Facie" was written in Plutarch's prime rather than in his later years.

The dates most considered in modern times have been:

(1) 71 March 20 Already discussed under Pliny's eclipse pair. Adopted as the eclipse of Plutarch in Ginzel 1882 (706), 1884 (539), see also 1899 (202, 205). Also suggested by Lynn, Obs. 19, 1896, 205, and consequently calculated by Stockwell, ibid. 330. Adopted in Fotheringham, MN 81, 1920, 112, see also 69, 1908, 29, and in P.V. Neugebauer 1930 (B23, Schoch; B27). Recently treated sympathetically, though excluded from final analysis, in Muller & Stephenson 1975 (475, 522); actually computed in Muller 1975 (8.35f., A3.3), and treated somewhat more favourably than (4) below (75 Jan. 5). See also W.H. White, JBAA, 56, 1946, 73.

(2) 75 January 5 Ginzel 1882 (706), 1884 (539), thinking of Chaeronea or Delphi, regarded this with disfavour. But Sandbach 1929 (loc.cit.), thinking of Rome, considered this identification possible, as did Cherniss 1957 (loc.cit., pp.8-12). A further possibility mentioned in Muller 1975 (8.36), is that this eclipse was seen by Lucius at Carthage. Ginzel 1899 (Map X) shows the band of totality passing during the afternoon from Tunisia via Southern Italy to a sunset point in Romania.

(3) 83 December 27 Calculated but rejected by Stockwell (AJ 11, 1891, 57); considered as possible by Sandbach and Cherniss, thinking of Alexandria, which lies just inside the band of totality shown in Ginzel 1899 (Map X).

Newton 1979 (214) does not see any reason to believe that the observation was real and, in our table, we have shared 6 points between the alternatives of AD 71, 75 and 83.

80 S.80 March 10 ANNULAR ECLIPSE NOT RECORDED

The track of annularity for starting at sunrise in NW Africa, traversing the Middle East, to a noon point in Central Asia, and ending at sunset in Siberia. The part shown in Ginzel 1899 (Map X) runs from Algeria via Libya, just north of the Nile delta, Syria and Mesopotamia to the Caspian. The true path thus traversed the south-eastern parts of the Roman empire during the morning, but we have encountered no Western record of this eclipse, which occurred during the short reign of Titus (AD 79-81).

88 S.88 Oct. 3 ANNULAR ECLIPSE NOT RECORDED

Annular in Oppolzer from the Atlantic to the Indian Ocean, with noon point in the Western Sudan. The path in Ginzel 1899 crosses Southern Spain and Algeria during the morning. We have encountered no Western record.

95? INVISIBLE 'SECOND' APOLLONIAN 'ECLIPSE'

Second solar occurrence of Philostratus (about AD 95). Besides the doubtful reference to the solar eclipse of (probably) AD 67, a second solar occurrence is mentioned in the Life of Apollonius of Tyana by Philostratus, a work whose possibly fictional nature has already been pointed out. The reference occurs in Book viii, Chap.23, also 25 (Loeb ed., tr. Conybeare, Vol.II, 1912, pp.386-9, also 388-391). Chap.23 says that, while Apollonius was in Greece, "The following remarkable portent overspread the heavens. The orb of the sun was surrounded by a wreath which resembled a rainbow, but dimmed the sunlight". Chap.25 mentions the assassination of Domitian (on the usual chronology, 96 Sept. 18).

Ginzel 1899 (205) found no suitable eclipse round this time, and regarded the description as referring to something other than an eclipse. He was followed in this view by Boll 1909 (2361).

The passage has, however, several times been taken, if hesitantly, as referring to S.95 May 22. Since the track of totality starts in the Indian Ocean and ends in the Pacific, the magnitude in Greece must have been much too small. As an identification for the passage quoted from Philostratus, S.95 May 22 seems absurd, and we know of nothing better. It seems safe to conclude that the passage does not refer to an eclipse.

97/
98

## GHOST 'ECLIPSE' OF NERVA'S DEATH

Speaking of the emperor Nerva, Sextus Aurelius Victor, in the section of his Roman History entitled 'De vita et moribus imperatorum Romanorum', often called the Epitome, Chap.xii (Nerva), §13, says: "On the day on which he died, an eclipse of the Sun took place". This cannot be true, for Nerva is known to have died in the second half of January; the year is AD 98 on the usual chronology, though AD 97 and 99 have sometimes been mooted; and there was no solar eclipse in January in any of these three years. Assuming, as one probably may, that the mention of an eclipse is not merely honorary, the solar eclipse which makes the best of the situation is undoubtedly S.98 March 21, whose track of annularity in Oppolzer runs from California, via a noon point in the North Atlantic, to Iceland (approximately) and the North Cape. Ginzel 1899 (205) finds magnitude of only 0.3 at Rome, and not much more than 0.7 even in Roman Britain.

Of some interest is a calculated eclipse (unobservable at Rome) mentioned in Struyck 1740 (102-3). As with the eclipse about the time of the death of Augustus, Struyck put forward for consideration a North American solar eclipse, in this case that of 97 Oct. 23, calculable by astronomers but invisible at Rome. This date, Struyck argues (from chronological information given in Suetonius, Cassius Dio, Victor and Eutropius), was about the time when Trajan was made co-emperor by Nerva, and may imply that Victor erred in associating an eclipse with the death of Nerva rather than with the beginning of the rule of Trajan. The conjecture deserves mention, because it provides the only way in which Victor's 'on the day' could possibly be justified. The actual dates of the stages of Nerva's adoption and elevation of Trajan appear to be somewhat uncertain. See Camb. Anc. Hist. 11 (1936), 196-8, and E. Mary Smallwood, Documents Illustrating the Principates of Nerva, Trajan and Hadrian (Cambridge, 1966), and further references there given.

The Chinese astronomers, already by the first century BC, had begun to predict eclipse-dates and it becomes increasingly difficult to distinguish in the Dynastic Histories eclipses that were observed from those that were predicted. Most of the Chinese eclipses in the standard list of Hoang 1925 were certainly calculations only. The specific eclipses selected for inclusion in our text will be limited to the few with some claim to be observed and scientifically significant.

In the first century AD there is only one, or possibly two, which meet our requirement, as we now explain.

AD 2 2 Nov.23 SOLAR ECLIPSE NEARLY TOTAL AT CAPITAL

This is of scientific importance because the Han-shu 'Five Elements' Chapter 27 specifies that it was total, and as an amnesty was granted it is clear that the eclipse was observed. Stephenson writes (Feb.84): "Calculation indicates that this eclipse was only barely total on the Earth's surface; in the vicinity of the capital Ch'ang-an it was neither total nor annular. Hence the probability of it being observed as a central eclipse is minute. The track passed well to the NE of Ch'ang-an (computed magnitude there 0.94). After about 28 BC, the eclipse records in chapter 27 of the Han-shu become very deficient; before that date the R.A. of the Sun is reported regularly with the eclipse record while afterwards it is very rare. As the same situation occurs towards the end of the lists of eclipses in the Hou-han shu (chapter 28), it seems that the original records in both periods were lost and had to be replaced by material from secondary sources". (cf. maps in Introduction and in Stephenson and Houlden, in progress, and Newton 1979, 159 and 419.)

65 65 Dec.16 TOTAL SOLAR ECLIPSE

In the History of the Later Han Dynasty, the Hou-han-shu, the astronomical treatise regarded as Chapter 28 refers to an eclipse as total and connects it with the suicide two years later of the king of Kuang-Ling. Modern calculations confirm that the date, day and solar position given in the text are correct, and Muller and Stephenson 1974 (and Muller 1975, p.8.35) accepted it as an observation made at the provincial court of king Ching at Kuang-Ling, 32.42°N, 119.45°E. Nevertheless, Stephenson and Houlden (letter Feb.1984) now consider (see their map in our Introduction from Ginzel's original calculation) that it was 'on the verge of totality at the Imperial capital (Lo-yang)' but the '$\Delta$ T at this period is not as well known as we could wish' (cf. also Newton 1979, 159, 165 and 419).

## LATER HAN ECLIPSES

The astronomical records from the Hou Han Shu or History of the Later Han Dynasty (AD 24-220) are scattered in its chronicles, its astronomical chapters and its five-element chapters and all three have been examined by K. Saito, who has published the results in Kagakusi Kenkyu or the Journal of History of Science, Japan, Series II, Vol.21, Nos. 141-142, pp.27-36 and 70-80. He lists 88 solar eclipse records, and considered that 65 could have been visible in Lo-yang and 17 from distant prefectures. Stephenson (personal communication) has, however, found that many of the eclipses in both the Han dynasties were calculated not observed. The eclipse list is given on pp.30-35, with 11 lunar eclipses (from the calendric chapter) used for improving the calendar on p.36. No translation is yet available. The Local Histories also are now being examined at the Academica Sinica at Beijing (Peking) but the observed eclipses in these are likely to belong to the second millennium AD.

# THE SECOND CENTURY

## INTRODUCTION

## THE AGE OF PTOLEMY IN EGYPT

The eclipses of the second century are less well documented than even those of the first. Such records as there are come from the Eastern Empire; chronicles were thus being kept (e.g. c.105/129) in Asia Minor. Astronomy flourished at Smyrna (127/131) in Asia Minor and especially at Alexandria in Egypt, then the scientific capital of the Western World: Ptolemy worked there in the second quarter of the century (127/151). Records of eclipses are very few in the second half of the century, when Latin literature flourished especially in North Africa (150/250). Dio becomes an eyewitness source for the period 193 to 219; he had come from Asia Minor and he wrote in Greek.

113 S.113 June 1 TOTAL SOLAR IN GAUL (UNRECORDED)

We have encountered this eclipse in the West (a Chinese mention exists) only as one of the less probable identifications (Stockwell's first) of the vague 'eclipse of Plutarch', on which see our discussion among first century eclipses. The central line of totality passed through France during the morning.

118 S.118 Sept. 3 TOTAL SOLAR RECORDED IN LATIN ANNALS

"Hadrian and Salinator. Under these consuls an eclipse of the Sun took place." This appears in the Fasti Vindobonenses priores (also known by other titles, e.g. Annals of Ravenna), ed. Mommsen, Chr.Min.1 = AA 9. 1892, 285. The second consulate of Hadrian, who had Salinator as colleague, occurred in AD 118 (on the usual chronology). The rough track of totality runs from the Atlantic to South-East Asia, with noon point in the Black Sea; his rough track illustrated crosses Europe from the English Channel to the Black Sea. Ginzel 1899 (Map XI) shows that part of the fully computed band of totality which passes over Bohemia, the northern part of the Danube delta, south Crimea, the south-east Black Sea (Batumi) and the south end of the Caspian. The eclipse was the greatest one visible in Europe around the time indicated. There seems no reason to doubt the identification, and we are not aware that it has ever been questioned. The only surprise is that no other certain Western record appears to survive. In the West, we have encountered the eclipse otherwise only as one of the less likely identifications (Stockwell's second) of the vague 'eclipse of Plutarch' (see our discussion among first century eclipses). Possibly bad weather may have had a hand in restricting observation of what should have been a striking eclipse. The eclipse is discussed briefly in Seyffarth 1878 (462). Newton 1972 (448, 451, 604) and 1979 (419) manages to allow the record a little weight, despite its vagueness about place and silence about time of day.

119 See Addendum for possible volcanic darkening in AD 119.

125/136 ECLIPSES OF PTOLEMY

125 (1) M.125 April 5-6 (Wed.-Th.)

133 (2) M.133 May 6-7 (Tu.-Wed.)

134 (3) M.134 Oct. 20-21 (Tu.-Wed.)

136 (4) M.136 March 5-6 (Sun.-Mon.)

We take together these four celebrated lunar eclipses given by Ptolemy as observed at Alexandria. They occur in Book 4 of the Syntaxis (Almagest), (1) in Chap.9 (8 Halma) and the other three in Chap.6 (5 Halma). The references in the most quoted sources are:

Halma (Greek with French translation) I, 1813, 267, 254-5,
Heiberg (Greek) I(I), 1898, 329, 314-5,
Manitius (German translation) I, 1912, 239, 228.

In addition, text and/or translation are set out in many standard astronomical discussions, e.g. in Newcomb 1878 (40), needing correction as given in Fotheringham, MN 69, 1909, 667.

The dates are not given by Ptolemy in the Julian calendar as used in our heading, but in his usual form. In principle, he uses years (each of exactly 365 days) of Nabonassar (i.e. Egyptian wandering years) and Egyptian 30-day months and days; but in the Roman imperial period he uses his own version of the regnal years of the emperors, the years of Nabonassar often needing to be hunted out from among subsequent calculations or supplied editorially. Ptolemy's dates in his original statements are given in the left-hand portion of the following table, in which the Roman numerals give the ordinal numbers of the Egyptian months (e.g. Pachon is the ninth month):

| | Had.yr. | Month | Day | Nab.yr. | Begins |
|---|---|---|---|---|---|
| (1) | 9 | Pachon (IX) | 17-18 | 872 | AD 124 July 23 |
| (2) | 17 | Payni (X) | 20-21 | 880 | 132 July 21 |
| (3) | 19 | Choiak (IV) | 2-3 | 882 | 134 July 21 |
| (4) | 20 | Pharmuthi (VIII) | 19-20 | 883 | 135 July 21 |

For the purposes of the present work, what matters mainly is that the dates are certain (having been confirmed by the calculations of numerous astronomers down the ages), and that the regnal years are of some interest. Indeed, without some explanation the reader may find it difficult to understand the years of Hadrian, considering that he began to reign in the first half of August, AD 117.

It is obvious that the regnal years of Hadrian given above do not tally with the years AD given in our heading; thus it is not the Roman consular year-beginning (Jan. 1) which is being used. By Ptolemy's time, Egypt had (besides the older one) an official calendar, in which the first day of the year (Thoth 1) fell on August 29 or 30 of the Roman Julian calendar. In Egypt, the first regnal year of a Roman emperor began on August 29 or 30 before his accession; see Bickerman 1968 (66, also 48).

It is natural to ask oneself whether Ptolemy is using a Nabonassar year-beginning (falling in his time in July or early August) or an August 29-30 year-beginning. The answer is that he uses the former, but keeps his regnal years as close as possible to the official ones.

Ptolemy's use of the Nabonassar year-beginning is clearly shown by his putting observations of

Venus on AD 140 July 29-30, Nab. 888 Thoth (I) 11-12
and Mercury " 141 Feb. 1-2, " 888 Phamenoth (VII) 18-19

both in the same fourth year of Antoninus Pius. If he had been using the official (Aug. 29) year-beginning he would have put the two observations in consecutive regnal years. The observations occur respectively in Book 10, Chap.1 and Book 9, Chap.7 and 8. The references are:

| | |
|---|---|
| Halma | II, 1816, 194, 167, 175, |
| Heiberg | I(II),1903, 297, 263, 273, |
| Manitius | II, 1913, 156, 131, 139. |

Ptolemy's keeping as close as possible to the official regnal years is shown in the following table:

| | DP | First regnal year begins Egypt (official) | Ptolemy | Thoth 1 of Nab. |
|---|---|---|---|---|
| Domitian | 81 Sept.13 | 81 Aug. 29 | 81 Aug. 3 | 829 |
| Nerva | 96 Sept.18 | 96 Aug. 29 | (96 July 30) | 844 |
| Trajan | 98 Jan. 25 | 97 Aug. 29 | 97 July 30 | 845 |
| Hadrian | 117 Aug. 8 | 116 Aug. 29 | 116 July 25 | 864 |
| Ant. Pius | 138 July 10 | 137 Aug. 29 | 137 July 20 | 885 |

The column headed DP gives date of death of the predecessor of each emperor. Errors should not exceed about one day; these do not affect year-beginnings, nor does the fact that Hadrian counted his accession as Aug. 11 (when news of Trajan's death reached him). The two columns of regnal years are self-explanatory. The last column gives the year of Nabonassar which began on the date given in the penultimate column.

The effect of Trajan's dying in early August is that the first year of Hadrian, in official terms, began almost a year before his accession; while the bizarre effect of Ptolemy's keeping as close as possible to the official regnal years is that Ptolemy's first regnal year of Hadrian began more than a year before his accession. The reader will have noticed that all four eclipses lie in Hadrian's reign.

Ptolemy's observations (gathered quickly from the index in Manitius) include none under the name of Nerva, but they do include two occultations during the reign of Nerva (98 Jan. 11 and 14, counted as the first year of Trajan; Book 7, Chap.3).

We have dealt mainly with the relevance of the four eclipses to chronology. We shall look only briefly at the detailed magnitudes and times given by Ptolemy. His maximum magnitudes and mid-eclipse times at Alexandria are:

| | | | | |
|---|---|---|---|---|
| (1) | Southern 1/6th of diameter eclipsed at | | | 8.24 p.m. |
| (2) | Total | | " | 11.15 p.m. |
| (3) | Northern 5/6ths of diameter | " | " | 11.00 p.m. |
| (4) | Northern ½ of diameter | " | " | 4.00 a.m. |

The magnitudes throw light on the Moon's node at that period, and the times on the mean elongation of the Moon from the Sun. Both have frequently been compared with the principal tables and theories. A few remarks on the times may be made.

The genuineness of the observations reported by Ptolemy has long been suspect. R.R. Newton (The Origins of Ptolemy's Astronomical Parameters, 1982, 55-56, Technical Publication No.4 of the Center for Archaeoastronomy, University of Maryland) has examined the matter in detail. He concludes that Ptolemy "fudged" the observations of the last three eclipses, passing off calculated quantities as observed quantities. However, this investigation, like earlier ones, is concerned with errors of minutes of time; the days and nights remain as we have stated them. Since the times given by Ptolemy, whatever their origin, seem never to be found more than 50 minutes in error, they are as accurate as any times one could hope for from a chronicler at this period, and their chronological usefulness is hardly affected at all.

The time quoted by Ptolemy for (1) is usually found to be too early, a fact which led Cowell (MN 66, 1906, 526) to treat it as referring to the start (not the middle) of the eclipse. But, as pointed out by Fotheringham (MN 80, 1920, 579-80), it is not easy to accept this, when Ptolemy explicitly states that the time refers to mid-eclipse.

For earlier astronomical discussion, see the works of Newcomb, Cowell, and Fotheringham already quoted, also Ginzel 1884 (538), 1899 (229), Cowell MN 66, 1905, 5, Nevill MN 67, 1906, 2, and Newton 1970 (152, 154, 215, 228).

Having treated together four lunar eclipses of 125-136, we resume our chronological order at AD 125

125 S.125 April 21 ANNULAR TOTAL IN EGYPT (NO RECORD)

We have encountered no Western record. Oppolzer's rough track as illustrated crossed the Sahara, the Suez Canal, Palestine and Syria. The fully computed band in Ginzel 1899 lies more to the south-east (Arabia, Lower Mesopotamia, Persia), and the absence of a record is no surprise.

131 S.131 June 12 TOTAL SOLAR IN SPAIN (NOT RECORDED)

We have encountered no Western record. Both Oppolzer and Ginzel 1899 make the track of totality pass over Spain and end at sunset in Algeria.

138 S.138 Jan. 28 TOTAL SOLAR IN GAUL (NO RECORD)

We have encountered no Western record. Oppolzer's rough track of totality illustrated has noon point in the Atlantic, then runs by France and Germany to a sunset point in the Baltic. The short stretch within the map area of Ginzel 1899 just misses North-West Spain, and runs towards Western France.

143 S.143 May 2 ANNULAR TOTAL IN N. EUROPE (NO RECORD)

We have encountered no Western record of this annular-total eclipse. Oppolzer's central line starts in the East Pacific, has noon point in mid-Atlantic, and is shown approximately as running via the English Channel and Germany to a sunset point in the Sea of Azov. The short stretch illustrated in Ginzel 1899 also ends in the Sea of Azov, approaching it from the Ukraine.

145 S.145 Sept. 4 TOTAL SOLAR IN SPAIN (NO RECORD)

We have encountered no Western record. Oppolzer's track of totality has sunrise and noon points in the Arctic, and is shown running roughly along the Pyrenees (too far north) before ending at sunset near Tripoli (Libya). The fully computed band in Ginzel 1899 runs via North-West Spain, Chaves (Portugal), Central Spain, Cartagena, Algeria and Tunisia to an end point also near Tripoli.

164 S.164 Sept. 4 ANNULAR ECLIPSE OF SOSIGENES

Neugebauer 1975 (1, 104) takes this to be the annular eclipse said by Proclus to have been observed (in Greece) by Sosigenes the Peripatetic (2nd century AD); see Proclus Hypotyposis, IV, §98 (ed. and tr. C. Manitius, Leipzig, 1909, pp.130-1, 295). Oppolzer's track of annularity was too far south, crossing Lower Egypt. The fully computed band in Ginzel 1899 (Map XI) crosses Central Spain, the foot of Italy, Greece, Western Cyprus, Palestine (just north of Jerusalem) and Arabia; the annularity at Athens speaks well for Ginzel's track.

172 S.172 Oct. 5 TOTAL SOLAR NEAR S. LIMITS OF EMPIRE (NO RECORD)

We have encountered no Western record. Oppolzer's track of totality starts in mid-Atlantic, is shown approximately as crossing North Africa, has noon point in South-West Egypt, and ends in the Bay of Bengal. But the fully computed stretch shown in Ginzel 1899 passes over Southern Spain as well as Northern Algeria and Libya.

174 S.174 Feb. 19 STRIKING ECLIPSE NOT RECORDED

We have encountered no Western record. As Oppolzer's track of totality starts at sunrise north-west of Sicily, and proceeds via approximately Greece and Asia Minor, this is one of the more surprising second century omissions. The initial stretch shown in Ginzel 1899 is rather similar to Oppolzer's. This should have been the most striking eclipse of the century at Athens and in the N.E. Mediterranean.

176 S.176 July 23 TOTAL SOLAR IN SPAIN (NO RECORD)

We have encountered no Western record. Oppolzer's track of totality has noon point off Labrador, and is shown approximately as crossing Spain and Tunisia before ending at sunset in Libya. The fully computed stretch shown in Ginzel 1899 crosses Southern Portugal and Spain, Algeria and Libya.

185 S.185 July 14 TOTAL SOLAR IN BLACK SEA AREA (NO RECORD)

We have encountered no Western record. Oppolzer's track of totality has noon point in Greenland, and is shown approximately as crossing the middle of the Black Sea (too far north as we now know) before ending in Eastern Iraq. The fully computed stretch shown in Ginzel 1899 runs via the Ukraine, the Danube delta, North-Eastern Asia Minor, and Northern Mesopotamia to an end in Eastern Iraq. Herodian, a Syrian Greek, later wrote on the period 180 to 238 but he did not refer to this eclipse.

(185) SUPERNOVA AND (191) COMET OF COMMODUS

<u>186</u> S.186 Dec. 28 ECLIPSE OF COMMODUS

Two sources have been claimed as evidence for a sizable solar eclipse visible at Rome in the reign of Commodus, who succeeded in 180 and was assassinated on 192 Dec. 31.

On the one hand Herodian (c.165 c.250), in his first book (devoted to Commodus) states, among portents, that "some stars shone continuously by day, others became elongated and seemed to hang in the middle of the sky" (Loeb ed., C.R. Whittaker, vol.I, 1969, pp.88-9; 1.14.1). The first part of this statement is sometimes considered to refer to an eclipse (total or nearly so), the second to a comet.

The word <u>continuously</u> suggests that the supernova of 185/187 rather than the eclipse of Dec. 28 is referred to, whereas the 'elongated stars' referred to the comets of 188 and 191 (not AD 192. See Ho 1962, 128 and 153-4). R. Stothers in 'Is the Supernova of AD 185 recorded in Ancient Roman literature?', <u>Isis</u>, 1977, 68, 443-448, reaches a similar conclusion (cf. also Stothers and Rampino 1983, 6368-9).

The second hint is contained in the unreliable Historia Augusta, in the life of Commodus, compiled perhaps in the fourth century by Aelius Lampridius, xvi, 1-2 (Loeb ed., Script. Hist. Aug., ed. D. Magie, vol.I, 1922, pp.300-3). "The prodigies that occurred in his reign ... were as follows. A comet appeared ... On the Kalends of January a swift coming mist and darkness (repentina caligo ac tenebra) arose in the Circus". Magie interprets "the Kalends of January" as 193 January 1; in that case, the sudden mist and darkness cannot refer to an eclipse.

But the author is explicitly listing portents of the reign of Commodus. Ginzel 1899 (205) points out that the only solar eclipse visible in Rome around January in the reign of Commodus was S.186 Dec. 28 (the month and day are 5. Kal. Jan.). The track of annularity in Oppolzer runs from Mexico to a point near Sardinia. Ginzel finds magnitude 0.79 at sunset at Rome, so that the eclipse could probably have been noticed there. The identification, which Ginzel attributes to Struyck 1740 (103), is also adopted by J.K. Fotheringham and by P.H. Cowell (MN 69, 1908, 29 and 69, 1909, 617 resp.) and in Boll 1909 (2361).

Newton 1970 (74, 121) accepts the identification, but regards the record as an inadequate basis for numerical calculation.

On the whole, it does seem that a faint record of the solar eclipse of 186 Dec. 28 may have survived in the Historia Augusta. But the record is almost useless astronomically, for lack of sufficient detail, and completely useless chronologically, since the eclipse (if such is indeed referred to) is linked with no other specific event whatever.

## 197 FALSE DATE FOR THE ECLIPSE OF TERTULLIAN

It is practically certain that the eclipse of Tertullian took place, in the region of Utica, near Carthage, on 212 Aug. 14 - see our account below under third century eclipses. Former second century identifications can be dismissed as erroneous.

S.197 June 3 was proposed in Struyck 1740 (103), and adopted, albeit with two question marks, as the nominal date under which the eclipse was discussed in Ginzel 1899 (206); however, Ginzel also calculated S.211 March 2 and S.212 Aug. 14, and in the end wisely left the decision to the historians. They in fact favour a date around AD 212 - see our discussion under the third century. Various references to S.197 June 3, under the impression that it was the eclipse of Tertullian, have been made, e.g. by Cowell, Nevill, Newcomb, and Crommelin, see MN 66, 1906, 35, 414, 471, 523; 68, 1907, 18; 69, 1909, 460; also Obs. 30, 1907, 137. But,as the identification is certainly false, all such work, even though by eminent astronomers, now lacks interest. Lynn, Obs. 28, 1905, 387 deserves mention for his clear recognition that AD 197 is too early.

199 S.199 Oct. 7 SUNSET SOLAR IN GAUL (NOT RECORDED)

We have encountered no Western record. Oppolzer's track of annularity has two sunset points, an initial one in the Arctic and a final one in Central France (the second is shown also just within the limits of Map XI in Ginzel 1899). Oppolzer's very rough track in between passes close to Iceland and South-West Ireland.

As a recent example of a solar eclipse with centrality running from one sunset point to another, the reader may find illuminating the map and tables for the total eclipse of 1968 Sept. 22 in the Astronomical Ephemeris for that year. The phenomenon cannot occur unless part of the track of centrality lies in high latitudes; a track entirely confined to moderate latitudes always begins at a sunrise point and ends at a sunset point.

120 120 Jan.18 NEAR TOTAL SOLAR ECLIPSE AT LO-YANG

The History of the Later Han Dynasty, the Hou-han-shu, includes a significant solar eclipse from the capital Lo-yang with details that proved it was observed. "6th year (of the Yüan-ch'u reign period), 12th Moon, day wu-wu, the first day of the Moon, there was an eclipse of the Sun and it was almost complete (chi-chin). On the earth it was like evening. It was 11 degrees in (the constellation of) Hsü-nu. The woman ruler showed aversion from it. Two years and three months later, Teng, the Empress Dowager, died". The date, the day and the solar position are all correct. In contrast to the eclipse of AD 65, the observation was apparently made in the capital Lo-yang and the prognostication therefore related to the Imperial family. The preceding and subsequent entries in this section of the treatise state that the respective observations were not made by the Imperial astronomers but were reported from the provinces stated. The inclusion of the statement "like evening" confirms the attribution of 'near-totality'. (The commentary adds that "the stars were all seen") and Stephenson finds this report the most valuable eclipse-report of the period AD 1-800 (cf. Newton 1979, 165 and 169. See also Stephenson and Clark 1978, p.23). Meeus and Mucke find (personal communication) that maximum magnitude was 0.989 at $6^h04^m$ UT, that is shortly after local noon, the Sun's latitude thus being 41°. Newton 1979 (pp.159, 83, 419, 422) however claims that this is not consistent (p.125 fig.iv 6) with other eclipse observations of this period and suggests (p.129) that the observation may have been made in Southern China.

Stephenson's latest computation (letter Feb.84) makes the eclipse total at Lo-yang but adds that $\Delta T$ at this period (cf. Map in Introduction) is not as well known as we would wish.

179 179 May 24 (ANNULAR) ECLIPSE OBSERVED IN CAPITAL

In the History of the Later Han, the Biography of Han Yüeh reports that he predicted a solar eclipse. "He begged the emperor to have officials don their most formal regalia. The emperor followed these instructions, and events transpired exactly as Yüeh had predicted." (tr. K.J. DeWoskin 'Doctors, Diviners and Magicians of Ancient China: Biographies of Fang-shih'. Columbia Univ. Press, New York, 1983, 73 and cf. 170 n.65.)

## THE THIRD CENTURY

### EAST MEDITERRANEAN SOURCES

Greek sources predominate again in this century. In the first half of the century some eyewitness history still comes from the Eastern Empire in Syria, Palestine and Asia Minor. Dio's history is followed by that of Herodian, a Syrian who again wrote in Greek. Julius Africanus flourished (c.220s) in Palestine and later in Alexandria, and like Dexippus (260s), who wrote a Universal History, his work, used as a source by Eusebius in the next century, has since been lost.

In the second half we have records from Greece and the Rhineland, but little else. Persian sources under the Sasanids must once have been detailed, but only their coinage survives.

212 S.212 Aug. 14 (Fri.) SOLAR ECLIPSE OF TERTULLIAN (N. AFRICA)

Although various identifications have been given for the eclipse of Tertullian at Utica near Carthage (including, wrongly, the eclipses of 197 and 200 mentioned at the end of our section on eclipses of the second century AD), the above identification is now almost definitive (8 points in our table).

Tertullian, in a letter of protest to the Roman governor Scapula, about persecution of Christians, says (§3):

"Nam et sol ille in conventu Uticensi, extincto paene lumine, adeo portentum fuit, ut non potuerit ex ordinario deliquio hoc pati, positus in suo hypsomate et domicilio. Habetis astrologos."

"That sun, too, in the metropolis of Utica, with light all but extinguished, was a portent which could not have occurred from an ordinary eclipse, situated as the lord of day was in his height and house. You have the astrologers, consult them about it."

We quote the Latin from E.F. Leopold's edition (Vol.I, 1839, p.249); much the same is quoted by various astronomers, e.g. by Ginzel 1899 (206) from Oehler's edition.

The somewhat expanded English version we quote from A. Roberts and J. Donaldson, The Writings of Tertullian, Vol.I (Ante-Nicene Christian Library, Vol.XI), Edinburgh, 1869, p.49.

Tertullian's contention that the Sun in any particular position is exempt from ordinary eclipse is nonsense. Nevertheless, the technical terms hypsoma and domicilium will be briefly considered (mostly from Boll) in the discussion below, as they are clues to the season at which the eclipse occurred.

Ginzel 1899 (206) briefly described proposed identifications, and regarded three as worth computing. His results for maximum magnitude at Utica are:

(1) S.197 June 3: 0.92 about 12.54 p.m., apparent time.

211 (2) S.211 March 2: 0.45 at sunset (5.37 p.m.).

212 (3) S.212 Aug.14: 0.93 at 5.56 a.m. (42 minutes after sunrise).

In Oppolzer

(1) is annular-total; the track begins in South America, has its noon point in Algeria, and proceeds via Syria to India.

(2) is total: the track begins in the Pacific and ends in Algeria.

(3) is total; the track starts in the Western Mediterranean, runs by Central Italy, the Balkans, the Black Sea, Siberia (noon point), and China, and ends in the Pacific.

Ginzel confirms that all three eclipses were visible in the region in question (Plate XII and p.82), the comparatively modest magnitude of (2) being compensated by its occurrence at sunset. He wisely leaves the decision to the historians.

The most useful literary-historical reference we have encountered is P. Monceaux, Histoire littéraire de l'Afrique chrétienne ... Tome I, Tertullien et les origines (Paris, 1901; reprinted Brussels, 1963), p.199. Among other points, Monceaux

(a) quotes Pallu de Lessert, Fastes des provinces africaines, 1, 252, as giving AD 211-3 for the years when Scapula was proconsul of Africa;

(b) deduces from a passage in §4 of the Ad Scapulam that Tertullian's letter was written after the death of Septimius Severus (211 Feb.), and even after the death of Geta (212 Feb.);

(c) says that the proconsuls held their sessions in the morning.

These indications evidently point to the eclipse (3) above.

The crucial statement is (b), which seems to us a reasonable deduction from the text of §4. This, as Monceaux explains, refers to (Septimius) "Severus, father of Antoninus" (Caracalla) in the past tense, and several times mentions "the emperor" in the singular, whereas the sons of Septimius Severus, namely Caracalla and Geta, were co-emperors until the elder murdered the younger.

In the statement which we have labelled (c), Monceaux evidently regards "in conventu Uticensi" as referring to routine proconsular business. It has sometimes been taken to refer to a particular assembly; e.g. Seyffarth 1878 (462) mentions a "Council of Utica" held in AD 200, but this is irrelevant if (b) is correct. "Conventus Uticensis" can also mean the whole region or province of which Utica was the metropolis, and is not necessarily limited to the city of Utica itself.

It may be noted that Boll 1909 (2361) also prefers eclipse (3) for quite different reasons, of an astrological and astronomical nature. He says that domicilium is unambiguous, as it refers to the house of the Sun, which astrologically lies in the sign of Leo; only (3) fits. Hypsoma, on the other hand, can have two meanings. In astrology, it implies that the Sun was in Aries, which is true for none of the three eclipses. In astronomy, it can mean the north declination of the Sun;

although greatest north declination occurs when the Sun is at the beginning of Cancer, the declination of the Sun in Leo is north, so that (3) partially fits. It may be added that (1) fits even better, but its date seems historically impossible. Ginzel 1899 (206) mentions a meaning of hypsoma as meridian altitude, rather than north declination; again, (1) fits best, but (3) fits partially.

An aurora is mentioned in the same context and the eclipse date thus suggests a possible date for the sunspot maximum (see ed. Schove, 1983, Appendix B).

218 S.218 Oct. 7 (Wed.) DIO'S SOLAR ECLIPSE

The only ancient writer known to have mentioned what must have been this eclipse is the contemporary imperial official Cassius Dio, but the passage survives only in the epitome by Xiphilinus (eleventh century); Cassius Dio, Book lxxviii (or lxxix), Chap.30 (Loeb ed., E. Cary, Vol.IX, 1927, 406-9). Speaking of the end of the 14-month reign of the usurper Macrinus (217 April-218 June), he says:

"It seems to me that this also had been indicated in advance as clearly as any event that ever happened. For a very distinct eclipse of the sun occurred just before that time and the comet was seen for a considerable period; also another star, whose tail extended from the west to the east for several nights, caused us terrible alarm."

Ginzel 1899 (206) rightly finds any other identification impossible. He finds maximum magnitude 0.87 at Rome about 7.04 a.m. apparent time, and 0.89 at Pergamum about two hours after sunrise. The track of annularity in Oppolzer starts in France, and runs roughly the length of Italy and through Palestine (i.e. too far south) to the Indian Ocean (noon point) and the Far East. The fully computed band in Ginzel 1899 (Map XII and p.83) runs fairly directly from France and Northern Italy to Mesopotamia.

Although the identification cannot reasonably be questioned, the report as it has come down to us is faulty. The date 218 Oct. 7 was after, not before, the fall, and probably the death, of Macrinus. The phrase translated as "before that time" could mean "about that time", but the sense of "indicated in advance" (or "foreshown") is certainly present. For this reason Ginzel (following Hofmann) accepts that chronological order is inverted by the error of either Dio or Xiphilinus. Boll 1909 (2362) regards the error as illustrating a tendency to relate eclipses to deaths of rulers.

The eclipse is not mentioned in Herodian, but his Loeb editor, C.R. Whittaker, has useful comments on chronology (Loeb ed., Vol.II,

1970, pp.24-5 and 34-5, being notes on Herodian 5.3.11 and 5.4.11 respectively). He appears to consider that it must be an error for the epitome of Dio to imply that the eclipse preceded the defeat of Macrinus in battle (218 June 8), but he refrains from claiming as erroneous that the eclipse should precede the death of Macrinus, who after his defeat was on the run for some time (at least a month in any case, and later tradition suggests still longer). He also thinks this may explain why Elagabalus was so long returning to Rome. (This return occurred in 219 July, according to D. Magie's commentary on the Life of Antoninus Heliogabalus, in Scriptores Historiae Augustae, Loeb ed., Vol.II, 1924, p.110, also 105).

Astronomers have been undecided about where to suppose that the eclipse was seen. Ginzel says Dio was chief official in Pergamum and Smyrna at the time, hence his calculation for Pergamum as well as Rome. Fotheringham and Cowell (MN 69, 1908-9, 29, 617 resp.) opt for Rome. Newton 1970 (117, 118, 121) accepts 218 Oct. 7 as the nearest possible eclipse in point of time, but regards it as not worth calculating, on account of lack of information about location. He states that Dio was administrator of Pergamum and Smyrna in the reign of Macrinus, but probably returned to Rome after the successful revolution of Heliogabalus.

The record may in fact warn us not to attempt to alter historical chronology on the basis of slender eclipse evidence. If we had regard only to the eclipse, we might wonder whether Macrinus did not fall in 219 rather than 218. But this is one of those cases in which good cometary evidence outweighs poor eclipse evidence. There is no doubt that the comet and star mentioned by Dio were both Halley's comet. Both Chinese records and modern perturbation calculations show that this was visible in the East in 218 April before, and (following an interval of some 20 days) in the West after, its perihelion passage of 218 May. See Lynn, Obs. 24, 1901, 165, Ho 1962 (128, 154) and T. Kiang, Mem. R.A.S., 76 (2), 1972, 48, 55. Thus Halley's comet strongly confirms AD 218 as the year of the fall of Macrinus.

228 S.228 March 23 UNRECORDED SOLAR ECLIPSE

We have encountered no Western record. This was total in England, but, as far as eclipse reporting goes, England might just as well not have been in the Roman Empire.

234 S.234 June 14 SUNRISE ECLIPSE AT ROME (NO RECORD)

We have encountered no Western record. As Ginzel's track of annularity starts south-west of Rome, and crosses Central Italy on its way to Russia, this is one of the more surprising apparent failures of record in the third century.

240 S.240 Aug. 5 ECLIPSE OF GORDIAN III, TOTAL IN ASIA MINOR

We may consider together five eclipses, one in each of the years AD 237-41 inclusive. Maximum magnitudes and times for Rome are from Ginzel 1899 (207 or 83).

| | | Rome | |
|---|---|---|---|
| (1) S.237 April 12 | (Total) | 0.87 | 5.46 p.m. |
| (2) S.238 April 2 | (Partial) | 0.65 | 8.40 a.m. |
| (3) S.239 Aug. 16 | (Total) | - | - |
| (4) S.240 Aug. 5 | (Total) | 0.78 | 5.20 a.m. |
| (5) S.241 Jan. 29 | (Annular) | 0.84 | 4.20 p.m. |

Except (3), these all arise explicitly as proposed identifications of an eclipse about the time of the youthful emperor Gordian III, who reigned from 238 (July or Sept.) to 244. The fourth century "Julius Capitolinus", in his Life of the Three Gordians, xxiii, 1-2 (Script. Hist. Aug., Loeb ed., D. Magie, Vol.II, 1924, pp.422-3) says:

"And an end of the civil strife was made when the boy Gordian was given the consulship [for 239]. There was an omen, however, that Gordian was not to rule for long, which was this: there occurred an eclipse of the sun [probably that of 238 April 2 DM], so black that men thought it was night and business could not be transacted without the aid of lanterns." Magie's footnotes have been incorporated, in square brackets; the "end of the civil strife" occurred in AD 238 on the usual chronology, and the region of action stretched from North Italy to North Africa. However, here we confidently allocate 8 points to the eclipse of AD 240.

We have merely run through the eclipses in chronological order, remarking however beforehand:

(a) that the passage quoted seems to refer to a total or near-total eclipse, yet in Oppolzer's Canon, which suffices for approximate preliminary orientation, only 237(1), 240 (4) and 241(5) were total, annular, or otherwise nearly total, anywhere in the Roman Empire;

(b) that there has been a shift of opinion among astronomers, if not among historians; whereas for a long time, from the early 17th century, the choice was believed to lie between 237(1) and 238(2), it is usually believed to lie between 238(2), 240(4) and 241(5), though 238(2) is only possible if the record describes the eclipse in untrue terms. As the eclipse of AD 240 was total in Asia Minor we now prefer solution (4).

237 (1) S.237 April 12. This identification goes back at least to Bunting 1590 (500). It is a good enough eclipse, with track of totality starting in the Pacific, having noon point in the Atlantic, and passing during the afternoon from France to near the Lower Danube. Ginzel 1899 (Plate XII and p.83) shows the stretch from France to Romania. The drawback is that it seems too early. With five emperors (Gordian I and II, Maximinus, Pupienus, and Balbinus) slain within 15 months of this eclipse, it could hardly have appeared in retrospect as a portent of the fall of Gordian III, who survived until 244.

238 (2) S.238 April 2. This identification goes back at least to Calvisius 1605 (538), where (1) is also mentioned but (2) is preferred. The posthumous editions (1620, etc.) have fuller accounts. This eclipse was not total or annular anywhere on the Earth; the shadow cone missed the Arctic cap. Ginzel 1899 rightly rejects this eclipse as of insufficient magnitude, as he finally calculates only the three "more considerable" eclipses (1), (4) and (5).

Historians still mention this eclipse, e.g. recently C.R. Whittaker in his edition of Herodian, Loeb ed., Vol.II, 1970, pp.186-7. The chronology of the emperors of April/Sept. 238 can now be given, from papyri and ostraca (J.R. Rea in Zeits. f. Papyrol. u. Epigraphik 9, 1972, 18). Probably the identification (2) stems from a feeling that the text tends to imply AD 238; but if Capitolinus does mean that, then his description of the eclipse is a travesty of what actually occurred. Still, the account is not a contemporary one. To justify this we could argue that Capitolinus found simply a bare record of an eclipse in the critical year, and, failing to appreciate the vast difference between a total and a moderate partial eclipse, have embellished his description for literary purposes with the characteristics of some eclipse he had read about, or even seen. This seems unlikely.

239 (3) S.239 Aug. 16. We can say little about the original mention of this eclipse, as we have seen neither the 'ancient reports' nor the 18th century 'identification'. The 19th century references we have are Seyffarth 1878 (463); Stockwell, Obs. 18, 1895, 58, 162; Lynn, Obs. 18, 1895, 198. They allude to a statement by one Nicolaas Struyck in vol.i, p.353 of a publication variously called 'Ruperti's Magazine' and 'Commentationes Theologicae, edited by Jo. Casp. Velthusen, C.F. Kuinoël, Ruperti, and others (Leipzic 1794-99)'. It may be added that the Nicolaas Struyck (Struijck) whose book of 1740 is often mentioned in eclipse literature lived from 1686 to 1769. This eclipse was conspicuous in Siberia, but not sufficiently important to be a portent in the Roman Empire.

240 (4) S.240 Aug. 5. Stockwell, Obs. 18, 1895, 57-8, after preferring (1) for the Historia Augusta eclipse, mentioned (4) for Struyck's second eclipse. But Lynn, Obs. 18, 1895, 92-3, preferred (4) to (1) for the Historia Augusta eclipse. The discussion was continued in Stockwell, Obs. 18, 1895, 162, 19, 1896, 56 and Lynn, Obs. 18, 1895, 197. Ginzel 1899 (207) adopted, with a question mark, Lynn's identification of the Historia Augusta eclipse, and Boll 1909 (2362) did likewise; Fotheringham, MN 69, 1908, 28 even accepted AD 240 without question mark, though also without including the eclipse in his list of those deserving further calculation. Ginzel 1899 (Map XII and p.84) plots the initial stretch from near Tripoli through Asia Minor to the Caspian. The eclipse was total in the early morning in some Eastern parts of the Roman empire in Asia Minor. (At Rome itself the magnitude of only about 0.78, even though at dawn, would hardly justify the language of Capitolinus.)

241 [240] (5) S.241 Jan. 29. Ginzel 1899 (207) quotes Hofmann (1885 and earlier) for this identification, also mentioned in Boll 1909 (2362). The track of annularity starts in the Pacific and has noon point in the Atlantic. Ginzel 1899 (Map XII) maps the final stretch from Southern Spain to near Vienna. The identification is plainly possible; calculators vary as to whether they find (1) or (5) to have the greatest magnitude of all five eclipses at Rome. However, (5) is rather late; (4) is a little earlier in the reign of Gordian III, and thus a little nearer to the year which the Historia Augusta rather implies. Moreover, (4) was total at some places in the Mediterranean region, whereas (5) was not more than annular anywhere. We have allocated 1 point to the eclipse of 241, but we feel that Asia Minor was the probable place and 240 almost certainly the correct eclipse.

265 S.265 April 3 SUNSET IN S.W. MEDITERRANEAN (NO RECORD)

We have encountered no Western record. The fully computed final stretch shown in Ginzel 1899 (Map XII) runs from Southern Portugal and Spain to a sunset point between Tunisia and Sicily.

266 S.266 Sept. 16 TOTAL IN SYRIA (NO RECORD)

We have encountered no Western record. The fully computed initial band shown in Ginzel 1899 (Map XII) starts near the east coast of Tunisia towards India, and proceeds through Syria, but well north of Babylon and Susa.

272 S.272 Nov. 8 ANNULAR IN N. EGYPT (NO RECORD)

We have encountered no Western record (a Chinese mention exists). Oppolzer's track of annularity starts near Sardinia and passes via (roughly) Libya, Egypt and Aden to a noon point in the Indian Ocean and an end between Borneo and Celebes. The fully computed initial band shown in Ginzel 1899 (Map XII and p.85) starts near Sardinia and passes via the Nile delta coast (not Libya) and Gaza towards Arabia. Papyrus evidence might possibly be found for this and especially for the 291 eclipse.

291 or 292 S.May 15 TOTAL SOLAR IN N. AFRICA AND EGYPT
ALTERNATIVE TOTAL ECLIPSE (W. MEDITERRANEAN)

"Tiberiano et Dione. His conss. tenebrae fuerunt inter diem; et eo anno levati sunt Constantius et Maximianus Caesares, die kal. Mart." "In the consulate of Tiberian and Dio, there was darkness by day; and this year Constantius and Maximian were raised to the rank of Caesars on the first day of March." Hydatius, Fasti, PL, 51, 1861, 906, and PG, 92, 1865, 1091; similarly (except "Maximinus (sic)" for "Maximianus") in Mommsen's edition, Consularia Constantinopolitana ad annum 395 cum additamento Hydatii ad annum 468 ..., Chron. Min. 1 = AA 9, 1892, 230.

Until 1889 there appears to have been no problem, the eclipse was considered to be S.291 May 15, as indicated in Petavius 1627 (831).

There seems to be universal agreement that the consular year of Tiberian and Dio was AD 291. But the formal elevation of the Caesars, Flavius Valerius Constantius (Chlorus) and Galerius Valerius Maximianus, occurred on March 1 in a year formerly reckoned as 292, but for some decades now considered by historians to be 293; see Camb. Anc. Hist. 12, 1939, 328.

Petavius 1627 put the eclipse in the seventh year of Diocletian, which ought to mean some 12-month period within 290-291. Tycho-Curtius 1666 (XXIV) says "Petavius in Chronol." put it in the eighth year of Diocletian [i.e. within 291-292]. The copy of Petavius 1627 used by us says seventh year.

If one accepts that the eclipse occurred in 291, it must be S.291 May 15. The track of totality starts in the Pacific, has noon point in the Atlantic, proceeds along the whole length of North Africa from Morocco to Egypt, and ends in Arabia. For Carthage, Petavius found maximum magnitude 0.93 and Ginzel (Plate XII) confirms that it was total in Egypt but not in Carthage.

But Ginzel 1899 (208) quotes Seeck (Jahrb. f. klass. Philol. 1889, p.631) as going deeply into the connection between the various sources, concluding that the annal occurs in Hydatius under the wrong consulate,

292 and consequently proposing S.292 May 4 as the true eclipse identification. The track of totality starts in the Atlantic, proceeds via N.W. Europe to a noon point in the Arctic. The fully computed band shown in Ginzel 1899 (Map XII) goes by Southern, Central and Eastern Spain and Central France. As Hydatius was bishop of Chaves in the extreme North of modern Portugal, it has been argued that he incorporated a record from the Western Mediterranean. However, there is nothing in the context to suggest such a source before AD 379.

Newton 1972 (529, 533) considers the identification S.291 May 15 as certain, but does not use the eclipse because the place of observation is unknown. We suspect a N. African or Egyptian source and have allocated 6 points to 291 and only 3 to 292.

Those who suppose a local N.W. Mediterranean source will follow Ginzel 1899 (207-8) and Boll 1909 (2362) in preferring Seeck's S.292.

For Rome, Ginzel 1899 (86) finds maximum magnitudes and times:

S.291 May 15, 0.73 at 4.32 p.m.; S.292 May 4, 0.93 at 7.24 a.m.

Fotheringham, MN 69, 1908, 29 accepts S.292, with Rome as "probable place of observation". The eclipse is recorded also in the Byzantine chronicle of Malalas.

295 ANNULAR IN MEDITERRANEAN (NO RECORD)

We have encountered no Western record. The track of annularity starts in South America, has noon point in N.W. Africa, and then passes approximately via the Mediterranean to an end in Russia. The fully computed stretch shown in Ginzel 1899 (Map XII) runs by Northern Algeria, Southern Sardinia, Central Italy (with Rome just inside the northern rim of the band of annularity), Yugoslavia, Romania, and the North-West of the Black Sea (with Odessa near the southern rim). In view of this, the apparent absence of record is rather surprising.

The quotation, under this date, from Malalas in Newton 1979 (385) is derived from a report for AD 291 (in Hydatius) discussed above.

Only one Chinese eclipse in this century is considered central and usable by Newton 1979 (Table V 4, p.166) and that is given a low reliability because of the uncertainty of the place of observation.

## THE FOURTH CENTURY

### BETTER RECORDS FROM THE EASTERN EMPIRE

The Mediterranean seaways still provide the basis for Roman culture and prosperity in the fourth century, when documentation of eclipses and comets becomes more reliable with the development of the chronicles of Eusebius (300/323) and his continuator, Jerome (333/378). As the Codex replaced the Roll the writing of annals became more frequent.

In the first half of the century we have chroniclers from Palestine, Sicily, Cyprus and Syria and in the second half from Antioch, Sicily, Italy, southern France and Constantinople. The Eastern or Greek portion of the Empire provides us with the most records, although the new capital of Constantinople is less important culturally than the Levant. Ammianus Marcellinus from Antioch nevertheless wrote in Latin and was in Persia for part of this time (350/365).

There are very good astronomical records from China (Stephenson and Clark 1978 and Schove and Ho 1967, in Schove 1983, 51-64 and 109-111 respectively) in this century; as usual they supply reliable evidence for comets, but their eclipse records were derived largely by calculation before insertion in the Dynastic Histories.

301 S.301 Apr. 25 TOTAL SOLAR IN EGYPT (NO RECORD)

We have encountered no Western record. The track of totality runs from West Africa to the Pacific, with noon point in Central Asia. The fully computed band in Ginz. 1899 passed over Egypt, Mesopotamia and Persia during the morning.

303-4? 303 or 304 GHOST LUNAR OF FELIX

Lunar eclipse (AD 303 or 304) about the time of the martyrdom of St. Felix? "And he was led to the place of martyrdom (when even the Moon itself was turned to blood) on the third of the kalends of September" (August 30). Before trying, with the help of Ginz. 1899, 208, to decide between the lunar eclipses in the evenings of 303 Sept.12 and 304 Aug.31, the reader should consult the discussion and references relating to the martyrdom of St. Felix of Thibiuca in Butler's Lives of the Saints, ed. H. Thurston and D. Attwater, 4, 1956, 188, or the briefer but useful account in D. Attwater, The Penguin Dictionary of Saints, 1983, 127. Relying mainly on Fr. Delehaye, Analecta Bollandiana, 39, 1921, 241-276, it is stated that St. Felix was executed at Carthage on 303 July 15 or 16, and that the account of his transport to Italy (in which the eclipse passage is embedded in those manuscripts in which it occurs at all) is fictitious. It is also mentioned that there has been hagiographical confusion between various persons named Felix. Indeed, it is a curious coincidence that Ginzel tended to favour M.304 Aug.31 for the now apparently discredited eclipse present in some manuscripts of the Passio S. Felicis Tubzacensis, while 'Butler' (3, 1956, 446) gives Aug.30 of the year '304(?)' as the date of the martyrdom of Saints Felix and Adauctus at Rome.

It may save trouble to add that one early 'identification' may be quickly disposed of. Scaliger 1598 (Prolegomena), 1629 (xviii) searched for a suitable lunar eclipse in AD 301. But both the martyrs Felix mentioned above suffered in the persecution under Diocletian, which in 301 had not yet begun.

306 S.306 July 27 ANNULAR OF CONSTANTINE

The approximate track of annularity starts at dawn in N.W. Africa, passes via the Mediterranean to a noon point in Siberia, and ends at sunset in the Pacific. The fully computed initial band in Ginz. 1899 starts near the Moroccan-Algerian border, and then passes over Tunisia, the foot of Italy, the Balkans and the Crimea.

Riccioli 1653, 293, in his list of astrologically significant eclipses, says "306 Jul.27. Solis Eclipsis. Stellae in Caelo visae. Constantius Imper. moritur. ex Fastis antiquis." Without more information about the old annals referred to, it is not clear whether the

Fasti noted the death and the eclipse, or only the former, leaving Riccioli to supply the latter. Certainly the death at York of Constantius Chlorus, father of Constantine the Great, is usually taken to have occurred on (or at least very near to) 306 July 25. We may also be sure that astrologers of the period noticed the near coincidence in the times of the death and the eclipse, even should no contemporary record of their awareness now survive. We have found no (later) Latin annals, but Riccioli was presumably citing Byzantine sources.

Recte 306

The solar eclipse (first half of the fourth century) of George the Monk ("Hamartolos") and others. George the Monk lived in the ninth century, and is also known by the self-deprecating name of Hamartolos. He says that, apparently during the reign of Constantine the Great, "an eclipse of the Sun occurred about the third hour of the day, so that stars appeared in the sky". Ed. de Boor, Vol.II, 1904, 502, or PG 110, 1863, col.611. The same information, except that the hour is lacking, was later given by Cedrenus (I, 1838, 499).

If the passage in Hamartolos is read as part of a continuous narrative, the date is around 325, and the standard suggestions are S.324 Aug.6 and S.326 Dec.11, discussed at some length in Ginz. 1899. 87, 209-210. S.324 is not satisfactory, as this annular eclipse occurred in the afternoon throughout the whole Mediterranean world; if we disregard "third hour of the day", in the general region of 9 a.m., we throw away our principal clue. S.326 is equally unsatisfactory; it was a morning annular eclipse, but the central line, in Africa, near which alone stars might appear in the sky, ran too far south.

Newton 1972, 530, 533-4, discovered that the passage occurs in "a collection of various small notes about events ... not in chronological order", and pointed out that "the only eclipse of Constantine's reign that would have been large anywhere in the Empire at the third hour is the eclipse of 306 July 27". This eclipse did occur in the first few days of the reign of Constantine the Great. Newton's suggestion, made à propos of Hamartolos, is accepted here, and we suspect a source from near Byzantium, already a region of Christian chronicles, and the eclipse would have been noteworthy there.

316-20 Solar eclipses of AD 316-320. There are three distinct possible allusions to one or other of the five eclipses, namely:

| | | | |
|---|---|---|---|
| (1) S.316 July 6 | Annular | Ginzel's Plate XIII & p.87 |
| (2) S.316 Dec.31 | Annular | p.87 |
| (3) S.317 Dec.20 | Partial | |
| (4) S.319 May 6 | Total | Plate XIII & p.87 |
| (5) S.320 Oct.18 | Total | Plate XIII & p.87 |

There were no solar eclipses visible in the Roman Empire in 315, 318 or 321. We now consider the three allusions, which we shall denote by (a), (b) and (c).

(a) The 'three Caesars' eclipse of Aurelius Victor. In his De Caesaribus (various editions by F. Pichlmayr and revisers), Chap. xli, Sextus Aurelius Victor mentions the simultaneous elevation of Crispus and Constantine (sons of the emperor Constantine the Great) and Licinianus or Licinius (son of the emperor Licinius), "which was shown by an eclipse of the Sun perpetrated on a day in the same months (iisdem mensibus die) to be destined to be hardly long-lived nor happy for those who were elevated". Dates between 315 and 317 have been mentioned for the elevation; the statement in Camb.Anc.Hist., 12, 1939, 693 that it occurred "at Serdica on 317 March 1" seems to represent modern opinion. In these circumstances it is not surprising that (1), (2) and (3), being all within ten months of the event, have all been proposed.

(1) goes back at least to Calvisius 1620, 389, and is mentioned in Ginz. 1899, 87, 208-9 and Boll 1909, 2362. The track of annularity in Oppolzer starts at sunrise in the Red Sea and proceeds eastwards through Arabia and the Persian Gulf to China and the Pacific. Ginz. found magnitude about 0.25 at Rome, about 0.52 at Athens, 0.77 at Memphis and 0.83 at Babylon.

(2) goes back at least to Struyck 1740, 139. The track of annularity starts in Central Africa and proceeds to a noon point in the Indian Ocean, then eastwards. This track is so far south-east that Ginz. 1899. 87, 209,found the eclipse invisible at Rome, but he found magnitude 0.31 at Athens, 0.40 at Memphis, and 0.25 at Babylon. Except for Struyck-Ferguson, Seyffarth 1878, 464, gives the only other support for this identification which we have noticed, although the eclipse occurred only two months before the date apparently usually accepted for the elevation.

(3) goes back at least to Petavius 1627, 834, and after being mentioned in Ricc. 1653, 368 and Tycho-Curtius 1666 (XXIV) seems to have fallen out of favour as an identification of (a). According to Oppolz., (3) was not central anywhere on the Earth's surface. Petavius found the magnitude at Paris in the early afternoon to be 0.6, still a defensible figure. All three eclipses were too inconspicuous to explain the references.

Ginz. 1899, 209, mentions both (4) and (5), finding magnitudes about 0.83 and 0.53 respectively at Rome, but he does not dwell on these eclipses, doubtless as too far removed from the assumed date of the triple elevation. In identifying (b), Ginz.1886 had already discussed (4), and despite the 319 date we consider this the only plausible solution.

(b) Eclipse of the Consularia Constantinopolitana (about AD 318). Mommsen, under the title Consularia Constantinopolitana ad annum 395 cum additamento Hydatii (Chron.Min. 1 = AA 9, 1892, 232) gives, under editorial year AD 318, "In the consulate of Licinius (fifth time) and Crispus Caesar there was darkness by day at the ninth hour". The same is found in Hydatius, Fasti, PL 51, 1861, 907.

AD 318 is still normally reckoned as the consulate of Licinius V and Crispus; the difficulty is that no solar eclipse was visible in the Roman Empire in that year.

The fact that the hour of day is stated enables astronomy to limit the possibilities. (1), (2) and (5) cannot possibly, despite occasional apparent support for (1), have been afternoon eclipses anywhere in the Roman Empire. Thus (3) and (4) are the only possibilities.

Authority undoubtedly favours (4), which was proposed in Struyck 1740, 104, and has been supported in Ginz.1886, 965, 969, Ginz.1899, 87, 209, Boll 1909, 2362 and Newt.1972, 529, 534; Newton considers there is no serious question about the identification. The track of totality of (4) in Oppolz. begins in the Pacific, crosses the United States, has noon point in the Atlantic (43° West) and proceeds via approximately Germany to an end in the Black Sea; about the ninth hour of the day, totality would be somewhere in the general region of the English Channel. Ginz.1899 found magnitude 0.83 at Rome, and states that Struyck, to justify a larger magnitude than that computed for Spain, suggested that Hydatius used a Frankish source. Ginz.1886 tabulated both Oppolz. and Ginz. tracks for Western Europe (S.W. Ireland, Cardiff, London, Brussels, Mainz), and found magnitude 0.73 at Chaves at 3.23 p.m.

It cannot be pretended that (3) would be noticed, but it could have been calculated, and it is more satisfactory to find an eclipse on 317 Dec.20, when Licinius and Crispus were doubtless already consuls designate, than on 319 May 6, when their consulate had been some months finished. We do not feel able to rule out (3) entirely, but, in assessing probabilities, we have assumed that both sources refer to No.4, the only eclipse (319) which could have been seen by the public. We have allocated 6 and 8 points respectively for identification to

(a) Aurelius Victor and (b) the Latin Annals.

(c) The conjunction of Pappus (AD 320?). Pappus, in his commentary on Ptolemy's Almagest (Book VI, Chap.4, 180-1 in the edition by A. Rome, Studi e Testi 54, Rome 1931), works out the time of mean conjunction of Sun and Moon on Nab.1068, Tybi 17 (= AD 320 Oct.18), without carrying the computation any further. Nevertheless, it has been thought that the date may have been chosen as one of particular interest, and even that Pappus may have seen the eclipse at Alexandria, where its magnitude, according to modern calculations, was 0.7 or 0.8. The dates of birth and death of Pappus are highly uncertain, but A. Rome (ibid., p.xi) wonders whether Pappus chose the day of this eclipse because S.291 May 15 (magnitude 0.95 at Alexandria) was too much out-of-date, while astronomers had not yet got round to predicting S.346 June 6 (total or nearly total at Alexandria). This would place the composition of the commentary of Pappus within 10 or 15 years of AD 330. (See Neugebauer 1975, 2, 966).

Pappus's time for mean conjunction is $8^{h}3^{m}.6$ p.m. at Alexandria (= $6^{h}4^{m}$ p.m. at Greenwich). This seems to us just over half an hour too early; Rome finds an error (apparently for true conjunction) of "a good three-quarters of an hour". We accept the suggestion and have allocated 6 points for identification.

The track of totality of S.320 Oct.18 starts the brief initial stretch illustrated in Ginz.1899, Map XIII, at sunrise in Algeria, and proceeds via approximately the Northern Sudan to a noon point in the Indian Ocean.

## c.331 GHOST ECLIPSE IN ARMENIA

Non-eclipse about AD 331. (cf. 320 or 346)

A Byzantine record describes what was almost certainly merely a meteorological darkness due to a cold front. Darkness in Iberia (north-east of Armenia) is mentioned somewhere around AD 331 in the ecclesiastical historians Socrates Scholasticus (born about 379) and Sozomen (born about 400). See The Ecclesiastical History of Socrates, Engl. trans., London 1880, Book I, Ch.xx, 54, under page heading AD 331, and The Ecclesiastical History of Sozomen, Engl. trans. by E. Walford, London 1855, Book II, Ch.vii, 62. There was no striking eclipse in Iberia around the year in question (cf. however AD 346).

334 S.334 July 17 ANNULAR IN SICILY

This is one of the most certainly dated of eclipses. It is mentioned in the Mathesis (an astrological work) of Julius Firmicus Maternus. Ginz. 1899, 210, quotes the Latin from Book I, Chap.IV, §10 in the Teubner edition of W. Kroll and F. Skutsch (Leipzig, 1897), p.13. Seventeenth century (and some later) writers refer the same passage to Book I, Chap.II (p.5) of the 1533 Basle edition.

Firmicus Maternus was a Sicilian, and Ginz. quotes authority for his work being already completed before the death of Constantine (AD 337). Firmicus, describing the fear caused by eclipses "in the middle of the day", says, "speaking of the more recent", that such a one was predicted by astrologers (mathematici) in the consulate of Optatus and Paulinus (AD 334). The Latin has also been held by some to hint at annularity, but Johnson (Obs.28, 1905, 174) is right in saying that "no mention of the actual ring-form occurs". However, the annular eclipse S.334 July 17 fills the bill uncommonly well, and the identification is universally accepted. All tracks of annularity (Oppolz. 1887, Ginz. 1899, 210, and Zech and Hofmann there quoted) go so close to the Straits of Messina in the first hour after noon as to leave no doubt. It is very unlikely that the consular year is even one unit out, and it would need to be much more in error to upset the identification, for there is no eclipse nearly as well-fitting in the years around 334. Even the independent-minded Seyffarth 1878, 464-5, mentions S.333 July 28 only to dismiss it (rightly so).

Ginz. 1899, 210, says "date from Petavius". Certainly Petavius 1627, 835, gives the date, but the identification goes back at least to Calvisius 1605, 568, briefly alluded to in Ginz. 1883, 645.

Riccioli 1653, 368 and Tycho-Curtius 1666 (XXIV) mention the eclipse as seen at Byzantium; compare Ginz. 1883, 646, on S.346 June 6 and Ginz. 1899, 210, on S.324 Aug.6. Petavius computes for Rome, which is one of the standard cities for which Ginz. 1899, 88 also gives a result. Struyck 1740, 139 and Struyck-Ferguson 1773, 175 list the place of observation as Toledo. Newcomb 1878, 34, Chambers 1902, 133 and Foth. (MN 69, 1908, 29) accept Sicily, and Cowell (MN 69, 1909, 617) finds this rather better fitted by Newcomb's formulae than by his own.

335 M.335 Dec.16

See additional note re Neugebauer, 2, 1052-3.

337 S.337 May 16? PREDICTED (UNOBSERVABLE) SOLAR ECLIPSE

Solar eclipse before the death of Constantine the Great (AD 337)? Ricc. 1653, 293 quotes the Life of Constantine for a solar eclipse a little before the death of Constantine, which he says occurred on 337 May 22, the feast of Pentecost. In 337 Whit Sunday did indeed fall on May 22. However, Ricc. wisely gives the eclipse only in his list of eclipses thought astrologically significant by others rather than by himself. There was in fact a total solar eclipse of 337 May 16, but it was quite invisible in the Mediterranean world; the track of totality in Oppolz. starts in the Far East, has noon point in Alaska, and ends in Eastern Canada. This is a particularly clear example of astrological consideration of a calculated, unobserved eclipse. The source has not been traced and is possibly based on the observed eclipse of 306.

346 S.346 June 6 TOTAL SOLAR IN EAST MEDITERRANEAN

This is the third of the only three eclipses (the others being in AD 45 and 59) in the first four centuries AD which have month and day in a Julian-type calendar explicitly stated in old (ancient or early mediaeval) records. "In the same year an eclipse of the Sun occurred, so that stars appeared in the sky, in the third hour of the day, on the sixth of the month Daisios." Theophanes, 9th C., Bonn, I, 1839, 57; Teubner ed. C. de Boor, I, 1883, 38.

Theophanes places this in the Year of the World (AM) 5838, which normally means some period of twelve months within AD 345-6; we leave the matter so, because we could quote high 'authority' for each of at least two different dates (month and day) of the year-beginning of the Annus Mundi of Theophanes. (He takes the first year of Constantius as AM 5829, meaning AD 336-7, so that effectively he puts the eclipse in the tenth year of Constantius. Constantine had died in 337 May.) Daisios means various months in various Eastern calendars, see Bickerman 1968, 50; scholars here take it in its most common meaning, namely June. This points to S.346 June 6, and in fact the identification is certain. "Third hour of the day" may pass muster. Calculators find mid-eclipse in the second or third seasonal hour, the result depending not only on the astronomical elements used but also on where the eclipse is supposed to have been seen.

In Ginz. the track of totality starts in North Africa, and proceeds in a more or less SW to NE direction towards a noon point in Siberia. Stockwell (AJ 13, 1893, 73) fancied he could push the central line far enough to the NW to cause totality at Constantinople, but Ginz. revealed that this was impossible. In any case, there is no reason why the record should refer to Constantinople. During the 340s Constantius was much engaged in operations in Mesopotamia, etc. against the Persians. Ginz. 1883, 645 computed a track which crosses the Constantinople-Jerusalem line in Cyprus, as does the track in Ginz.1899. Ginz. 1899, 211 suggested that this eclipse is the only post-Eusebian eclipse mentioned by Jerome (see our AD 346-8 section), and was so mentioned because Jerome used Palestinian sources.

The record by Theophanes (758-c.818) was echoed in almost identical terms by Cedrenus (early twelfth century?), I, 1838, 523. It is the identification of these Byzantine records which is certain. Although this eclipse is probably also to be regarded as the most favoured identification of the eclipse recorded more vaguely by Jerome (whence by Cassiodorus and various mediaeval sources), we relegate such considerations to separate discussion below. Identification of the Byzantine records goes back at least to Calvisius 1605, 573, who, however, mentions only Cedrenus; recognition of Theophanes as the earliest extant detailed source goes back at least to the same century. The useful Ginz.1883 quotes (of the Byzantines) only Cedrenus, but Ginz. 1899 recognises him as secondary to Theophanes. Ginz.1899 discusses Byzantine and Latin sources together, with sufficient references to earlier work. After 1899, there are brief mentions of the Byzantine aspect in Foth. (MN <u>69</u>, 1908, 29), Cowell (MN <u>69</u>, 1909, 617; S.346 by implication only), and Boll 1909, 2362. A much more notable discussion occurs in Newt.1972, 529, 531, 534-6, 604.

Theophanes wrote over 4½ centuries after the eclipse. Whatever the process of transmission was, a correct message got through to him, at least in respect of month, day, and approximate hour. This is a little more than can be said of the next eclipse, S.348 Oct.9, to which we now proceed. After that, we shall return to the years around 346-8 for a discussion of the debatable identification of an eclipse mentioned by Jerome and later writers.

<u>348</u> S.348 Oct.9 (Sun.) MORNING ECLIPSE OF JEROME

In the same decade, Theophanes has another eclipse. "In the same year the Sun again became impoverished (αὐχμηρότερος) in the second hour of the Lord's Day." Bonn ed., I, 1839, 58; Teubner ed., I, 1883, 39. There is now neither month nor any suggestion of totality. Unfortunately the information about year is faulty. As we shall see, the identification here (and commonly) adopted is almost certain, but Theophanes puts the eclipse in Annus Mundi 5839 (AD 346-7); he should have put it in AM 5841 (AD 348-9).

348 Oct.9 was a Sunday, and on that day the Sun was eclipsed at about the second hour of the day in Asia Minor. The track of totality in Oppolz. ran from Denmark via approximately the Crimea to a noon point in India. This gave totality nowhere in the Eastern Roman Empire,

and a partial eclipse of magnitude which may be estimated at about 0.7 at Constantinople. Stock. (AJ 13, 1893, 73) found magnitude 0.81 at Constantinople at $2^{h}12^{m}$ after sunrise, thus within the third hour of the day. The fully computed band shown in Ginz.1899 (Map XIII) passes just North of the Sea of Azov, i.e. more to the North-East than Oppolzer's rough central line shown in Blatt 75. At Constantinople Newt. 1972, 536 now finds a magnitude "near 0.9". Such a magnitude there would imply a track of totality still further South-West, crossing the Empire not only in Pontus but also in a militarily strongly contested area of the Mesopotamian frontier, say somewhere within 100 miles or so of Nisibis.

There was another Sunday eclipse around the years in question, namely S.345 June 16. But this was undoubtedly an afternoon eclipse in the Eastern Mediterranean, and may be ruled out. The comment in Newt.1972, 536 that the second eclipse of this pair in Theophanes is "almost surely that of 348 Oct.9" seems well justified. For, in the first place, the content of an annal (Sunday, second hour) seems likely to be more reliable than its positioning, several centuries back, in a chronological scheme (AM). Secondly, we have seen that the gradual refinement of eclipse formulae appears to be leading to an eclipse in 348 of greater magnitude in parts of the Byzantine Empire than formerly seemed probable; this must enhance the status of the identification.

The identification of the eclipse with reference to Theophanes goes back at least to Petav.1627, 839. Calv.1620, 397 quotes only Cassiodorus, whom we consider, along with his source, Jerome, and later Latin writers, in the following section.

:.346-8

## AMBIGUOUS ECLIPSE OF JEROME

Solar eclipse (about AD 346-8) in Jerome's Chronological Canons. The chronological tables of Eusebius, which end about AD 327 with the 'vicennalia' (twentieth anniversary celebrations) of Constantine, were translated into Latin by Jerome, and also continued by him to the death of Valens in AD 378. Thus the last half-century or more in the Latin is due solely to Jerome, and in this continuation there is, as we shall see, one and only one eclipse mentioned. It would be a waste of space to discuss at length this vague and much copied record, but the principal facts may be stated. The record is a mere "solis facta defectio" (an eclipse of the Sun occurred - no month, day, hour, or place) in a debatable year. It is well known that it is often difficult to be sure in which year a Eusebius-Jerome manuscript places an event.

We mention two well-known editions of Eusebius-Jerome. A. Schoene, II, Chron.Can., 1866, 1967, 193, places the eclipse in the 2nd year of the 281st Olympiad, and the 9th year of Constantius. J.K. Fotheringham, Chron.Can., 1923, 318, places it in the 3rd year of the 281st Olympiad, and the 10th year of Constantius (variants 11th, 12th). The table in Schoene II, 231-6, translates Ol.281/2 as AD 348. Foth. gives no conversion table, but the most usual assumption, namely that Ol.195/1 = AD 1-2 (which implies that the first year of the first Olympiad began in the summer of 776 BC) makes Ol.281/3 = AD 347-8. (The same process applied to Schoene's Ol.281/2 would give AD 346-7.) Both editors also give years of Abraham (2362 and 2363 respectively), but these are related to Olympic years in the same way in both editions, and we have chosen to use the Olympic years.

The two chief candidates have usually been S.346 June 6 and S.348 Oct.9, both of which we have already discussed as established identifications of a pair of eclipses mentioned by Theophanes. Opinion on the Jerome eclipse seems, on the whole, fairly evenly divided between the two, which are the two greatest eclipses of the time in the region mainly concerned, say between Constantinople and Palestine inclusive. When Jerome wrote his continuation, perhaps about AD 380 in Constantinople, or soon after, he had recently resided much in Antioch and the desert of Chalas, some 50 miles away (his long residence at Bethlehem was yet to come), and he was translating and continuing the tables of Eusebius Pamphili, bishop of Caesarea in Palestine. Writing only some 35 years after the eclipse, Jerome may have had oral evidence from Western Syria; or, as Ginz.1899, 211 says, he may have had post-Eusebian records from the diocese of Caesarea at his disposal. The truth of either hypothesis would strengthen the case for S.346, which (as remarked in relation to Theophanes) may have been total somewhere in the region of Syria or Palestine, as against S.348, which could hardly have been total further South-West than Mesopotamia.

As Jerome sets the eclipse among historical events, one might expect help from these, but unfortunately the Persian campaigns of Constantius are not well documented. With reference to the second siege of Nisibis by Sapor, the eclipse is placed in the same year by Schoene and in the following year by Fotheringham. Both editors agree, however, in placing the eclipse in the year before the battle at Singara. But when the Cambridge Medieval History, 1, 1924 ed., 58, is reduced to dating this battle as "probably 344 (possibly 348)", we see how little historical chronology can help. Rather historical chronology tends to draw on astronomy at this point. Thus H.F. Clinton, Fasti Romani, 1.

1845, 408, dates the second siege of Nisibis to 346, as preceding S.346 June 6, with the perilous support of Jerome and the more detailed (and hopefully more reliable) help of Theophanes.

Other eclipses which occurred around the years in question are the afternoon eclipses of S.345 June 16, S.347 Oct.20, and S.349 April 4. One common property which immediately enables these eclipses to be ruled out in connection with Theophanes is that all three were afternoon eclipses as far as the Eastern Mediterranean is concerned, whereas both eclipses in Theophanes were definitely morning ones (in fact both S.346 and S.348 took place before 9 a.m.). In relation to Jerome, this reason for exclusion is not available. All three, however, are for various reasons less probable than the eclipses of 346 and 348 as identifications of the Jerome eclipse.

S.345 June 16 was mentioned in relation to Theophanes in Seyff.1878, 465, and has been mentioned by us under S.348 Oct.9 in the same context. The track of totality in Oppolz. starts in South America, has noon point in West Africa (Mauretania), and proceeds via approximately Somaliland to an end in the Indian Ocean. It can hardly have been a large eclipse at Constantinople; only Seyffarth's dubious empirical corrections enable him to say that it must have been nearly total there. As far as we know, this eclipse has never been proposed as an identification of the eclipse of Jerome; its date is rather far removed from the year 347 about which Jerome's chronological indications seem to centre. Yet it is curious that in one manuscript (M) Fotheringham's critical apparatus (318) reports the addition by a later hand, after "solis facta defectio", of "xvii kal. aug.". For this is July 16, and the only eclipse on the 16th of any month in the years we are considering is S.345 June 16 (admittedly one month out). The eclipses of 345 and 347 may be eliminated as too small in the Roman Empire.

S.349 April 4 has its track of annularity in Oppolz. going from the Atlantic via about Nigeria to a noon point on the Libya-Sudan border and then approximately via Upper Egypt and the head of the Persian Gulf to an end in Asia (Sinkiang Uighur). The short stretch of the fully computed band in Ginz.1899 lying within the area of Map XIII lies further North-West, passing within some 230 km. of Jerusalem on its way from the southern end of the Sinai Peninsula to the southern end of the Caspian Sea. Newt.1972, 536 finds a magnitude "near 0.7", apparently at Constantinople, and considers that the eclipse cannot be ruled out as a possibility. One must agree with this judgement and 1 point has been allocated in our table. The magnitude in Palestine would be considerably greater than that at Constantinople. It is only the year which makes one hesitate. The eclipse itself may or may not, according to the astronomical elements used, have been, in Palestine, the largest of the five which we have considered.

Readers may be warned against a tendency to credit Jerome's continuation of Eusebius with two eclipses instead of one. Ginz.1899, 211 alludes with surprise to a reference in Seyff.1878, 466 to the discovery by Petavius, in a Jerome manuscript, of mention of a solar eclipse in the 12th year of Constantius (and apparently in Ol.282/1). In fact, such a manuscript does exist. It is the one denoted by the letter F in Fotheringham's edition, described by him (xviii) as used by Scaliger, and shown in his critical apparatus (318) to be the only manuscript which reads 12th year (two others read 11th year, and eight others read 10th year). This manuscript F is rather late and not of great authority. Inspection of the variants in Foth. shows that in the reign of Constantius, the regnal year of an event in F is usually one or two units greater than that in the majority of manuscripts. There is thus no case for supposing Jerome to have mentioned two eclipses. It is simply a matter of a systematic difference in dating between one manuscript and the majority. At most the difference may perhaps provide an argument (of small weight, however) for preferring the eclipse of 348 to that of 346.

Finally, it should be mentioned that Jerome was a principal chronological authority in Western Europe during the Middle Ages, so that Ginz.1883, 1899 can quote various references in chronicles probably derived from Jerome. The earliest he mentions is in the Chronica of Cassiodorus. In Mommsen's edition (Chron.Min. 2 = AA 11, 1894, 151) we find the eclipse (the usual laconic "solis facta defectio") placed in the consulate of Philip and Salia, and dated (by Momm.) AD 348, in accordance with the usual consular lists. A similar fairly early borrowed report, not mentioned by Ginz., occurs in Fredegar, Book II, Chap.42 (SrM 2, 1888, ed. B. Krusch, 67, 1.2-4). For the rest, we refer our readers to Ginzel.

355 S.355 May 28 NEAR TOTAL AT CONSTANTINOPLE (NO RECORD)

The 350s contain two solar eclipses whose central lines fell so much in the region of classical antiquity (Italy, the Balkans, etc.) that we are somewhat surprised to have encountered no Western record.

The first is the total eclipse of 355 May 28. The fully computed initial stretch shown in Ginz.1899 (Map XIII) starts at sunrise off the East coast of Tunisia and runs via Eastern Sicily, the toe of Italy, Northern Greece and the Western Crimea towards a noon point in Siberia.

359 S.359 Mar.15 ANNULAR TOTAL (NOT RECORDED)

This is the second Roman Empire eclipse of the same decade to lack record. Ginz.1899 (Map XIII) shows the track passing through Southern Spain, Central Italy (Rome!), Romania and the southernmost tip of the Crimea to an end in the Black Sea. The eclipse must have been an afternoon one in the Roman part of its course.

360 S.360 Aug.28 ANNULAR ECLIPSE IN PERSIA

There is virtually no doubt about this identification of a solar eclipse inadequately described in Ammianus Marcellinus, xx, 3 (Loeb ed., vol.II, 1937, J.C. Rolfe, 6-9): "At that same time, throughout the regions of the East (per Eoos tractus) the heaven was seen to be overcast with dark mist (subtextum caligine obscura), through which the stars were visible continually from the first break of day until noon ... men thought that the darkening of the sun lasted too long, but it thinned out at first into the form of the crescent moon, then growing to the shape of the half moon, and was finally restored."

The standard identification, since at least Bunting 1590, 500 and Scaliger 1598, 479; 1629, 511, has always been as above. In view of difficulties which we shall shortly explain, Stock. (AJ 10, 1890, 35) suggested S.348 Oct.9 as an alternative; but the historical chronology cannot bear such a large displacement, and the memory of Ammianus, who was probably in Syria at the time, can hardly have been quite so faulty. Bury, in his edition of Gibbon (vol.II, 1898; third ed., 1901, 537), queried whether Ammianus, by a smaller mistake, did not really refer to S.364 June 16. Newt.1972, 537, 597, doubts this identification, and accepts the usual one, in our view rightly. Ammianus has the eclipse under the consular year 360 (Constantius X and Julian III), and the eclipse of 360, properly understood, fits better than that of 364. We deal with the undoubted reports of the eclipses of 348 and 364 in their proper places.

There are two difficulties about the standard identification, but neither is serious. The description in Ammianus implies a total eclipse, while S.360 was certainly only annular. The difficulty was noted early. Thus Kepler 1604, 293-4 was criticised in Ricc.1653, 369, and hence in Tycho-Curt.1666 (xxv), for appearing to entertain the truth of the account of Ammianus in this respect. Nowadays, it is accepted that the account is 'exaggerated' or 'confused'; the author inappropriately used a stock literary description of a total eclipse in the case of an annular eclipse.

The second difficulty seems to have taken longer to realize. The account of the eclipse occurs in a Mesopotamian setting, with the Romans and Persians fighting near the Tigris, somewhere east of Nisibis. Ginz.1899, 212 takes the position of the Roman army to be in longitude 41°.8 East, latitude 37°.4 North. Yet the sunrise point at the beginning of the central line of the eclipse is given in Oppolz. as longitude 55° East, latitude 32° North, nearly 400 miles east of Susa and about 800 miles from the Roman army. This distance is not quite far enough to ensure that the Roman army saw no eclipse at all, but it is enough to make quite certain that any partial eclipse which it did see at sunrise was of very small magnitude.

One looks in vain for any recognition of this difficulty in early authors. Such notes as we have made of authors from Bunting 1590 to Tycho-Curt.1666 inclusive show no such recognition. The reason is alluded to in Ginz.1899; the main computation was by Petavius, to whom the identification must have seemed satisfactory; for he was assuming the Tigris, in latitude 36°, to be some 16° (or nearly 900 miles) East of where it really is! Early chronologists operated not only with inadequate astronomical theories, but also with poor longitudes. It may be added, however, that Petavius had longitudes of other key places (Compostela, Carthage, Rome, Constantinople) less than 6° in error. In the 18th century, the difficulty seems to have been better appreciated; Struyck 1740, 140, see also Struyck-Ferguson, found sunrise annularity near Ispahan (Isfahan, capital of Persia in the days of Struyck and Ferguson).

There appears to be no significant error in Oppolzer's sunrise point. Ginz.1899, 212 accepted it, and quoted Zech 1853, 38, 54, as finding a point even somewhat further East; it may have been slightly further West, and Newt.1979, 385, 420, 469 accepts this eclipse as Syrian (34°/37°N and 36°/42°E).

The explanation which seems to have met with tacit acceptance, and may probably be regarded as definitive, is set out rather fully in Ginz.1899, 212-3. For the assumed position of the Roman army he finds a small partial eclipse (magnitude 0.20) at sunrise. It is known from records of other eclipses that an obscuration of such small magnitude, which would stand little chance of being noticed in the middle of the day, can be observed near sunrise (or sunset). Ginz. proceeds: "Since the magnitude of the eclipse rapidly increases in the direction of Persia, and centrality commences in the enemy's land and was probably accompanied by the appearance of stars, therefore the way in which the account of Ammianus arose is probably to be explained as follows: the observation by the [Roman] army of a small eclipse, which in itself would scarcely have been worth mentioning, gained in importance through reports soon afterwards arriving from regions further East (through Persian prisoners?), that the darkness there had been much greater. From Persia itself news may finally have arrived, that there the Sun had been <u>completely</u> darkened, and stars observed to come out. The element of the miraculous in the event may have caused Ammianus to speak of the appearance of stars 'from early morning to noon'."

Ginz. notes that A. von Humboldt's Cosmos lists this eclipse merely under observed 'darkenings' of the Sun. The English translation by E. Sabine (vol.III, 1852, xcix) says: "The description is quite that of a true solar eclipse; but what is to be done with the length of time and 'caligo' in all the Eastern provinces?" The answer seems to be that Ammianus ran together, without discrimination, accounts from various places, as though they all related to the army on the Tigris; the phrase "per Eoos tractus" gives a fair hint of the conflation.

Muller 1975 (8.39) concurs with a thesis (1972) by Stephenson in regarding the place of observation as unknown; it does indeed seem uncertain by several degrees of latitude and longitude, but it cannot have been west of Nisibis.

364 S.364 June 16 MINOR ECLIPSE OF THEON AT ALEXANDRIA

This total eclipse was seen at Alexandria as a moderate partial eclipse (magnitude about 0.4). According to A. Rome 1952 (a lecture given in 1950), Theon of Alexandria mentioned it in three works (one unpublished). The published ones are:

(1) Theon's Commentary on Ptolemy's Almagest. For Book VI of the Commentary (which relates to Ptolemy's Book VI) the only available edition is the poor Basle one of 1538, in much-contracted Greek. The eclipse is mentioned chiefly on p.332, but see also p.277.

(2) Theon's Small Commentary on Ptolemy's Handy Tables; Theon-Halma 1822, 77 ff., 161.

The gist of the passage from p.332 is translated into English in Foth., MN 81, 1920-1, 114, and into German in Ginz.1899, 213; the latter needs correction, according to Rome 1952, 211. (2) contains a French translation.

There is no doubt about the identification of this afternoon eclipse, which occurred on Nab.1112, Thoth 24 in the Egyptian vague year of 365 days used by astronomers and others (in 364, Thoth 1 fell on May 24), and on Nab.1112, Payni 22 in the Egyptian civil Julian-type calendar (in 363, Thoth 1 fell on Aug.30). Identification did however get off to a bad start. The only identification we have encountered in the classic 84 years from Scaliger 1583 to Tycho-Curt.1666 is wrong. Ricc.1653, 369, followed by Tycho-Curt.1666 (XXV), gave 365 March 10, though taking Payni 22 to refer to the astronomical instead of the civil calendar, and failing to check that a solar eclipse occurred on the day proposed (in fact, none did).

Correct identification goes back at least to Struyck 1740, 107-8 and Dunthorne 1749, 164. Discussions in Zech 1853, 55, Ginz.1899, 213 and Foth. (loc.cit.), ignoring Struyck, agreed with Dunthorne's identification, but rejected his interpretation of the observed hours, which found only occasional favour (e.g. in Stock., AJ 10, 1890, 35) until the researches of A. Rome brought it strongly to the fore again.

Theon in both (1) and (2) gave times after noon at Alexandria for beginning, middle and end of the eclipse. For simplicity, and to avoid the complication of a few minutes difference between greatest phase and 'half time', we shall consider only first and last contact. The equation of time being only about 3½ minutes, we shall disregard in our brief account the distinction between mean and apparent time. The important distinction, which in this case may make a difference of about half an hour, is between equinoctial (ordinary) and seasonal hours. Near mid-summer, in the latitude of Alexandria, the Sun sets about 7 p.m.; thus

12 seasonal hours = interval from sunrise to sunset
= about 14 equinoctial hours.

Theon's observed times of contact, 2.50 and 4.30 p.m., were usually interpreted up to 1950 as time in equinoctial hours, because equinoctial hours are clearly and repeatedly mentioned on p.332 of the 1538 edition. But modern astronomers calculated times varying at least from 3.00 and 4.54 p.m. (Hofmann, quoted by Ginz.) to 4.10 and 5.56 p.m. (Stock., AJ 10, 1890, 35); Ginz.1899, 213 found about 3.31 and 5.16, and Foth. 1920-1, 114 about 3.27 and 5.08. Rome 1952, 211 found 3.12 and 5.20 from P.V. Neugebauer's tables, and 3.28 and 5.19 from Oppolzer's Canon. On the whole, Theon appeared to give observed times about half an hour too small, compared with the general trend of modern tables.

But Rome, in the course of editing Book VI of Theon's Commentary on the Almagest, found the Greek word for 'equinoctial' sometimes missing in the best manuscript when present in others, so that interpolation could be suspected. Working through Theon's arithmetic in detail, he found that Theon calculated times in equinoctial hours and reduced them to seasonal hours for comparison with the observations. Working to about the nearest 5 minutes, Theon gives the following contact times p.m., except that equinoctial hours in brackets are deduced from his seasonal hours by applying a factor $\frac{7}{6}$ (amply accurate enough):

| | First contact | | Last contact | |
|---|---|---|---|---|
| | Equin.hrs. | Seas.hrs. | Equin.hrs. | Seas.hrs. |
| Observed | (3.18) | 2.50 | (5.15) | 4.30 |
| Calc. from Alm. | 3.20 | 2.50 | 5.15 | 4.30 |
| " " H.T. | 3.15 | 2.50 | 5.10 | 4.30 |

For the last line (Handy Tables) we rely on Rome 1952. Clearly Theon's times agree tolerably, after all, with modern computations by astronomers like Ginzel and Fotheringham, who took an extract from the only available printed edition of Book VI of Theon's Commentary on the Almagest at face value. To Rome 1952, which depends on critical discussion of manuscripts, is now due the chief credit for establishing what appears, as far as one can judge, to be the true situation. But the times 3.18 and 5.15 approximately, obtained by applying the $\frac{7}{6}$ factor to 2.50 and 4.30, appear already in Struyck 1740, 108 and Dunthorne 1749, 184.

For recent work on this eclipse, see Newt.1970, 126, 152-4, 247-9 and 1979, 218-9. Newton appears to interpret the hours in the Zech-Ginzel-Fotheringham sense, rather in the Struyck-Dunthorne-Stockwell-Rome sense.

364 M.364 Nov.25-26 TOTAL LUNAR ECLIPSE OF THEON AT ALEXANDRIA

This total eclipse is mentioned by Theon in the same works quoted above for S.364 June 16. The lunar eclipse is mentioned in (1) Theon's Commentary on the Almagest (1538 ed., 284, 319) and (2) Theon-Halma 1822, 90-1, 162-3. The eclipse occurred entirely after midnight at Alexandria, and might be described as M.364 Nov.26 (a.m.), but it is well to state the two days explicitly, since Theon times it from noon on Nov.25. Oppolz. gives magnitude 1.10 at 1.47 UT, equivalent to 3.47 a.m. Alexandria Mean Time, with the eclipse lasting altogether from 2.05 to 5.29 a.m. Correct identification goes back at least to Scal.1598, 466; 1629, 496, and other references are fairly numerous; most recently Neuge.1975 (2, 965).

The date given corresponds to Nab.1112, Phamenoth 6-7 in the astronomical 365-day reckoning, and to Athyr 29-30 in the Julian-type year commencing 364 Aug.29. It may be noted that Theon places S.364 June 16 and M.364 Nov.25-26 respectively in the 80th and 81st years of Diocletian. This is consistent with the date AD 284 Aug.29 usually given for the era of Diocletian, see e.g. Bickerman 1968, 72; it was in this context that Scal. handled the lunar eclipse.

We can mention only briefly the hour at which the eclipse was observed, as we have inspected (1) and (2) only under limitations of time, and in any case expert reference to the manuscripts is needed. (1), page 319, appears to give mid-eclipse as $16^h45^m$ after noon on Nov.25, and Halma's note on page 162 of (2) appears to mention $17^h12^m$, but is unclear to us. Prima facie, these times are 4.45 and 5.12 a.m., and are later than the time calculated by Oppolzer and others.

The so-called second solar eclipse of Theon has proved to be a ghost. The early edition (Theon-Halma 1822, 74) led Seyff.1878, 467 and 529, to consider two ill-fitting eclipses, S.374 Nov.20 and S.378 Sept.8. However, it is clear from the superior modern edition (Theon-Tihon 1978, 262 and 331-2), as well as Neuge.1975 (2, 966) that the numerical work, whether text or scholium, which mentions a conjunction in astronomical Phamenoth in the 94th year of Diocletian relates to the new moon of AD 377 Nov.17 (Astronomical Phamenoth 1, civil Athyr 21) at which no eclipse occurred.

380-465 On a lunar eclipse (associated with Maximus of Turin), unidentifiable within these wide limits, but probably fifth century, see the beginning of our account of western eclipses of that century.

386 S.386 April 15 TOTAL ECLIPSE (UNRECORDED)

The central line of this annular-total eclipse in Ginz.1899 starts at sunrise in Central Algeria, and proceeds via Libya, the extreme south of the Peloponnese, the Sea of Marmara, slightly west of Constantinople, and the western tip of the Crimea, into Russia towards the Arctic. Clearly only bad weather could have prevented a striking morning eclipse in parts of the Roman Empire, but we have encountered no Western record.

393 S.393 Nov.20 (Sun.) IMPORTANT ECLIPSE IN ITALIAN ANNALS

This is the long-established identification of the eclipse mentioned in both Western sources and the Eastern author Marcellinus Comes as having occurred in the consulate of Theodosius (third time; variant sixth time erroneous) and Abundantius, i.e. AD 393. Leaving some minor references to be mentioned later below, the chief modern references are:

(1) Fasti Vindobonenses priores cum excerptis Sangallensibus, in Mommsen's Consularia Italica section, Chron.Min. 1 = MGH AA 9, 1892, 298;

(2) Marcellinus, Momm., Chron.Min. 2 = MGH AA 11, 1894, 63.

Both say darkness occurred (tenebrae fastae sunt). Only (1) mentions month and day, "die Solis VI kal. Nov.", i.e. Sunday, Oct.27; but Oct.27 was not a Sunday, whereas Nov.20 was a Sunday, and a great eclipse occurred on that day; consequently VI kal. Nov. is usually taken (as by Mommsen) to be a mistake for XII kal. Dec. (= Nov.20), a mistake accepted in Newt.1972, 452 as "not too hard to make".

As to the hour of day, "third" is given in the St. Gall manuscript (9th century) and in Marcellinus, "second" in the Vienna manuscript (15th century). I.e., the eclipse was rather after or rather before 9 a.m., and anyway not too far from that hour. This is reasonable for France or Italy, but about noon at Constantinople. Perhaps Marcellinus, a Byzantine, used a Western source for at least the hour of day.

Newt.1972, 452 reasonably considers "most of Italy" as eligible. The eclipse would have been more conspicuous at Ravenna than at Rome.

Another Western source, Prosper's Chronicle, has been quoted since at least Calv.1605, 597. This means the chronicle, arranged by consular years, composed by St. Prosper of Aquitaine. The text in PL 51, 1861, 587 mentions the third hour, but not the day, and the editor warns that the annal may be a later interpolation from Marcellinus. Mommsen, Prosper Tiro, Epitoma Chronicon, Chron.Min. 1 = AA 9, 1892, 463, accepts interpolation from Marcellinus, and relegates the whole annal to the critical apparatus in the footnotes.

There is also a shorter work, arranged by regnal years, formerly attributed to Prosper. The modern edition of this Chronicon Imperiale, or Pithoeanum (so-called because first edited at Paris by P. Pithou in 1588), is also by Mommsen, under the title Chronica Gallica a. 452, Chron.Min. 1 = AA 9, 1892, 631-662. In this we find (pp.648-9), under editorial AD 393, "Terribile in caelo signum per omnia simile apparuit"; this is copied word for word in Sigebert (SS 6, 1844, 304); but the sentence appears to be abbreviated from a parallel non-eclipse (auroral) passage found elsewhere.

The two events chiefly chronicled for AD 393 are the elevation of Honorius by his father Theodosius at Constantinople, and the eclipse. In a very concise chronicle they are necessarily mentioned in close juxtaposition; but the two events were well separated in time (Jan.10 and Nov.20 respectively).

Zosimus (c.475) mentions an eclipse during the battle in which Theodosius defeated and killed Eugenius. The battle is usually placed in 394 Sept. or Oct. The eclipse made Stockwell wish to place the battle in 393, and change the year of the death of Theodosius from 395 to 394. But it is usually considered that Zosimus erred in mentioning an eclipse where other writers mention only a storm. There was much discussion about this in 1896 between Stockwell and Lynn, see Obs.19, 57, 91, 238, 331 and AJ 16, 89, 118, 175, 17, 6 and cf. Newton 1979, 386 and 420.

S.393 Nov.20 was the only striking European solar eclipse of its decade. The central line passes via Southern England and Northern Italy to a noon point not far West of Constantinople, and proceeds via approximately Northern Asia Minor to an end in Siberia. This is a strongly curved track, consequently the band of totality was wide; Stockwell in 1896 (Obs.19, 57; AJ 16, 89) found over 100 miles.

The principal post-Oppolzer information is contained in Stock., Ginz., 90, 213-4 and Newt.1972, 448, 452, 530, 537-8, 604. There are brief references in Foth., MN 69, 1908, 29, Boll 1909, 2363 and Cowell, MN 69, 1909, 617.

Ginz.1899 found the magnitude to be 0.96 at Rome, and very great in the whole Byzantine Empire, Northern Italy and Southern France.

Early chronologists mention, often under the heading of this eclipse, a solar obscuration alluded to by Jerome in one of his letters; see our 'circa 398' section.

395 S.395 April 6 TOTAL SOLAR IN ARABIA

For this also see our section 'circa 398' (Jerome), below.

c.398 GHOST ECLIPSE (AURORA?) OF JEROME

Solar obscuration in a treatise of Jerome (about AD 398). Jerome, in a treatise addressed to Pammachius against John, bishop of Jerusalem, appears prima facie to speak of an eclipse: "a few months ago, about the days of Pentecost (circa dies Pentecostes), when, the Sun being obscured, everyone feared that the day of judgement was already at hand (jamjamque venturum judicem)". PL 23, 1883, 411. The passage is on p.446 of W.H. Fremantle's English translation in Select Lib. of Nicene and Post-Nicene Fathers ..., 2nd ser., vol.6, 1893, reprinted Grand Rapids, Michigan, 1954.

Ginz.1899, 214-5, in a rather extended discussion, quotes literary-historical opinion dating the book about 398 or 399; at any rate, during the quarrel between Jerome and John, which he says began in 394 and was composed about 400. Fremantle seems to date the work 397-9.

One may notice, as the old chronologists did, that no eclipse, either solar or lunar, can occur on the day of Pentecost itself. The fact that

Whit Sunday is 7 weeks after Easter Sunday, which lies within about the week after Full Moon, ensures that Whit Sunday always falls in the first or second quarter of the Moon, and always at least a day or two away from both the preceding New Moon and the following Full Moon.

This does not greatly matter in principle, for an eclipse about the time of Pentecost might well signify one several days away from Whit Sunday. Unfortunately, when one looks at specific eclipses of the period, the problem hardens; at Bethlehem there was no total or almost total eclipse of the Sun, such as Jerome's language might suggest, about the time in question.

Ginzel can find nothing better than S.400 July 8 (Pentecost May 20), magnitude at Bethlehem about 0.45 at sunrise. He also mentions a number of others. Seyffarth's S.392 June 7 (Pentecost May 16) seems too early; S.393 Nov.20, magnitude about 0.9 at Bethlehem, too early and not near Pentecost; Struyck's S.395 April 6 (Pentecost May 13), magnitude about 0.65 at Bethlehem, possibly too early; one may think of some of the lunar eclipses of Claudian (see our western eclipses of the fifth century), such as M.401 June 12 (Pentecost June 2) and M.402 June 1 (Pentecost May 25), but it is hardly conceivable that Jerome should be so confused about a recent eclipse, except through a slip of the pen.

The report of the obscuration is too vague to be of use to astronomers, and can hardly be of much use to chronologists; it is too probable that any solar obscuration could refer to sunspots and the main phenomenon to an aurora. An auroral maximum is known to have occurred about 396 (cf. Schove 1983a, Appendix B) and displays reached as far south as 32°N in China in 395 and 400.

400 M.400 Dec.17 (p.m.) See 'The eclipses in Claudian, about AD 400-402' in our account of the fifth century western eclipses.

360 360 Aug.28

"4th year of the Sheng-p'ing reign period, 8th month, day Hsin-ch'ou, the 1st day of the month, the Sun was eclipsed and it was almost total; it was in Chueh." This was a wide belt of annularity; presumably only the crescent phase was observed at the capital. My computations make the eclipse annular to the south of Nan-ching, magnitude there 0.88. Source: Chin-shu Astronomical Treatise. (Contribution by R. Stephenson.)

# THE FIFTH CENTURY

## A CHRONICLE FROM 'PORTUGAL'

The Latin and Greek halves of the Empire are now separate, but the conventional collapse of the Roman Empire in the fifth century is not matched by a collapse of the eclipse records. Indeed, the decline of the land routes through Gaul is offset by an increase in the use of the Western Seaways from Iberia to Ireland, and a good record of eclipses in this century comes from Galicia in what is now Portugal. The Mediterranean sources of Marcellinus (420/450) and Hydatius (440/450) were later used to fill up gaps in the Irish Annals. The same growth of the sea routes in the north-west helps to explain the archaeological finds in Denmark in this century; at the same time increased use of rivers in what is now the USSR allowed artistic influences to spread from the prosperous cities around the Black Sea as far as the Scandinavian world. The fifth-century dating of the Danish eclipse record we discuss below has nevertheless to be confirmed by the dendrochronologists and the art historians.

In the first half of the century eyewitness history is available from Constantinople in the East and Galicia and Aquitaine (now southern France) in the West. Astronomy was important further east, notably in western India at Ujjain from AD 400-650, but the extant testimony lies in calculations rather than observations. Eclipse observations were recorded in this century by the Maya, judging from the astrological dates in the medieval Dresden Codex (Schove 1982/4).

400-1 400 Dec.17 THREE LUNAR ECLIPSES OF CLAUDIAN
401 June 12
401 Dec.6-7

The comet and eclipses of Claudian (M.400 Dec.17, M.401 June 12, M.401 Dec.6-7).

"Constant eclipses of the Moon alarmed us and night after night throughout the cities of Italy sounded wailings and the beating of brazen gongs to scare the shadow from off her darkened face. Men would not believe that the Moon had been defrauded of her brother the Sun, forbidden to give light by the interposition of the Earth ... Then ... signs of the past year ... a comet - ne'er seen in heaven without disaster." Claudian, Gothic (or Pollentine) War, lines 233-6, 238, 243 (e.g. Loeb ed. and tr. M. Platnauer, 2, 1922, 142-5).

The eclipse passage may sound vague, but with the help of the comet there is in fact, as we shall see, no difficulty in making reasonably certain identifications.

The passage refers to a time during Stilicho's war against Alaric. The poem is sometimes called The Pollentine War, because it goes only as far as the battle of Pollentia; the later battle of Verona is mentioned in (and only in) another of Claudian's panegyrics (On the sixth consulate of the emperor Honorius [AD 404], line 201; Loeb 2, 1922, 88-9). Claudian is very explicit that the campaign up to Pollentia lasted only one winter (op.cit. 11.151-3; Loeb 2, 1922, 136-7). The usual chronology puts the Gothic invasion of North Italy in the second half of 401, and the battle of Pollentia in the first half of 402. But winters 400-1 and 402-3 have also been suggested; see J.H.E. Crees, Claudian as an Historical Authority, 1908, 175ff. (See A. Cameron 'Claudian' 1970 Oxford, 138-145 for a chronology of his life.)

Regarding the path of the comet, Claudian mentions explicitly the constellation Cepheus, and in mythological disguise Andromeda and Ursa Major. The Chinese records of the comet, 400 March 19 until May, give more detail, and mention Chinese asterisms wholly or partly in, among other classical constellations, Andromeda and Ursa Major. The Chinese records are independent of Western ones, so there is no doubt that Claudian was referring to the great comet of AD 400, March to May.

There is also no doubt that in considering AD 400-3 one is working in the correct quadrennium. There was no lunar eclipse in 403, and the lunar eclipses available are:

| | Rome MT | Mag. | Visibility |
|---|---|---|---|
| (1) M.400 Dec.17 (p.m.) | $19^h45^m$ | 1.07 | All visible |
| (2) M.401 June 12 (a.m.) | 4 35 | 1.55 | First half visible |
| (3) M.401 Dec.6-7 | 0 38 | 1.32 | All visible |
| (4) M.402 June 1 (p.m.) | 20 19 | 0.86 | Second half visible |
| (5) M.402 Nov.25-26 | 0 01 | 0.05 | All visible |

Oppolzer's UT times of mid-eclipse have been converted to Rome mean times; Goldstine 1973 agrees within 11 minutes. It will be seen that the first three eclipses were total. The visibilities in the last column apply approximately to Italy generally.

At what time did the Romans become worried about "frequent eclipses" and, among "anni signa prioris", which seems to us to mean "portents of the previous year", a comet? Clearly it was in winter.

The winter of 400-1 can be ruled out. As Platnauer's translation reads "signs of the past year", the comet may perhaps be regarded as indecisive. But of eclipses, only (1) would have occurred! Powerful panegyric tends to exaggerate, but Claudian may be acquitted of having turned one single eclipse into "frequent eclipses".

On the other hand, consider the winter of 402-3; i.e. suppose that the Romans became worried after eclipse (5). This was such a small eclipse that it would not be noticed. Moreover, the comet would then have to be counted among "portents of an earlier year" (namely, the year _before_ the previous year), which seems strained. Crees admitted (p.180) that the argument from eclipses is the hardest to refute; it seems to us that the combined evidence of comet and eclipses is even harder to resist.

J.B. Bury, History of the Later Roman Empire, _1_, 1923, 137, 160, adopted the usual chronology and referred (correctly, in our view) to simply the first three eclipses. A number of further references will be found in a full discussion in Ginzel 1899, 215-6, where also 401-2 appears to be favoured for the troubled winter between the original invasion of North Italy and the battle of Pollentia. Ginzel quotes the outstanding critical edition by T. Birt (AA _10_, 1892); we have quoted the Loeb edition for the convenience of English readers.

Claudian seems to us not to mention the considerable _solar_ eclipse of 402 Nov.11 (q.v., magnitude almost 0.9 in Italy), though Ginzel suspects an allusion.

We conclude that 402 is certainly the correct year for the battle of Pollentia. As the battle is known from other sources to have occurred on Easter Sunday, we believe the usual 402 April 6 to be the correct date.

400-465?? M.c.400-465 UNDATED LUNAR ECLIPSE OF MAXIMUS

Lunar eclipse of Maximus of Turin. See Ginz.1899, 219-221, also Boll 1909 (col.2363). Ginzel discusses at length a sermon of Maximus, bishop of Turin, in which he tried to enlighten members of his flock who had cried aloud to assist the Moon when it was labouring under an eclipse one evening. The dates of the activity of Maximus appear to be very uncertain. Ginzel is not able to do more than pick out all the lunar eclipses visible at Turin in the evening in the years 380-423 and 430-465. These number 22, and Ginzel marks seven with asterisks as particularly worthy of attention (these seven all lie in the years 400-465).

402 S.402 Nov.11 (Tues.) NEAR TOTAL SOLAR IN N.E. PORTUGAL

The Portuguese bishop Hydatius mentions this eclipse in both Chronicon 1894, 16 and Fasti 1892, 246:

Chronicon (Mommsen, Chr.Min. 2 = AA 11, 1894, 16):

"In the eighth year of Arcadius and Honorius, Ol.295/2, an eclipse of the Sun occurred on the third of the ides of November."

Fasti (or Consularia Constantinopolitana):

"In the fifth consulate of both Arcadius and Honorius, an eclipse of the Sun occurred on the third of the ides of November."

3.id.Nov. = Nov.11, and all three year indications imply AD 402. A manuscript of the Chronicle gives "second day of the week" (Monday), but this is wrong. The eclipse is also mentioned (without month or day) in the Gallic Chronicles of 452 and 511 (Chr.Min. 1 = AA 9, 1892, 652, 653 resp.). The track of totality enters North-East Spain from the Bay of Biscay, and proceeds via Tunis to East Africa. In the region from the Bay of Biscay to near Malta, Ginz.1886, 970 tabulates Oppolz. and Ginz. tracks, which differ little, and finds magnitude 0.87 at Rome about 9.10 a.m. (the earlier eclipses of Hydatius seem to be derived from an Italian source). See also Ginz.1899, 216 (Map IV) and Newt.1969, 1970 (71, 75, 263), 1972 (508, 538).

Identification goes back at least to Petavius 1627, 843, who found magnitude 0.875 at Rome, but Portugal is the usual place of observation in the fifth century part of the chronicle and a reference to Toledo, given in AD 399, seems to justify this.

410? Xc.410 FICTITIOUS SOLAR ECLIPSE OF ALARIC

Solar eclipse around the time of the sack of Rome by Alaric (410, late August)? References to such an eclipse appear not in Zosimus but in the secondary literature, and they do not have any basis in fact, since all we have encountered mention stars seen in the daytime, but there was no suitable eclipse between 402 and 418. The annular eclipse of 410 June 18 is mapped in Ginz.1899 (Map XIV), who shows a short stretch running about E.S.E. through Morocco and Algeria, with the central line passing approximately through the point on the Greenwich meridian in 30° North latitude.

The 14th-century Byzantine writer Nicephorus Callistus Xanthopulus, in his Ecclesiastical History, Book 13, Chap.36, claims "Then there was such an eclipse of the Sun, that stars shone in the middle of the day" (PG 14B, 1865, cols.1047-8). This is the origin of the mention in Ricc.1653 and Tycho-Curt.1666. The primary source for this period, Eunapius, is no longer extant except as summarized by Zosimus.

412-3 M.412 Nov.4 (p.m.) THE GOLDEN HORN (LUNAR) ECLIPSE
S.413 Apr.16 THE GOLDEN HORN (SOLAR) ECLIPSE

These eclipses have been identified by Willy Hartner as underlying runic inscriptions on two golden horns found in Northern Jutland in 1639 and 1734. See W. Hartner, Die Goldhörner von Gallehus (Wiesbaden, Franz Steiner Verlag, 1969), and A. Beer, 'Hartner and the Riddle of the Golden Horns', Journal for the History of Astronomy, 1(2), 1970, 139-143. The case is unusual, in that the evidence is archaeological rather than historical; we leave the English reader to begin by consulting Dr. Beer's article for further information.

The lunar eclipse was a deep total one. Oppolzer's data make its magnitude 1.62 around 9 p.m. Danish mean time.

The solar eclipse, an afternoon one in Denmark, was also total. Oppolzer makes its track run from Cuba via Southern England and approximately the Kiel Canal to Russia and Siberia.

Thus both eclipses would (granted fine weather) be very striking in Denmark.

Cdr. Knud Alsen has kindly sent me the recent Danish details on Golden Horns (Aarbøker for Nordisk Oldkyndighed og Historie for 1972 and 1979, pub. 1974 and 1980 respectively). He notes that archaeologists differ about the interpretation of the horns but agree that their dating is c.AD 400-450.

418 S.418 July 19 (Fri.) IMPORTANT SOLAR ECLIPSE FROM PORTUGAL TO ASIA MINOR

Among Western sources the best known is the Chronicle of Hydatius (Chr.Min. 2 = AA 11, 1894, 19). This gives the correct date (14.Kal.Aug.), but wrongly says fifth day of the week (Thursday) instead of Friday, and gives no hour.

There are in fact Western sources which give more information, but, except perhaps for Prosper (see below), they are too late to have much independent authority. We have encountered three which say "third hour", besides giving the correct date:

(i) Some "adnotationes antiquiores" to the Easter cycles of Dionysius Exiguus, edited by Mommsen (Chr.Min. 1 = AA 9, 1892, 755);

(ii) Annales chronographi vetusti (378-768), SS 13, 1881, 716;

(iii) Annals of Lund (or Esrom), SS 29, 1892, 191 (Late Medieval).

In addition there are:

(iv) Annales Blandinienses (Ghent), SS 5, 1844, 25 (Late Medieval); these give the same daily information (July 19, third hour), but under the year 951. This is 533 years out, and Newt.1972, 228 has suggested that the compiler made the mistake of putting the eclipse in the wrong Dionysian 532-year Easter cycle period, and also was an additional year out through poor legibility or error in his original.

Three Byzantine records carry more weight than the dozen or more Western records. These Byzantines all give the date correctly, and time the eclipse at or about the eighth hour of the day. The sources are Philostorgius (born about 364), Marcellinus Comes (died about 534?), and the Chronicon Paschale (7th C). The details are:

(i) Philostorgius, Ecclesiastical History (primary but extant only in the ninth-century Epitome by Photius), PG 65, 1864, 616; Engl. trans. (of Sozomen and Philostorgius) by E. Walford, London 1855, p.517. "When Theodosius had entered the years of youth, on July 19, about the eighth hour of the day, the Sun was so deeply eclipsed that stars appeared ..." Walford weakly has "a little after noon-day". Philostorgius (of 425) goes on to describe a drought, a comet (i.e. 418 June), etc. Theodosius was born in 401.

(ii) Marcellinus Comes, Chr.Min. 2 = AA 11, 1894, 74. "Honorius XII and Theodosius VIII. An eclipse of the Sun happened in the month of Panemos (here July) on 14.Kal.Aug. (July 19) on the day of preparation (Friday) at the eighth hour".

(iii) Chronicon Paschale, ed. L. Dindorf, Vol.I, Bonn 1832, 574. As in Marcellinus, prefaced by "First indiction. Tenth year of Theodosius".

Correct identification goes back at least to Kepler 1604, 294 (Ges. Werke, 2, 1938, 255). The eclipse confirms the usual AD 418 as the correct year of the twelfth consulate of Honorius and the eighth of Theodosius II. Calculators agree that the band of totality, on its way from, approximately, the Caribbean to the Bay of Bengal, ran not far from Northern Spain, Central Italy, and Asia Minor, with a noon point in the Adriatic region. Ginz.1883, 1899 queried whether Philostorgius observed at Constantinople or in his home town of Borissus (in Cappadocia, perhaps near Armenia). This is an important point, because only by supposing the Byzantine record to originate so far East as Borissus can one easily accept "eighth hour of the day" and a consequent difference of five hours local time between Western and Middle Eastern records. At Constantinople, calculators find mid-eclipse nearer the seventh than the eighth hour of the day. Moreover, the difference in longitude between the West coast of Portugal and Spain on the one hand, and Constantinople on the other, accounts for less than three hours time difference, while the actual travel time of the Moon's shadow accounts for only about another hour. Consequently, one cannot accept as accurate a Portugal-Constantinople time difference of five equinoctial hours (still less five summer seasonal hours).

Later records may be found in Ginz.1883 or Ginz.1899 (some in both). Several sources mention also the great comet of 418. Muller 1975, 840 takes account of Stephenson's 1972 thesis and considers that the Byzantine observation "could have come from anywhere in Asia Minor at least". Newt.1979, 420, 469, uses this famous eclipse for scientific calculations.

See also S.421 May 17, where at least one record refers to S.418 July 19.

21 ecte 18

S.418 July 19 MISDATED ECLIPSE OF PROSPER

Possible records formerly ascribed to S.421 May 17. Scal.1598, 573; 1629, 611, discusses records from South Gaul which we may call (i) Prosper, (ii) Pseudo-Prosper.

(i) is based on consular fasti (arranged by pairs of consuls); a standard edition is now Mommsen's, under the title 'Prosper Tiro, Epitoma Chronicon' (Chr.Min. 1 = AA 9, 1892, 341-499).

(ii) is a chronicle (arranged by regnal years), sometimes called Chronicon Pithoeanum, because edited by P. Pithoeus at Paris in 1588; it is not now attributed to Prosper, and a standard edition is again Mommsen's, under the title 'Chronica Gallica a.452' (Chr.Min.1 = AA 9, 1892, 631-662).

In Mommsen's edition, (ii) does give, under the 26th year of Honorius (about 421) "This year an eclipse of the Sun occurred". Scaliger noted this, and, calling on both (i) and another manuscript of (ii) for ancillary chronology, identified the eclipse as S.421 May 17. Oppolzer shows this as approximately annular in West Africa (near Dakar), the South Sahara, and the southern part of the Red Sea - suspiciously far from Gaul. Scaliger's argument reads cogently and might be valid if his sources were reliable, but they are not. The inaccuracy of (i) is notorious, while, in the immediate vicinity of the eclipse, (ii) has errors of more than ten years in both directions. It is evident that S.418 July 19 is meant (Muralt, 30), as said in Ginzel 1899, 218, quoting Struyck 1740, 111.

Scaliger's identification, copied in Ricc.1653 and Tycho-Curt.1666, must be rejected. We consider this may be an independent Italian record of the 418 eclipse.

421 or 418? S.421 May 17 SOLAR ECLIPSE IN SYRIA

A Syriac source for the 421 eclipse is more promising. The 10th-century writer Agapius, 1912, 408, says, after the death of Yezdegerd I, "Then Warahran (Bahram), his son, who succeeded him, persecuted and oppressed the Christians. That year there was an eclipse of the Sun. The same year there was a battle between the Greeks and the Persians, and many were killed on both sides; the Persians were routed, and the persecution of the Christians ceased". It seems from Camb.Med.Hist., 1, 1924, 464 that Warahran V succeeded in late 420, and that the battle took place in 421 Aug. or Sept. If this is so, the report is more likely to refer to S.421 May 17 than to S.418 July 19, even though the latter would be somewhat larger in Syria. The observation could have been made in Arabia or conceivably the year of the 418 eclipse could have been misdated to fit the battle.

443 S.443 March 17 SUNSET SOLAR IN S.W. EUROPE (UNRECORDED)

The track of annularity starts in the Pacific, has noon point in the Atlantic, and proceeds via Spain to a sunset point near the south coast of France. Ginz.1899 (Map XIV) plots the European end. We have encountered no Western record.

445 (See 447 S.445 July 20 FALSE DATE FOR IRISH ECLIPSE

This former interpretation of the solar eclipse "in the ninth hour" mentioned under 444 in the Annals of Inisfallen is of long standing (e.g. C. O'Connor in his 1825 edition), but is evidently improbable; see under the revised identification, S.447 Dec.23. The eclipse of 445 July 20 was nowhere central, and O'Connor 1952 calculated for central Ireland (8°W, 53°N) a maximum magnitude of only 0.65 at $17^h44^m$ local apparent time; beginning of eclipse $16^h48^m$, end $18^h36^m$. The eclipse of 445 did not begin until the tenth seasonal hour of the day; this is consistent with "in nona hora", or local apparent time before $16^h00^m$.

447 S.447 Dec.23 (Tues.) TOTAL SOLAR IN N.E. PORTUGAL

This eclipse is mentioned most clearly by Hydatius, Chronicon (Chr. Min. 2 = AA 11, 1894, 25): "In the 23rd year of Valentinian, and the 306th Olympiad, an eclipse of the Sun occurred on 10.Kal.Jan. (= Dec.23), which was the third day of the week (Tuesday)". Valentinian III began to reign 425 Oct.23. Some older editions wrongly have 9.Kal.Jan.

A Montpellier manuscript (Excerpta Montepessulana), included by Mommsen among the critical apparatus at the foot of the page, contains an extraordinary hotch-potch, including "from the fourth hour to the sixth and from the eighth hour to the ninth"!

The track of totality goes from a noon point in the East Atlantic, via approximately Portugal, France and Germany to the Baltic. Ginz.1886, 971 tabulates the Hispanic and French parts of the band of totality according to both Oppolz. and himself. He finds that Chaves, believed to have been the episcopal seat of Hydatius, lies within this band. Ginz.1899, 219 also finds totality at Chaves.

Mid-eclipse at Chaves and in Central France occurred about one and two seasonal hours, respectively, after local apparent noon. "From the eighth hour (of the daytime) to the ninth" is reasonably acceptable; "from the fourth hour to the sixth" is clearly to be discarded, as far as this eclipse is concerned.

There are weak reports, perhaps derived from Hydatius, in later writers (Fredegar, etc.). Identification goes back at least to Calvisius 1620, 433. The eclipse is used in Newt.1969, 1970 (71, 75, 263), 1972 (207-9, 507-9). This brings us to consideration of the Irish and Welsh records.

Of these, the chief one is in the Annals of Inisfallen (e.g., ed. S. MacAirt, Dublin 1951, 58-9), where under 444 we read "An eclipse of the Sun in the ninth hour". MacAirt viewed 444 as emendable to 445, and followed C. O'Connor (in his 1825 edition) and Anscombe in identifying the eclipse as S.445 July 20 (q.v.), an identification we have suggested is impossible, see L. Bieler, Irish Hist. Studies, 6, 1949, 252, citing a lecture in Dublin by D.J. Schove. Schove, JBAA, 65, 1954, 38-9, rejected the earlier eclipse as too late in the day, and too small in Ireland. The revised identification, S.447 Dec.23, fits better. For Central Ireland, O'Connor 1952 calculates maximum magnitude 0.90 at the reasonably acceptable time of $13^h00^m$. Anyone who follows Bieler in being dissatisfied with this may note the suggestion that the eclipse report was borrowed from S.W. France or the Mediterranean.

There is also a probable reference in MS. B of the so-called Annales Cambriae; e.g. in the edition by J. Williams ab Ithel (London 1860, p.3) we find under editorial 447 "Dies tenebrosa sicut nox", "A day as dark as night". The identification is fairly certain, but there is a difficulty, for the statement implies totality (or almost so), which is impossible in either Wales or Ireland. Thus the Welsh record, if literally true, comes from S.W. Europe, perhaps via Ireland. Newt.1972, 208, rightly deals leniently with the editorial 447. The eclipse is placed in the fourth year of the Welsh Era of the Annales Cambriae, and normally AD - WE = 444, so that WE 4 would normally mean AD 448; but the year beginning is uncertain, and about Christmas was common. Newt. appears to accept totality at face value (place St. David's) but with low weight.

The late Annals of Lund (or Esrom) have a milder statement on similar lines, under 447 (ed. J. Langebek, Script.ver.Dan., 1, Copenhagen 1772, 221) or 448 (SS 29, 1892, 191). They simply say "Hic dies tenebrosa fuit", "In this year a dark day occurred"; this falls a little short of implying totality. Ginz.1899, 219 mentioned that the eclipse must have been very striking in Denmark, but regarded the ultimate source as not yet determined, which seems to be still the case; Schove (loc.cit.) suspected Western France or the Montpellier manuscript.

449 S May 8 SUNRISE ECLIPSE (?) IN SYRIA
or
450 X Aug.25 or DARKNESS IN CONSTANTINOPLE

John of Asia (in his 6th century Ecclesiastical History) says "On the day on which Marcian assumed the crown, the sky was dark until the evening". Marcian was crowned at Constantinople on 450 Aug.25. No solar eclipse in 450 or 451 was visible in Europe or the Near East. The darkness, if genuine, must have been meteorological. Newt.1979, 386, from the parallel passage in the 13th century version of Michael the Syrian, suggests that the eclipse of 449 May 8 was probably large at sunrise in Syria, but the translation of the original, as given by F. Nau in Revue de l'Orient Chrétien, 1897, 445-493 (see p.457) specially states "Jusqu'au soir" and we have allowed only 1 point for the possibility that the solar and meteorological events of 449 and 450 had been conflated by John's predecessor.

450-3 FICTITIOUS ECLIPSE OF ATTILA

Discredited solar eclipse in the time of Attila.
The first eclipse mentioned by Gregory of Tours is a solar eclipse placed in Book 2, Ch.3 (SrM 1(1), 1884, 66 or 1951, 45; Latouche 1963, 84). This chapter deals entirely with the Vandals in Africa, and in due course we shall discuss the eclipse (AD 485 or 497). Much misdirection of effort appears to stem from Scaliger's imagining that the eclipse related to the famous invasions of France in AD 451 and Italy AD 452 by Attila and the Huns (the subject of Book 2, Ch.5 onwards).

In any case, the identification S.450 April 27, proposed in Scaliger 1598, 575; 1629, 612, and copied in Ricc.1653, 293, 369 and Tycho-Curt. 1666, xxvi, is impossible if an observed eclipse is meant. Meeus and Mucke show the track of totality as beginning in the Pacific, crossing Central America, and ending in mid-Atlantic. Equally invisible in Europe were the other three solar eclipses of 450-1, namely S.450 Oct.21 (South America), S.451 April 17 (far southern hemisphere), and S.451 Oct.10 (North America).

It may be added that the lunar eclipse of 451 Sept.26 (q.v.) is well authenticated.

451 M.451 Sept.26 (p.m.) LARGE LUNAR IN N.E. PORTUGAL

This eclipse is clearly mentioned in Hydatius 1894, 26: "In the 28th year of Valentinian, in the 307th Olympiad, the Moon is darkened from the East (a parte Orientis) 5.Kal.Oct. (= Sept.27)". The 28th year of Valentinian III normally means 452. The eclipse took place early in the evening of Sept.26; Sept.27 in the record may be explained by ecclesiastical usage (taking the evening with the next day), but some editors correct to 6.Kal.Oct. (= Sept.26). Oppolz. gives magnitude 0.83 with mid-eclipse at 6.45 p.m. mean time at Greenwich, or soon after 6 p.m. mean time at Chaves, where the eclipse would already be in progress as the Moon rose in the East, though Ginz.1899, 219 says the whole eclipse was visible there. Identification goes back at least to Petav.1627, 845. The comet mentioned by Hydatius as appearing on June 18 was Halley's (451 June-August).

The eclipse is taken over, much more vaguely, in the History of the Goths, Vandals and Sueves by Isidore of Seville (7th C), and in the Historia Pseudo-Isidoriana (Chr.Min. 2 = AA 11, 1894, 278 and 384 resp.), and probably in Sigebert, s.a.452 (SS 6, 1844, 309); Sigebert's year numbers at this point tend to be one unit too large, so that his 452 is probably our 451. One original source for the Hun invasion of this period was Priscus of Panium, whose work has not survived.

452 recte 451 M.452 Sept.14-15 FALSE YEAR FOR LUNAR ECLIPSE AND COMET

This is the identification given in Calv.1620, 432-6; 1650, 551-4, and copied in Tycho-Curt., xxvi, of an eclipse mentioned by Trithemius as occurring in the eighth year of Meroveus (after whom the Merovingian Kings are named). Meeus and Mucke do indeed give a deep total eclipse of the Moon, of maximum magnitude 1.54 at 1.55 a.m. UT (Goldstine 1973 gives full moon at 1.48 a.m. UT). The identification is reasonable, given Tritheim's work - but reference to this latter reveals a fatal flaw. The work is Joh. Tritheim (Tritemius), Compendium sive Breviarium primi voluminis annalium sive historiarum ..., Mainz 1515, and it mentions (F.vi, recto), among other things:

(i) In the 8th year of Meroveus, "An eclipse of the Moon; a comet of fearful size appeared".

(ii) In the 9th year of Meroveus, AD 453, roman indiction 6, Attila invaded France.

In view of Tritheim's 453 for the Hun invasion, it was natural that the (posthumous) editions of Calvisius should give an eclipse of 452. But Attila's invasion is nowadays dated 451, in spring. One might look for an eclipse in 450, but no lunar eclipse occurred. Tritheim was referring, whether or not he knew it, to Halley's comet of 451 and presumably M.451 Sept.26 (q.v.); as some of Tritheim's material is derived, directly or indirectly, from Hydatius. The year of the accession of Meroveus is uncertain.

453/4 SUNSPOT IN PORTUGAL ?

It may be added that Hydatius has, under editorial AD 453 or 454 (29th or 30th year of Valentinian), a curious "in sole signum in ortu", "a sign in the Sun at rising" (1861, 884, 74; 1879, 731-2). The words "in ortu" seem to rule out any reference to the sunset eclipse S.453 Feb.24 mentioned above, and indeed the statement is worlds away from the form of detailed statement normal for eclipses in Hydatius. It seems best to follow Newt.1972, 509, and ignore this portent, which could refer to a sunspot, as auroral activity was evident about 451 (Schove 1983a, Appendices B and C).

458 S.458 May 28 (Wed.) PARTIAL SOLAR ECLIPSE IN PORTUGAL

This eclipse is clearly referred to in the Chronicle of Hydatius (Chr. Min. 2 = AA 11, 1894, 30): "In the first year of Majorian in Italy and Leo in Constantinople, in the 309th Olympiad, on 5.Kal.Jun. (= May 28), the fourth day of the week (Wednesday), from the fourth to the sixth hour, the Sun appeared reduced in the light of his orb to the figure of the Moon five or six days old". This description implies a magnitude of 0.8 (Newt.1979, 459). Leo reigned from 457 Feb.7, Majorian from 457 April 1, and some editors give second (not first) year. Muralt misdated the eclipse as 457.

The track of totality goes across the Atlantic via the British Isles to Asia; the noon point is in the North Sea. Ginz.1886, 972 tabulated the part of the Oppolz. track which runs between Ireland and the North Sea, and found magnitude 0.775 at Chaves (Aquae Flaviae) at 10.43 a.m. Ginz.1899, 221 found magnitude 0.78 at Chaves about 10.45 a.m., and pointed out that the magnitude corresponds to a six-day-old Moon (he referred to Isidore of Seville, Origines 3, 54, for terms describing sickle shapes of the Moon). The magnitude at Chaves was modest, but agrees tolerably with the statement of Hydatius, who, being interested in eclipses, may have been expecting this one. The eclipse is considered in Newt.1970, 76; 1972, 509. Identification goes back at least to Petav.1627, 847. The eclipse having been total in Britain, Schove, JBAA 65(1), 1954, 39, comments on its non-recording in the British Isles.

462 M.462 March 1-2 (Th.-Fri.) BLOOD-RED LUNAR ECLIPSE IN PORTUGAL

This eclipse also is clearly referred to in Hydatius (Chr.Min. 2 = AA 11, 1894, 32): "In the first year of Severus, in the 310th Olympiad, in the province of Gallaecia various signs of portents are seen. ⟨Aera D⟩ 6.non.Mar. (March 2), from sunset (to) cock-crow the full Moon is turned to blood; it was the sixth day of the week (Friday)". Severus began to reign on 461 Nov.19. "Aera D" means Spanish Era 500, hence AD 462 (the difference in the reckonings being 38 years). Oppolz. gives magnitude 1.02 with mid-eclipse at 2.05 a.m., mean time at Greenwich, and a total duration of $3^h20^m$(the duration of the eclipse is, of course, exaggerated in the record). Ginz.1899, 221 finds magnitude 1.01 at 1.41 a.m., mean time at Chaves. Identification goes back at least to Petav.1627, 846.

A probable brief reference in Fredegar, Book 2, Ch.7 (SrM 2, 1888, p.77, 1.7) would have been difficult to date if its wording had not suggested an abbreviated borrowing from Hydatius.

464 S.464 July 20 (Mon.) SOLAR ECLIPSES, PARTIAL IN PORTUGAL AND LARGE IN SYRIA

This is the last eclipse unmistakably mentioned in Hydatius (Chr.Min. 2 = AA 11, 1894, 33): "In the second year of Severus, in the 311th Olympiad (should be 310th), on 13.Kal.Aug. (= July 20), the second day of the week (Monday), from the third to the sixth hour, the Sun is observed to be diminished in its light to the figure of a 5-day-old Moon". Severus began to reign 461 Nov.19, and some editors put the eclipse in his third (not second) year. This description enables Newt. 1979 (354, 372, 459) to estimate its magnitude as $0.85 \pm 0.075$.

The track of annularity runs from the Atlantic to approximately Brittany, Northern France, Central Europe, South Russia to the Himalayas. The Western European part of the Oppolz. track is tabulated in Ginz.1886, 972, which gives the greatest magnitude at Chaves as 0.82 about 7.07 a.m. Ginz.1899, 222 finds the greatest magnitude at Chaves to be 0.91 at 7.01 a.m. The times are perhaps a little earlier than one would expect from the record. Identification goes back at least to Petav.1627, 848. See also Louis du Four de Longuerue, Disquisitio de annis Childerici I Francorum regis (Bouquet, 3, 1869, 681). The eclipse is mentioned vaguely in Fredegar (SrM 2, 1888, 77). It has recently been dealt with in Newt.1972, 510.

The Syrian writer, Agapius (8(3), 1912, 419) says "In the ninth year of the reign of Leo, there was an eclipse of the Sun and the stars appeared (in daytime)". As Leo I ascended the throne in 457 Jan., this is a reference to S.464 July 20, with an error of one year. Ginz.1899 (Map XIV) shows a brief stretch of the band of annularity running north of the Sea of Azov and through the North-Central part of the Caspian Sea. The source of Agapius may thus have been from Armenia.

467 M.467 June 3 (p.m.)? COMET (OR ECLIPSE GHOST) IN GAUL

The last certain eclipse in the Chronicle of Hydatius is S.464 July 20. But a damaged section (Chr.Min. 2 = AA 11, 1894, p.34, c.242) has something meteorological or astronomical, apparently among portents seen in Gaul and reported by an embassy which returned in the second year of Anthemius (whose reign began on 467 April 12). We read of "another sun seen after sunset". Newt.1972, 510 thinks that the passage may refer to the partial lunar eclipse of 467 June 3. No other lunar eclipse was visible in Western Europe until the autumn of 469, although the comet of 467 seems to account for the remark. The sunspot maximum was uncertain (Schove 1983a, Appendix B), so that an auroral corona is possible.

c.472 VOLCANIC DARKNESS

The darkness was ascribed by Count Marcellinus to an eruption of Vesuvius (see Stothers and Rampino 1983).

484 S.484 Jan.14 (Sat.) SOLAR ECLIPSE IN GREECE AND PERSIA

In the life of the Athenian philosopher Proclus, written, probably very soon after the death of Proclus, by his pupil and successor Marinus, an observed solar eclipse and a predicted eclipse are mentioned; Marini vita Procli, ed. J.F. Boissonade, 1814, Ch.37, p.29. "Portents occurred a year before his death, such as the solar eclipse, which was so considerable that night occurred in the daytime. For there was deep darkness and stars were seen. This happened in Capricorn near the rising point (of the Sun). The Almanac makers also noted another eclipse as due to occur about the end of the first year." Ginz.1899, 222 gives Greek text and German translation; Newt.1970, 119-120; 1972, 540, gives a first and a revised English translation, with comments; there is an English translation of the whole Life in L.J. Rosán, The Philosophy of Proclus, New York 1949, pp.13-35. Stephenson and Clark 1978, 4, give a revised translation and state that this is probably the most reliable of all solar eclipses reported in the Classics, adding "It is a pity that there is no precise mention of totality". We have credited it with 9 points for identification and 6 for the information contained. This eclipse is used by Newt.1979, 420, by Muller and Stephenson 1975, and by Muller 1975.

In Ch.35, Marinus says that Proclus died on April 17 in the 124th year after the rule of Julian. Here Marinus is counting his years for ideological reasons from the reign of Julian the Apostate, who became sole emperor on 361 Nov.3. We are aware of no reckoning which would make Marinus put the death of Proclus outside the triennium AD 484-6; AD 485 is most commonly accepted. Marinus also gives the archon of Athens for the year as the younger Nicagoras, who appears to have functioned in 484-5.

In that period, or even one extended at both ends, the only solar eclipse which occurred in Capricorn was S.484 Jan.14, which did occur in Capricorn and around sunrise at Athens. As far as we know, the identification has never been challenged. The discussion in Ginz.1899, 222 and the re-discussion by Neugebauer (1931) argue respectively for totality actually at, and only near, Athens.

The earliest identification we have ourselves inspected is given in Ricc.1653, but this refers back a few years to the catalogue of Reinerius (Vincenzo Reinieri, d.1648).

The other eclipse, which is merely predicted, is usually considered to be S.486 May 19 (q.v.).

S.484 Jan.14 is best known in relation to Proclus, as above. But the track of totality, beginning at sunrise in or near Greece, travelled via approximately Southern Asia Minor, Syria and Mesopotamia to a noon point in Central Asia, and there is a correctly dated (though not contemporary) record from the Near East. "An 795. En lequel le soleil s'éclipsa le samedi 14 Kanūn II, à trois heures de la journée, et les étoiles apparurent. En ce temps-là, Piruz, roi des Perses, fut tué. (Hist. ecclés. de Barsohède de Karka)". The extant source is Elias 1910, 74. Seleucid 795 is AD 483-4, second Kanūn is January. According to Bury, Hist. Later Rom. Emp., 1, 1923, 397, Piruz fell in battle in 484 January. "At three hours of the day" sounds late for an eclipse which probably occurred an hour or so after sunrise, but no great accuracy was intended; the third, sixth and ninth hours may be regarded as a canonical division of the day into four parts. Barsohedes is described by Delaporte (p.xi) as a Nestorian writer of the commencement of the eighth century.

485 S.485 May 29 USUAL (FALSE) DATE FOR GREGORY'S ECLIPSE

See 497.

486 S.486 May 19 SOLAR ECLIPSE IN SYRIA OR ARABIA

This eclipse is mentioned probably by Marinus and certainly by Elias. Apart from the observed solar eclipse of 484 Jan.14 (q.v.), the Vita

Procli mentions only a predicted eclipse (solar not stated, but probably meant). There is doubt, if only slight, about the identification of the predicted eclipse. The following possibilities arise.

The death of Proclus may have occurred on April 17 of (A) 484, (B) 485, or (C) 486. Strict regnal years of Julian point to (B). Year-beginnings on Jan.1 (consular) or in spring (before April 17), both unlikely, could give (A) or (B); summer or autumn year-beginnings could give (B) or (C).

The predicted eclipse may have been (1) S.485 May 29, (2) S.486 May 19, or (3) S.487 Nov.1. For (1), see above, under Gregory's first eclipse; Ginzel found it invisible at Athens, but it may still have been predicted. For (2), total in west Africa, Libya, Arabia, etc., Ginz.1899 gave 0.68 as the magnitude of the partial eclipse seen at Athens. (3) was an annular eclipse, shown as traversing the length of the Mediterranean from the Pyrenees to Palestine (Oppolzer's rough track) or Lower Egypt (Ginz.1899, Map XIV); it probably had a somewhat greater magnitude than (2) as a partial eclipse at Athens.

The time intervals to be considered are (i) between the observed eclipse and the death of Proclus, (ii) between the death of Proclus and the predicted eclipse. With regard to (i), what we have translated (with Rosán) as "a year before his death" is πρὸ ἐνιαυτοῦ τῆς τελευτῆς ; this would presumably allow an interval differing from one year by a few months either way. Other translations are "for a year before his death" (Newt.1970, 1972) and "before the year of his death" (Ginz.). With regard to (ii), what we have translated as "about the end of the first year" is πληρουμένου τοῦ πρώτου ἐνιαυτοῦ; this is usually taken as meaning after the death of Proclus, and does strongly suggest an interval fairly close to one year.

The most likely combination is B2, i.e. death of Proclus on 485 April 17 and predicted eclipse S.486 May 19. On account of the comparative precision of (ii), the minor possibilities, in order of decreasing likelihood, appear to be A1, C3, A2. The likelihood of C3 is somewhat enhanced if "about the end of the first year" may be taken as referring to a year 'of Julian' beginning on 486 Nov.3 and ending on 487 Nov.2. See Newt.1972, 526, 540-1.

The above relates to eclipses within a year or two of the death of Proclus. There are also a queried horoscopic date of birth and an age at death to be accommodated. See Rosán (loc.cit. under S.484, 34) and references there given. Neugebauer 1975 (2, 1032ff.) discusses the horoscope in detail and confirms 412 Feb.8 for the birth of Proclus.

Elias 1910, 74 says: "An 797. En lequel le soleil s'éclipsa le lundi 19 'Ijar, à neuf heures de la journée, et les étoiles apparurent". Seleucid 797 is AD 485-6 (autumn to autumn in Elias), and 'Ijar is May, so that the date is correct. "At nine hours of the day" in Elias implies little more than that the eclipse occurred in the afternoon; this is doubtless true, as the noon point is in Libya (Map XIV in Ginz. 1899). The track continued approximately from West to East through Sinai and Northern Arabia. For this eclipse it happens that Elias, uncharacteristically, gives no source; but his sources in general include Arabic as well as Syrian authors.

493 recte 496/7 S.493 Jan.4 SUNRISE ECLIPSE IN SYRIA (UNLIKELY RECORD)

The 12th century Michael (ix, 7, t.II, Paris 1901 = Brussels 1963, p.154) mentions an eclipse of the Sun early in the reign of Anastasius I (491-518). This might refer to S.493 Jan.4; Oppolzer's track of totality begins at sunrise in Arabia, and on this basis the eclipse would be visible as a partial one at sunrise in Syria. Newt.1979, 386 assumed that it was the first year of the reign and thus considered 492 Jan.15 as a possibility.

Michael's eclipse is probably that of 497 (or 496), recorded in the earlier source Marcellinus Comes. Probably Vasiliev, in his translation of Agapius (PO 8(3), 1912, 425) does not really mean to imply identification as S.512 June 29; he refers to "Mich. le Syr., II, 154", but in his context this appears to be an inadvertent reference to the wrong eclipse passage; for 154 read 168.

The solar eclipse in CS under "493" really belongs to 497 (or 496).

496 S.496 Oct.22 SOLAR ECLIPSE IN S.W. ASIA (NO CLEAR RECORD)

The eclipse of this date would have been visible in S.W. Asia and a possible reference to it occurs in John of Asia (mid 6th century) (p.463), "In the year 811 (which should convert to AD 500), Saturday, October 23, the sun was obscured up to the 8th hour". However, as we explain under 497, the magnitude in even Armenia would be insufficient to cause darkness. A similar report in the contemporary Chronicle of Edessa (AD 499) suggests that the event was meteorological and three or four years later.

496-7 S.497 April 18 (or 496 Oct.22) SOLAR ECLIPSES OF MARCELLINUS (S.E. MEDITERRANEAN)

S.497 April 18 (more probable than S.496 Oct.22).

There is a reasonably contemporary record (prima facie under 497) from Byzantium, and from Ireland there are records (ostensibly under 496) which are derivative. We shall see that all probably refer to S.497, but that S.496 cannot be ruled out. We shall consider first the Byzantine record, then the Irish records.

(i) The Byzantine record.

This occurs in Marcellinus Comes, Chronicle, Chr.Min. 2 = AA 11, 1894, 94. The account is a mere "Solis defectus apparuit" (An eclipse of the Sun occurred"). It is placed in (editorial) year AD 497, in the fifth indiction (496-7), in the second consulate of the emperor Anastasius I (without colleague). Similarly in PL 51, 1861, 935.

The usual identification, S.497 April 18, goes back at least to Calv.1620, 448. For Constantinople he found greatest phase at 6.05 p.m. and greatest magnitude 0.66 (given sexagesimally as 7.57 digits, and misquoted in at least one edition of Struyck-Ferguson as 17.57). For the same eclipse Ginz.1899, 223 found, again for Constantinople, maximum phase at 5.42 p.m. and maximum magnitude 0.68. Boll 1909, 2364 adopted the same identification.

Newt.1972, 541 regards S.496 Oct.22 as a possible, if less probable, identification of the record in Marcellinus, which he consequently considers cannot be identified safely.

The central line for S.496 Oct.22 runs from Northern Scandinavia via Russia to a noon point in Central Asia. Oppolz. gives it as total in his tables but as annular on his map. Newt. estimates the magnitude as only about 0.7 in Constantinople or Illyria, the native land of Marcellinus.

The central line for the annular S.497 April 18 runs from a noon point in the Atlantic and then by North Africa and Egypt to Arabia. From the map, Newt. estimates magnitude perhaps 0.8 at Constantinople and somewhat less in Illyria. We have seen that Ginz. found less than 0.7 at Constantinople. He plots the final stretch from Algeria to Arabia.

Note that both S.496 Oct.22 and S.497 April 18 fell in the fifth indiction (so also did S.512 June 29, mentioned by Marcellinus in rather similar words). Thus the annalistic decision is left to the consular year. If Marcellinus put the eclipse in the correct consular year, and this is correctly equated with AD 497, then the eclipse was S.497 April 18. But if there is doubt on either score, then the eclipse identification becomes doubtful. However, there is a non-annalistic consideration. While the two eclipses had rather similar magnitudes (0.7, more or less) at Constantinople, under conditions about equally favouring visibility (that of 496 not long after sunrise, that of 497 not long before sunset), the central line in 496 ran far north of Constantinople and in 497 far south. Now (as references to Antioch that follow in Marcellinus confirm) the contacts of Byzantium were stronger with the south than with the north. Consequently, while allowing S.496 Oct.22 as a possibility, we consider the traditional identification of the Byzantine record, namely S.497 April 18, as correct, but consider the place of observation likely to have been Syria or Egypt.

(ii) The Irish records.

The chief surviving record is contained in the Annals of Ulster (1887, 32-3) under manuscript year 495 (which would normally mean true 496). The wording is "Solis defectus apparuit", just as in Marcellinus. The eclipse is indexed as S.496 Oct.22 (AU 4, 1901, 140). Similar brief mention occurs in AT and CS. In AT, 1896, 122, the annal has no explicit year number of its own, but the ferial "K.ii" (i.e. Jan.1 fell on Monday) is consistent with AD 496. In CS, 1866, 32, the mention appears under editorial "493"; but although Hennessy (p.xlvi) gives the general systematic error at this period as zero, his comments on this particular annal point to 496, and in fact the annal has similar content to AU "495" (496). J. O'Donovan, in his Introduction to Annals of the Four Masters (1, Dublin 1851, xlviii), quotes AU as 495 (496). A.O. Anderson (Early Sources of Scottish History, London 1922, p.1) identifies the eclipse as perhaps S.496 Oct.22, visible at Rome at 8 a.m.

It will be seen that S.496 Oct.22 was long the standard identification. But, as explained in Schove, 1954, 37-43, it is now thought that the eclipse report may be copied from Marcellinus. We have already seen that S.497 April 18 is the more probable identification of that Byzantine record.

The difficulty about regarding the report as really originating from Ireland is that both eclipses were so unremarkable there. S.496 Oct. 22 cannot, with centrality in Scandinavia, have been at all striking in Ireland. Eddington (see Schove 1954, 40) found maximum magnitude 0.10 for Ireland; even allowing for the helpful effect of possible mist at sunrise, the magnitude is not enough. For S.497 April 18, O'Connor 1952 gave, for mid-Ireland, maximum magnitude 0.31 at 2.40 p.m.; such an eclipse in broad daylight would not have been noticed.

497 or 485?

S.497 April 18 GREGORY'S FIRST SOLAR ECLIPSE (N. AFRICA?)

(or 485 May 29)

The first eclipse in Gregory of Tours (485? 497?). "Then the Sun appeared hideous (teter), so that scarcely a third of it gave light; I believe (this occurred) on account of such crimes and the shedding of innocent blood." Gregory (Book 2, Ch.3, SrM 1(1), 1884, 66, or 1951, 45; Latouche 1963, 84). Unlike the other eclipses in Gregory's famous work, this one took place well before his own lifetime, in a period about which his sources failed to inform him satisfactorily. His arrangement is not designed to be strictly chronological, and is even less so than he intended. Except for the eclipse, whose place of observation is not stated, Ch.3 of his second book deals entirely with the Vandals in Africa, and covers the whole period from their crossing to Africa in 429 to the extinction of the Vandal state in 534. Gregory's chronological weakness in this chapter goes beyond the omission of dates; he does not even know the correct order of succession of the Vandal kings.

Thus although the identification of the eclipse as S.485 May 29 has been standard, a revision is suggested below. Oppolz. shows S.485 May 29 as total in the far North, with central track going approximately by Kamchatka, the polar regions, North Greenland, and Central Scandinavia. Ginz.1899, 223 finds magnitude only 0.73 at Clermont, and the eclipse would agree well enough with the record almost anywhere in France (if that is indeed the region in which it was seen). Ginz. finds the same magnitude, 0.73, at Rome. The identification goes back at least as far as Calv.1620, 445. Ginzel's quotation of the Paschale Campanum in connection with S.485 May 29 is wrong; the passage refers to S.512 June 29 (q.v.).

Let us now consider the degree of validity of this common identification. Gregory appears to narrate the eclipse most closely in the context of

(i) miracles performed by African Catholic bishops Eugenius, Vindimial, and Longinus,

(ii) anger of Arian king Huneric,

(iii) deposition and exile of Eugenius,

(iv) martyrdom of Vindimial, Octavian, and many others.

(v) apostasy of bishop Revocatus from the Catholic faith,

(vi) death of Huneric.

Since Huneric is believed to have succeeded in 477 and to have died on

484 Dec.23, and one exile of St. Eugenius to have lasted from 484 to 488, and the persecution under Huneric to have occurred especially in 483 and 484, the standard identification of the eclipse appears, in these respects, reasonable. But after the persecuting king Huneric there followed a more tolerant king Gunthamund (omitted by Gregory), and then another king Thrasamund (seriously misplaced by Gregory), under whom there was further persecution about 498. Eugenius was finally exiled in 497. We notice that a considerable authority gives "about 498" for the martyrdom of Vindimial and Octavian (also, apparently, Longinus); P. Monceaux, Histoire Littéraire de l'Afrique Chrétienne, Tome 3, Paris 1905 (reprint Brussels 1963), pp.543-551. Such a date leads naturally to the consideration of the solar eclipse of 497 April 18, whose track of annularity runs the whole length of North Africa from south of the Canaries to Egypt. We regard S.497 April 18 as the probable identification of Gregory's first eclipse; this eclipse is mentioned (see under 497) also in Marcellinus Comes and in Irish Annals, but the primary source has not been traced.

429 429 Dec.12

"6th year of the Yuan Chia reign period, 11th month, day chi-ch'ou, the 1st day of the month, the Sun was eclipsed; it was not complete but was like a hook; at the time of the eclipse a star was/stars were seen; at the hour pu (3 - 5 p.m.) it was over; in Ho-pei the Earth was dark." This account is from the Treatise on the Five Elements in the Sung-shu (chapter 34). It would seem that the main observation was made in the Sung capital Nan-ching (computed magnitude 0.92); Venus would probably be the only object seen. The computed track runs to the north of Nan-ching; the wide belt of totality would cover most of the Ho-pei province. (Contribution by R. Stephenson.)

S.447 Jan.2

S.462 Mar.17

M.474 Jan.19

In Mesoamerica at this time Maya astronomers were able to predict lunar eclipses with great accuracy. Their medieval Dresden Codex includes a Lunar or Eclipse table for predictions up to 33 years ahead. In its extant form this is built up from three bases relating to the eclipse month beginning with 842 March 15 (AD 756 in the usual conventional chronology as we explain in our Appendix C). The table betrays evidence for a prototype and three 5th century dates in the Codex are structurally related to visible eclipses. One date is 462 Mar.14, 380 years before the first base, and 3 days before a calculated solar eclipse and 12 days after a visible total lunar eclipse. The other two dates are in 447 and 474, both years of visible solar eclipses; the dates are incorrect as they convert to Jan.23 and Jan.28, whereas the solar eclipses were respectively Jan.2 and Dec.24; however, the second date was only 9 days after the total lunar eclipse of 474 Jan.19 at node passage. Some of the Maya dates were adjusted to fit solar and lunar cycles (e.g. the 11,960 day cycle) and to avoid dates that were unlucky. These relationships are explained in Schove 1983 and the new correlation in Schove 1982/4

C.H. Smiley in ed. A.F. Aveni 'Archaeoastronomy' 1975, 253 (Austin, Texas) had a different correlation, but he pointed out that in the "interval AD 477 to 510, not a single solar eclipse occurred anywhere on earth without warning (in the table), nor was there a single false warning". The Maya were certainly successful in predicting solar eclipses, as 11th century dates in the same Codex confirm. Probably observations of solar and lunar eclipses in the 5th century helped to provide the empirical basis for the Dresden Table.

# THE SIXTH CENTURY

## THE AGE OF SYRIA: FIRST RECORDS FROM IRELAND AND FRANCE

The revival of the Mediterranean Empire under Justinian was cut short before mid-century by famine and plague, and cultural activity in Constantinople and in Italy about the 530s reflects this brief prosperity. However, even in the first half of the century Syria is important and in the second half the south-east part of the Empire becomes the 'tail that wagged the dog'. The influence of Syria extended to the British Isles in Art and in Religion. In religion this division within the Eastern Empire is reflected in the Monophysite heresy, which was opposed to Constantinople and which prepared the way for the transition to Islamic rule. The contemporary historians of Antioch (e.g. Malalas, Evagrius) are now more important than those of Constantinople.

In the West the detailed chronicle of Gregory of Tours in what is now north-west France gives eyewitness accounts of natural phenomena (cf. Schove 1983, 43-44) and paves the way for the growth of chronicles in Western Europe, a growth which gathers momentum over the next few centuries; already the Irish Annals provide an eyewitness record of an eclipse seen at Bangor (N.E. Ireland) or in Iona (Scotland) in the 590s.

In India astronomy was sophisticated but we have no records of observations, and our cometary chronology depends as usual largely on the Chinese Histories.

511 S.511 Jan.15 POSSIBLE SUNSET ECLIPSE

Possibly a sunset eclipse at Constantinople; hence a debatable alternative to S.512 June 29 as an identification of an eclipse report given below. Totality in Oppolz. runs from a noon point in the Atlantic to the Sahara, Tripolitania, and an end in the Mediterranean (between Benghazi and Sicily). This would give virtually no sunset eclipse at Constantinople. Meeus (letter 20 Oct. 1982) confirms that it was not large enough to be noticed at the capital itself. However, in view of Newton's comments in 1972, 542, 592 and 1979, 387, we have allowed 1 point for an observation from further to the south-west.

<u>512</u> S.512 June 29 (Fri.) LARGE MEDITERRANEAN ECLIPSE

This was primarily a Mediterranean eclipse. The approximate track of totality shown runs from the Atlantic to China, passing near Constantinople and the south coast of the Black Sea. A full calculation in Ginz. 1899 made the magnitude about 0.94 at Constantinople about 10.35 a.m. We shall consider first Latin and Greek accounts, and probable Irish copyings from Marcellinus Comes, and then some Syriac records.

The eclipse is fairly certainly (but see below) referred to in Latin sources edited by Mommsen:

(i) Paschale Campanum, Chr.Min. <u>1</u> (= AA <u>9</u>, 1892), 330, 747;

(ii) Marcellinus Comes, Chr. Min. <u>2</u> (= AA <u>11</u>, 1894), 98.

The title of the first source refers to Campania, the region round Naples; the work, which does indeed mention two eruptions of Vesuvius, is of the late 6th century. Count Marcellinus was a high Byzantine official of the early 6th century.

The year is given as 512 by Mommsen in each case, but arrives differently: in (i) as "post consulatum Felicis" and in (ii) as the consulate of Paulus and Muscianus. These indicate the same year, as Felix was consul in 511, and Paulus and Muscianus were consuls in 512.

The descriptions are:

(i) "Hoc anno in k.Iul. sol eclipsin passus est, et monte Besuvio ardente [i.e. eruption of Vesuvius] VIII id. Iulias tenebrae factae sunt per vicinium montis";

(ii) "His fere temporibus solis defectus contigit".

Though June 29 is not the kalends of July, it does lie within the Kalendic-reckoning period (June 14 to July 1 inclusive). Moreover, emendation of "in k.Iul" to "iii.k.Iul." (June 29) suggests itself, and has several times been proposed.

The same eclipse is mentioned by the contemporary Byzantine Greek author John Laurentus Lydus (c.490-c.565), in his work on portents (De Ostentis, Ch.6; ed. I. Bekker, Bonn 1837, 280; second Teubner ed., C. Wachsmuth, 1897, 13). He records that stars appeared; he mentions no month or day, but correctly says that the eclipse occurred "six years before the death of Anastasius" (518 July 8).

S.512 June 29 can hardly fail to be the correct identification of the Paschale Campanum eclipse, but the reader may be warned of a muddle in the literature. Ginz.1899 gives the reference to Paschale Campanum, 747, under both S.485 May 29 and S.512 June 29. For 485 he emends "in k.Iul." to "IV kal.Iunias" (= May 29); for 512, he tacitly emends "in k.Iul." to "III.kal.Iul." (= June 29). But the year 485 is calculated according to the cycle of Victorius (Chr.Min. 1 = AA 9, 1892, 686), which differs by 27 years from the AD reckoning; it corresponds with AD 512. The eclipse of AD 485 (q.v.) was conceivably mentioned by Gregory of Tours, but there seems to be no genuine mention of it in the Paschale Campanum.

This is relevant to the Irish records, which we now consider. The Annals of Ulster, 1887, 36-37, under AD 511 (normally meaning true AD 512, as also here implied by ferial number and epact), have "Solis defectus contigit", the very words of Marcellinus. The CS, 1866, 36-37, has "Defectus solis contigit" under 510, but the editor refers to a note by R. O'Flaherty mentioning 512.

As Marcellinus is known to have been among the sources of AU, the Irish records of this eclipse are usually thought to have been borrowed from Marcellinus (or other Mediterranean source); see Schove 1954, 39 and Newton 1972 (pag.cit.). The fact that Irish records do not mention the large solar eclipse of 507 March 29 now makes one realise that they were not recording eclipses as phenomena observed in Ireland at this period. The index to AU (4, 1901, 140) naturally adopts the traditional S.512 June 29.

S.512 June 29 is also mentioned in Syriac sources, which add rough information about time of day.

The 6th century Syriac Chronicle, Book vii, end of Ch.9 (English translation by F.J. Hamilton and E.W. Brooks, London 1899, 178) mentions, in connection with ecclesiastical events known to have occurred near Amida in Syria in 512, "an eclipse of the Sun, which took place in those days, and produced darkness from the sixth hour unto the ninth hour".

Agapius (10th C), 1912, 425, says: "In the 22nd year of Anastasius, in the month of Haziran (June), at midday, there was an eclipse of the Sun". As Anastasius was proclaimed Emperor in 491 April, the year is correct.

In Michael the Syrian, Chronicle, Book ix, Ch.xi (tr. J.B. Chabot, 2, Paris 1901 = Bruxelles 1963, 168) the account becomes "In the time of

Anastasius ... There was also a solar eclipse, on a Friday, from the third to the ninth hour: a portentous sign (signe prodigieux)". The editor refers to James of Edessa. The duration of the eclipse is obviously exaggerated. Newt.1979, 386 also accepts this identification.

The noon point on Oppolzer's track being near Tiflis, the three Syriac statements about time of day may pass as crudely correct. None strays outside the interval from the third to the ninth hour, so that all refer to what may be called the middle of the day (as opposed to near sunrise or near sunset). As far as the Syriac Chronicle is concerned, it should be noted that any eclipse at all near noon is liable to be described, under gospel influence, as "darkness from the sixth hour unto the ninth hour".

512 again LARGE MEDITERRANEAN ECLIPSE
NOT S.526 Sept.22 Not the African 526 eclipse

Oppolzer shows the central line of the 526 annular-total eclipse as starting off the north-west coast of Africa, passing via Mauretania to a noon point in Kenya. The true track was slightly further north and is supposed to be referred to in the early 11th century by Elias 1910, 75: "An 837. En lequel le soleil s'éclipsa au milieu du jour". With Elias, Seleucid 837 means AD 525-6, and the year begins on Oct.1; thus his year is correct. It would be a surprise to find a Syriac writer able to mention an eclipse, some five centuries earlier, with its noon point two thousand miles south of Syria. He frequently indicates his source, but unfortunately not here. His sources in general include both Christian and Islamic writers. The eclipse would be very large in Ethiopia. A clue is provided by the work of the Armenian historian Samuel of Ani (see traduit M. Brosset 'Collections d'historiens arméniens 1876, St. Petersburg, Vol.II, 391. Samuel d'Ani, Tables chronologiques) who placed beside the section dated c.527/530 the statement "A total eclipse of the sun, followed by a horrible famine". Ani's dates are all too late - he places this before Justinian's first year which he dates 533 instead of 527/8. Famine and locusts were reported in 512 in both versions of Bar Hebraeus (cf. Muralt 1855, 680). We conclude that the eclipse is probably that of 512 and we have allowed 1 point for 526.

534 S.534 April 29 (Sat.) ANNULAR ECLIPSE IN S.E. MEDITERRANEAN

Oppolzer's chart shows the approximate central line of this annular eclipse as starting at sunrise in West Africa, proceeding via the East Mediterranean to a noon point in the Northern Urals, and then on to N.E. Asia. Conjunction was about UT 7.05 (Oppolz.), 7.01 (Goldstine 1973). The accurate band of annularity in Ginz.1899 goes by North-West Egypt, Eastern Asia Minor, and the South-East Black Sea.

Newton 1972, 455-8 makes out a strong case for this as the eclipse referred to under "539?" in a defective annal edited by Mommsen, and sometimes mentioned in connection with Bede's eclipses of 538 and 540. Among a number of Italian sources, which he called collectively "Consularia Italica", Momm. edited the 6th century Fasti Vindobonenses with Excerpta Sangallensia (so-called from manuscripts at Vienna and St. Gall respectively). In Mommsen's edition, the St. Gall extracts state:

> "539? p.c. Bilisarii IIII et Stratigi IIII [sequitur litura sex litt.] tenebrae factae sunt ab hora diei III usque in horam IIII die Saturnis [sic]".

The day of the week is not fitted by any eclipse available in 536-541 (qq.v.). Belisarius was consul in 535-6-7 only; Strategus merely means General. The hour of the day (third to fourth) is reasonable, but the annalist seems to have put the eclipse in the wrong year.

Records of this eclipse seem fated to err. Agapius (10th C), 1912, 428, speaking of Justinian, says: "In the eighth year of his reign, there was an eclipse of the Sun, on the 29th of Nisan (April), at two hours in the afternoon". Justinian was emperor from 527 April 1, sole emperor from 527 Aug.1, so that the year is reasonable. But in the afternoon is quite wrong; no doubt the second hour of the day is meant.

536 Mar./537 June VOLCANIC DARKNESS

Cassiodorus, Variae, Book XII, Epistle 25, addressed to a certain Ambrose, has a passage which has been thought to refer fairly plainly to an eclipse of the Moon, and also apparently to a weak Sun for almost a year. The passage (in the author's usual wordy style) may be found in MGH AA 12, 1894, 381 (ed. Momm.) or in PG 69, 1865, 875, or ed. A.J. Fridh, Corpus Christianorum, Ser.Lat., 96, Turnhout (Belgium), 1973, 493. The passage in question refers to the weak sun and describes the moon as "full in its circle (orbe suo), and drained of its natural brightness". The Variae were published in AD 537, and since Ep.25 is, in position, towards the end of the set of official letters which Cassiodorus wrote as pretorian prefect (AD 533 on), one thinks first of the deep total lunar eclipse of 535 April 4, entirely visible at Ravenna in the early hours of the morning. However, Momm. dates the letter AD 533, the earliest possible year; Mommsen's name index (p.488, under Ambrose) gives a possible clue to his reasons. Fridh also gives 533, but we show that it must be 536.

Procopius, despite the length of his works, mentions no clear eclipse. He does however say (Wars, Book 4 = Vandal War, Book 2, Ch.14; Loeb ed., H.B. Dewing, Vol.2, 1916, 326-9):

"During this winter Belisarius remained in Syracuse and Solomon in Carthage. And ... during this year a most dread portent took place. For the Sun gave forth its light without brightness, like the Moon, during this whole year, and it seemed exceedingly like the Sun in eclipse, for the beams it shed were not clear nor such as it is accustomed to shed ... And it was the time when Justinian was in the tenth year of his reign". We may add the Bonn reference (ed. W. Dindorf, I, 1833, 469) and the Teubner reference (ed. J. Haury, I, 1905, 482).

Most contemporary reports are no longer extant, but they were summarized later by Michael (2, 26, 1963, 220-221), probably after John of Ephesus, whose work is partly lost. "... in the year of the Greeks 848 (i.e. AD 536) there was a sign in the Sun (Editor compares with Pseudo-Denys ad ann. 842, another late source, Rev. de l'Or. chr. 1897, 476). Such a thing had never been seen before and it is nowhere written that anything similar had happened in the world ... The Sun was dimmed, and the darkening lasted a year and a half, that is to say, 18 months. Each day it shone about four hours and even then the light was merely that of a weak shadow. Everybody declared that its original light would never return. The fruits did not ripen and wine had the taste of 'acid grapes' (the Chronography in the Bar Hebraeus ed. E.A.W. Budge, Vol.I, 1932, Oxford Univ. Press, says 'urine').

Theophanes (8th C) wrote AM 6026 (i.e. AD 535, 1839 ed., 313, 1883 ed., 202): "The Sun was without rays like the Moon and its brightness was dimmed for a whole year. For the most part it seemed to be in eclipse, not shining as clearly as usual. And it was the tenth period of the reign of Justinian. In this period neither war nor death ceased attacking Mankind".

The Syriac chronicle ascribed to Zacharias of Mytilene is a contemporary source and stated (with reference to the Constantinople visit of

Pope Agepetus who died there on 536 April 22) that "the earth (at Constantinople) with all that is upon it quaked; and the Sun began to be darkened by day and the Moon by night, while (the) ocean was tumultuous with spray (?) from March 24 in this year till June 24 in the following year (i.e. 536/7) fifteen". Book 9, Ch.19, trans. Hamilton & Brooks, 1899, 267 ("... And, as the winter (in Mesopotamia) was a severe one, so much so that from the large and unwonted quantity of snow birds perished ... there was distress ... among men ... from the evil things"). (Personal communication from Dr. R. Stothers; see now R.B. Stothers 'Mystery cloud of AD 536', Nature, (accepted) 1984.)

The solar observation is often ascribed to sunspot activity, but this is inaccurate (see Schove 1983).

An acidity layer in the Greenland ice-cores was dated 540±10 (Hammer et al. 1980, 235 note) and a volcanic eruption in 536 is almost certainly the explanation of all the reports.

538 and 540 S.538 Feb.15 (Mon.) BEDE'S MEDITERRANEAN ECLIPSES

The earliest known record of two morning eclipses, in 538 and 540, occurs in Bede's Ecclesiastical History, Book V, Ch.24 (in the Epitome, not the main narrative), ed. B. Colgrave and R.A.B. Mynors, 1969, 562. At 538 we read: "Anno DXXXVIII eclypsis solis facta est XIIII kalendas Martias ab hora prima usque ad tertiam", i.e., "In the year 538 an eclipse of the Sun occurred on Feb.16, from the first to the third hour". Notice that the date is one day out (the 14 days being normally counted inclusively at each end); see also under S.664 May 1. The account appears (apparently copied, either directly or indirectly) in many later English sources (e.g. Anglo-Saxon Chronicle, "In this year there was an eclipse of the Sun on Feb.16 from daybreak until nine o'clock in the morning", tr. D. Whitelock et al., 1961, p.12) and some European sources; all the more important such parallel passages are noted in Ginz.1883, 1899 and Newt.1972, 143-9.

The part of the band of totality shown in Ginz.1899 passes over Cyrenaica, Central Asia Minor (including Ankara), and Rostov-on-Don towards Siberia. Magnitude at London was not more than about 0.7, somewhere round 8 a.m. This could not be a genuine English record, and some Mediterranean source (e.g. Asia Minor or Italy) is necessary. The eclipses of 538 and 540 are discussed in Newt.1969 and in many places in Newt.1970, 1972.

Moreover, the eclipse of 540 was total or very large near Rome, and it is difficult to imagine a foreign place from which news might more easily percolate to remote monasteries. One must also remember that Bede has an acknowledged general debt to correspondents.

The hours of the eclipses of 538 and 540 are not worth prolonged consideration. Our translations of "hora prima" and "hora tertia" as "first hour" and "third hour" are literal; many translators either use "Prime" and "Terce", or interpret these conventionally as something like 7 a.m. and 9 a.m. In almost all places, seasons and centuries, those times do indeed lie within an hour or two of the times of the offices mentioned; on canonical hours and monastic horaria, see, for example, D. Knowles, Christian Monasticism, London 1969, 197-222. The time indications in Bede, even quite crudely considered, not only provide a measure of confirmation of identifications already tolerably clear, but also perhaps suffice (when taken in conjunction with the approximately known circumstances of the eclipses) to suggest the Latin West rather than the Greek East as containing the places of observation.

541 S.541 Dec.3 (Tues.) ANNULAR ECLIPSE (NO RECORD)

The band of annularity shown in Ginzel 1899 travels via Cyrenaica and Eastern Asia Minor. We are rather surprised to have encountered no record.

<u>547</u> S.547 Feb.6 ECLIPSES OF COSMAS (IN EGYPT?)
M.547 Aug.17

These two eclipses are easily identified as those mentioned in Book vi of the Topographia Christiana of Cosmas ("Indicopleustes") of Alexandria, apparently written close to the time of the eclipses. Cosmas refers to both eclipses as predicted at Alexandria by one Stephan of Antioch; he mentions that the solar eclipse actually occurred, but is silent on whether the lunar eclipse was seen. Ginz.1899, 225 gives both the Greek (PG, 88, 1864, 321) and a German translation. Greek text and French translation appear in the edition by Wanda Wolska-Conus, Tome III (= Sources Chrétiennes No.197, Paris, Les éditions du Cerf, 1973, 14-17). As Cosmas gives no specific time of day, the interest is primarily calendrical. The dates provide a good example of the working of the Egyptian calendar in its Julian form, with fixed year-beginning Thoth 1 = Aug.29, twelve months each of 30 days, and 5 or 6 supplementary days. Cosmas gives the dates as Mechir 12 and Mesori 24; on the scheme just stated, the 12th of Mechir (the sixth month) is Feb.6, and the 24th of Mesori (the twelfth month) is, when no leap year is involved, Aug.17. Since there were eclipses on those dates in AD 547, Ginz. has no hesitation in confirming the identifications made by J. Krall in 1890. In treating of this matter, J.B. Bury (Hist. Later Rom. Emp., 2, 1923, 319) has Mesori 14 by misprint for Mesori 24.

The solar eclipse was total in or near Nubia, India and China; Ginz. finds magnitude 0.50 at Alexandria about 8.35 a.m. The lunar eclipse was partial (magnitude about 0.44), and took place there entirely after Alexandria midnight. The solar eclipse is accepted by Newt.1979, 387, but he points out that Cosmas was famous for his travels and we should not necessarily assume that the observation was made in Alexandria. A solar eclipse of 0.5 is not likely to have been noticed.

560 M.560 Nov.18-19 DEEP TOTAL IN SWITZERLAND

In the Chronicle of Marius, bishop of Avenches (near Fribourg in Switzerland), edited by Mommsen, Chr.Min. 2 = AA 11, 1894, 237, we find:

> "a.560. p.c. Basili ann. xviiii. Ind. viii. Hoc anno serenitate caeli inter stellas splendidas obscurata est luna xvi, ut vix conspici posset."

Oppolz. gives a deep total eclipse of the Moon (magnitude 1.75) on the night of 560 Nov.18-19, with mid-eclipse about UT $1^h9^m$. Although month and day are lacking, there is only small doubt about the identification. The eclipse is in the stated year 560, which tallies with "post consulatum Basilii ann. 19", as Basil, the last consul, officiated in AD 541. In Byzantium the eighth indiction ran from 559 Sept.1 to 560 Aug.31, but Marius may well be using a Roman indiction, with eighth indiction from 560 Jan.1 to 560 Dec.31. The Moon would be in the part of the ecliptic which lies in a bright part of Taurus just north of Orion, and so would be very much "among bright stars".

Among alternatives, M.559 Nov.30 was only partial, and there would be no question of the Moon's being almost invisible. M.560 May 25 was invisible in Europe. M.561 May 14-15 was well visible, and just total (Oppolz. magnitude 1.02); the Moon would be in Ophiuchus, near Scorpio (including Antares). This last eclipse fits reasonably well, except for wrong year 561, but M.560 Nov.18-19 fits better.

563 S.563 Oct.3 PARTIAL SOLAR IN FRANCE

Gregory of Tours (Book 4, Ch.31; SrM 1(1), 1885, 167, or 1951, 165; Latouche 1963, 215): "But once on the kalends of October the Sun appeared so obscured that not even a quarter remained shining; it seemed hideous and discoloured, and looked like a sack. And a star, which some call a comet [AD 565 DJS], having a ray like a sword, appeared over that region for a whole year, and the sky was seen to burn [The aurora DJS], and many other signs appeared".

Following Struyck 1740, 113, the passage is generally accepted (by Gregory's numerous editors, by Ginzel 1883, 1899, and most recently by Newton 1972, 322) as referring to the annular eclipse of 563 Oct.3, not quite on the kalends of October (Oct.1). The comet is that of 565.

The track of annularity runs from near Iceland to Russia and China, so that the eclipse was not very large in France. Ginzel 1883 quotes Gregory's Latin from Bouquet 2, 1869, 218 and calculates magnitude 0.78 at 7.22 a.m. at Clermont. Ginz.1899 quotes Gregory's Latin from SrM 1(1), 1885, 167, and calculates magnitude 0.55 at 7.21 a.m. at Clermont, but 0.60 at Tours. Thin cloud or haze must have made the observation possible.

c.563-5 PORTENTS OF JUSTINIAN'S DEATH

Signs connected with the death of Justinian (565 Nov.). The Chronicle of Sigebert of Gembloux (SS 6, 1844, 318) says "Many signs appeared in the Sun and the Moon". There were only weak portents in 565.

It does not seem worth while to investigate in depth the sources for such a general statement by such a late author as Sigebert (died 1112). The editor in SS mentions Bede and Paul the Deacon, but whether as sources for substance or models for phraseology is not clear. The extant early sources (Muralt, 222) mention no signs, but the full version of Menander the Protector is now lost. Moreover, Sigebert used 6th century sources (the Historia Miscella and Jordanes) which may have been more complete than the extant versions. We suspect a lost Byzantine reference to the solar eclipse of 563 and the lunar eclipse of 564 March 13-14.

566 S.566 Aug.1 (Sun.) TOTAL SOLAR IN ARABIA

We have encountered no European reference to this eclipse. There is, however, a perfectly clear Syriac record. Agapius, 8(3), 1912, 435 says: "In the first year of his (Justin the Second's - Ed.) reign, there was an eclipse of the Sun, on Sunday the first of Ab (August)". Justinian I died on 565 Nov.14, and his associate Justin II then ruled alone; the statement quoted is entirely correct. Unfortunately, on page 434 the year of Alexander is given as 788; this should be 877 (AD 565-6, autumn to autumn in Syria). The approximate track of totality runs from West Africa by Middle Egypt, the Northern part of the Red Sea, and the Persian Gulf, to a noon point in North-West India. The observations may have come from Egypt or Arabia rather than from Syria itself.

567 M.567 Dec.31 (p.m.) DEEP TOTAL LUNAR IN S.E. EUROPE

Excerpta Sangallensia, ed. Momm., Chr.Min. 1 (= AA 9, 1892), p.335, has, in the reign of Justin II, about the year 567:

"in caelo luna XVI non conparuit II kl. Ian."

Ginz.1899, 226 takes "non conparuit" as meaning "disparuit"; II kal.Ian. is Dec.31. There was a deep total eclipse of the Moon in mid-evening on 567 Dec.31. Oppolzer gives magnitude 1.74 and mid-eclipse at UT 19.44 (7.44 p.m. at Greenwich). Thus the record is clearly correct. St. Gall was not founded until 614 and the source may have been the lost chronicle of Theophanes of Byzantium (a 6th century predecessor of the well-known 9th century Theophanes).

577 M.577 Dec.11 (a.m.) LUNAR IN FRANCE

There is a lunar entry in Gregory (Book 5, Ch.23, or 24 in some editions; SrM 1(1), 1885, 219, or 1951, 230; Latouche 1963, 285-6). The entry is for a year usually taken to be 577 (or rather the second year of Childebert II, running from 576 Dec.25 to 577 Dec.24). After mentioning a lunar phenomenon (near-by stars, etc.) on 3.Id.Nov. (= Nov.11), which is not said to be an eclipse date, Gregory continues his signs and portents with: "In that year we frequently (saepe) saw the Moon turned black (in nigredinem versam)". In fact M.577 Dec.11 is the only lunar eclipse listed by Oppolzer for 577; he gives magnitude 0.65, with mid-eclipse at 5.35 a.m., Greenwich mean time. Ginz.1899, 226 finds the same magnitude, with mid-eclipse at 5.46 a.m., Clermont mean time. (By 577, Gregory, born at Clermont, modern Clermont-Ferrand, was already bishop of Tours.)

Gregory's account is paraphrased in the later Chronicon Vedastinum (of Arras; SS 13, 1881, 688).

Why Gregory used the word 'frequently' is unclear. Struyck 1740, 141, while inserting M.577 Dec.10-11 in one of his lists, pointed in a footnote to (i) M.574 Feb.21 (p.m.) and (2) M.575 Feb.10-11 as two eclipses visible at Tours within the space of one year. He found for (1) a magnitude slightly more than one-half at 5.17 p.m., and for (2) totality, with mid-eclipse at 11.42 p.m. He suggested that these are the eclipses alluded to by Gregory. Whatever may be thought of this suggestion, Struyck did draw attention to the only available eclipses before M.577 Dec.10-11 itself.

581 M.581 April 5 (a.m.) LUNAR ECLIPSE & COMET IN FRANCE

There is a second lunar eclipse recorded in Gregory (Book 5, Ch.41, or 42 in some editions; SrM 1(1), 1885, 233, or 1951, 248; Latouche 1963, 305). The passage appears to relate to the year 580 (or rather the fifth year of Childebert II, running from 579 Dec.25 to 580 Dec.24). It says "the Moon was darkened and a comet appeared". There was no lunar eclipse in 580, and the passage is usually taken to refer to the partial lunar eclipse of 581 April 5. Oppolz. gives magnitude 0.55 at 2.33 UT, while Ginz.1899, 226 finds magnitude 0.52 at 2.46 a.m., Clermont mean time. The eclipse has been similarly identified since at least Petavius 1627, 849-850, where magnitude 0.56 is calculated. The comet is certainly that which the Chinese saw in 581 January, although Dr. Ian Wood tells me (letter 15 Feb. 1978) that our interpretation of Gregory's chronology is not the standard view.

582 (i) M.582 Mar.25 (a.m.)
or (ii) M.582 Sept.17-18 LUNAR ECLIPSE OF GREGORY

A third lunar eclipse appears in Gregory (Book 6, Ch.21; SrM 1(1), 1885, 262, or 1951, 289; Latouche 1963-5, 2, 37). Among the prodigies for, apparently, the seventh year of Childebert II (running from 581 Dec.25 to 582 Dec.24), he says "the Moon suffered an eclipse". This might refer to either of the above; Petavius 1627, 850-852 gave both. Some calculated magnitudes and mid-eclipse times are:

| | | |
|---|---|---|
| (i) | Petavius 1627 | 1.84 at 3.56 a.m. Paris time |
| | Oppolzer | 1.80 at 4.55 a.m. Greenwich mean time |
| (ii) | Petavius 1627 | 1.75 at 12.41 a.m. Paris time |
| | Oppolzer | 1.84 at 1.05 a.m. Greenwich mean time |
| | Ginzel 1899 | 1.82 at 1.22 a.m. Clermont mean time |

It will be seen that both eclipses were deep total ones. Calvisius gave only the first. Ginzel suspects that Gregory refers to the second, as the first was only partially visible owing to the dawn, and we have allocated

7 points to the September eclipse.

Gregory's information is taken over by later writers, Aimoin (Bouquet 3, 1869, 88), Chroniques de Saint Denis (Bouquet 3, 1869, 233), also weakly by Fredegar (Book 3, Ch.88 in SrM 2, 1888, 117). Calvisius 1620, 475 identified the eclipse in Aimoin as M.582 March 25.

590 S.590 Oct.4 PARTIAL SOLAR IN FRANCE

Gregory of Tours (Book 10, Ch.23; MGH SrM 1(1), 1885, 435, or 1951, 515; Latouche 1965, 301) has, apparently in the fifteenth year of Childebert (running from 589 Dec.25 to 590 Dec.24) and the twenty-ninth year of Guntram: "The Sun suffered an eclipse in the middle of the eighth month (mense octavo mediante): and its light so diminished, that what it had available was scarcely as much as the horns of a five-day-old Moon have". In Gregory "the eighth month" normally means October. In spite of "the middle of the eighth month", the eclipse occurred on 590 Oct.4. There is no doubt about the year, because it was one of the occasional years in which the Easter cycle of Victorius of Aquitaine mentions rival dates on the fifteenth and twenty-second days of the Moon, and Gregory alludes to the fact in the same chapter.

The eclipse was nothing like total in France; Oppolzer's track runs from Greenland to a noon point near Gothenburg (Sweden), then on to Persia and Kutch (India). Ginz.1883 computed magnitude 0.72 at Clermont at 11.20 a.m. Ginz.1899 computed magnitude 0.66 at Clermont at 11.17 a.m. and 0.63 at Tours. These agree well enough with Gregory's account. Newt.1972 considers the eclipse in a number of places. Correct identification goes back at least to Petavius 1627, 852.

590 S.590 Oct.4
592 S.592 Mar.19 SOLAR ECLIPSES AT CONSTANTINOPLE

The situation concerning the Byzantine records is at first sight quite simple if we restrict ourselves to a brief passage from the only record which was approximately contemporary. Theophylactus Simocattes (Simocatta, fl.c.610-40), Hist., Book V, Ch.16 (ed. I. Bekker, Bonn 1834, 236; Teubner ed., C. de Boor, 1887, 218) says of the Emperor Maurice (582-602): "He went out from the palace (at Constantinople) one and a half parasangs, to the Hebdomon, as the Byzantines called it. On that day indeed an eclipse of the Sun took place. It happened in the ninth year of the reign of Maurice". As Maurice succeeded on 582 Aug.14, this points to

S.590 Oct.4 as the eclipse, a view accepted by Newton 1972, 542.

The account by Theophanes (died c.817) makes one aware of a difficulty. In his annal for AM 6083 (AD 590-1) we find (ed. J. Classen 1839, 412; Teubner ed., C. de Boor, Vol.I, 1883, 268): "This year, as spring was beginning, Maurice collected his forces in Thrace and went out to see the ravages of the barbarians [the Avars]. But the Empress, the Patriarch, and the Senate urged the Emperor not to conduct the war himself, but to entrust its direction to a general; which he declined to do. When he had gone out to war, an eclipse of the Sun occurred." The information "as spring was beginning" fits ill with an October eclipse if, with Simocatta, we put the eclipse on the same day as the departure from Constantinople. Both S.590 Oct.4 and S.592 March 19 were large eclipses at that city. Ginz.1883, 652-4, 1899, 227-8 decided in favour of the first eclipse, without any apparent misgiving on account of the "spring" of Theophanes. Ginzel has been followed recently by H.W. Haussig in Byzantion, 23, 1954, 275-462, who wrongly quoted him as saying that S.592 was invisible at Constantinople. As his chronology is confused in years AM 6082/4 (cf. Muralt, 249-252) we consider that this was really the 592 eclipse.

The eclipse is also mentioned, but no more informatively, by the 12th century historian Zonaras, Book XIV, Ch.12 (Vol.III, 1897, 189-190; Teubner ed., L. Dindorf, Vol.III, 1870, 295).

The solar eclipses of "592" (now often 590) and 760 are ascribed by several 17th century authors (also Bunting 1590 in the case of the first eclipse) to "Annales Constantinopolitani", Books 17 and 22 respectively. It may be useful to say that by these annals they appear to mean Theophanes as exhibited in the Latin of the Historia Miscella; as printed by Muratori (Rev. Ital. Script., 1, 1723, 117 and 158), this does give the two eclipse passages mentioned in Books 17 and 22 respectively. For these passages (and those relating to several other Theophanic eclipses), the Latin of the Historia Miscella is almost identical with that of Anastasius the Librarian, in his abbreviation of Theophanes, as printed in the Bonn edition of the latter (ed. J. Classen, 2, 1841, 123-4 and 231) and in de Boor's Teubner edition (2, 1885, 164-5 and 283); but the narrative of Anastasius is not divided here into "Books".

590 M.590 Oct.18 (p.m.) PARTIAL LUNAR IN FRANCE

Fredegar, Book 4 (nowadays), Ch.11 (ed. B. Krusch, SrM 2, 1888, 127, or Wallace-Hadrill 1960, 10) has, in the 30th year of the reign of Guntram, usually taken to coincide mainly with AD 590, "In this year the Moon was eclipsed". There was a partial eclipse of the Moon on 590 Oct.18, with Oppolz. magnitude 0.77, and opposition at UT 18.13 (Oppolz.),

18.29 (Goldstine 1973). The eclipse is discussed in Ginz.1899, 228; also in Struyck 1740, 153, which refers only to Aimoin (see below). It is mentioned, with a question mark, for Fredegar in Newt.1972, 659, and discussed for China in Newt.1970, 144,215.

It may puzzle readers that the solar eclipse which Gregory of Tours puts in the 29th year of Guntram is usually identified as S.590 Oct.4, while the lunar eclipse which Fredegar puts in the 30th year of Guntram is usually identified as M.590 Oct.18. But Fredegar is known to number the years of Guntram's reign differently from Gregory. Common matter is slight, and there is not complete consistency, but, in general, Fredegar's year number in Guntram's reign is one more than Gregory's. Thus, Gregory's 29th year of Guntram and Fredegar's 30th year of Guntram contain common matter, namely a campaign of the Franks against the Bretons (Beppolen killed, Ebrachar disgraced). This year of Guntram's reign coincides largely with what we now call AD 590 (Gregory of Tours rightly mentions the election of Gregory I as pope), and it is in this year that Gregory puts an unusually roughly dated solar eclipse, and Fredegar an undated lunar eclipse.

Krusch (loc.cit.) refers also to M.590 April 25, but this lunar eclipse was completely invisible in Europe, and may be ruled out.

Fredegar's eclipse was earlier identified as M.589 May 5-6, a deep total eclipse, in Calvisius 1620 (477, where March is a mistake for May), on the strength of a quotation from Aimoin (Book 3, Ch.77 in Bouquet 3, 1869, 105), a writer several centuries later, and of no great consequence in the present instance.

Fredegar was writing in 658, but he evidently had access to earlier annals, a supposed 'Burgundian Chronicle' which no longer exists.

592 S.592 Mar.19 TOTAL AND PARTIAL PHASES OF SOLAR ECLIPSE

Oppolzer's track of totality runs from near the Cape Verde Islands to a noon point in Russia (36°W., 54°N.) and a sunset point in Siberia. The approximate line on his Map 86 goes through Morocco, South Italy, and the Balkans. The precisely computed band in Ginz.1899 runs more to the East during the morning - through Tripolitania, Northern Greece, Romania, and the Ukraine.

The 10th century Syriac writer Agapius, 8(3), 1912, 447, says, under Maurice: "Next year, which was the 903rd year of Alexander, in the month of Adar (March), in the middle of the day, there was an eclipse of the Sun". The 903rd year of Alexander was AD 591-2 (autumn to autumn in Syria), and the passage clearly refers to S.592 March 19. Michael,

X, 23, 1963, 373, states "there was an eclipse of the Sun, and darkness there, on March 10 from the third to the sixth hour. Everyone said that the Sun obscured itself because of the massacre of the monks, servants of Christ". This is evidently a Greek, not a Syrian, record.

Fredegar, Book 4, Ch.13 (ed. Krusch, SrM 2, 1888, 127, or Wallace-Hadrill 1960, 10) has: "In the 32nd year of Guntram's reign the Sun was eclipsed from dawn to midday to the extent that the third part of it was scarcely visible". As we have explained (under M.590 Oct.18), the 32nd year of Guntram is usually taken to coincide mainly with AD 592.

Ginz.1883 found maximum magnitude 0.72 at Dijon; Ginz.1899 found 0.76 at 9.30 a.m. at Avenches. Fredegar exaggerates the duration of the eclipse. Our identification of the Burgundian record goes back at least to Petav.1627, 853; Bunting 1590, 350 referred to the Byzantine record.

Krusch (SrM 2, 1888, 127) identified the eclipse as S.591 Sept.23 (q.v.). Ginz.1899 rightly rejected this, but Krusch had already retracted his identification (SrM 2, 1888, 576, 578).

We must agree with Newt.1972, 323 that S.594 July 23 is a possibility. but we feel that S.592 has much greater probability. We believe that Fredegar's chronology is just sufficiently good to make S.594 difficult of acceptance. The information in Theophanes that does not come from Simocatta (see 590 above) seems to come from a Greek Chronographer.

## 594 S.594 July 23 TOTAL SOLAR IN BRITISH ISLES

The track of totality starts at sunrise in the Atlantic, passes via Ireland and Britain to a noon point in Russia, then on to China where it would have been very striking. The eclipse is recorded in Irish annals.

There are certainly one or two solar eclipses mentioned in Irish annals around AD 590-4; all records refer to a morning eclipse, though in the case of Chronicum Scotorum the word is supplied by the editor. Before considering identifications, we shall list four which have been suggested, with their magnitudes and local apparent times for mid-Ireland as given in O'Connor 1952 (see also Schove 1954), and their maximum possible magnitudes for Ireland as given in Newt.1972, 593. We omit S.591 March 30 as a sunset eclipse, and note that S.596 Jan.5 was an afternoon eclipse.

| | O'Connor | Newton |
|---|---|---|
| (i) S.590 Oct.4 | 0.65 at 10.20 a.m. | $\leqslant 0.75$ |
| (ii) S.591 Sept.23 | $< 0.20$ | $\leqslant 0.34$ |
| (iii) S.592 March 19 | 0.58 at 8.18 a.m. | $\leqslant 0.63$ |
| (iv) S.594 July 23 | 1.00 at 5.32 a.m. | 1.00 |

(ii) occurred in the late morning, say 11 a.m., in Ireland.

The AU (1887, 72-5) has, prima facie, two eclipses:

(AU 1) Defectio solis .i. mane tenebrosum (An eclipse of the Sun, i.e., a dark early morning).

(AU 2) Matutina tenebrosa (A dark morning - with some hint of early morning).

AU 1 occurs under MS. year 590, but with feria (2) and epact (1) on Jan.1 indicating AD 591; AU 2 under MS. year 591, but with feria (3) and epact (12) indicating AD 592. Consequently, it is easy to see why the index to AU, 4, 1901, 140, identifies AU 1 as (ii) and AU 2 as (iii). Nevertheless, this includes the eclipse least likely to have been seen, (ii), and excludes the most likely (iv). Anderson 1922 (94f, 104) implies identifications (iii) and (iv), and consideration of related Irish annals certainly points to (iv); it is uncertain whether or not AU should be taken at face value, as referring to two different eclipses.

The CS, 1866, 62-3, has:

(CS) Defectio solis, [mane] tenebrosum (tenbrarum, etc.).

This occurs under editorial year 590, but under MS. feria vi, which implies AD 594.

Similarly, AT, 1896, 160, have:

(AT) Defectio solis ... .i. mane tenebrosum.

These annals have no year AD of their own, but the entry occurs under MS. feria vi, again implying AD 594.

Finally, the AI, 1951, 78-9, have:

(AI) Defectio solis in matutina hora.

The editorial year in this more modern edition is 594, though MacAirt did not fail to point out that the revised AU (i.e. "AU + 1") dates 591 as 592.

We agree with New.1972, 593 that at least three possible identifications must be kept in mind. But it seems nowadays that the great Irish total eclipse of the decade, namely S.594 July 23, is not merely present in Irish annals, but also occurs in some cases with a correct year-indicator. It would lead us too far to consider how the dating of Irish annals relates to non-astronomical events (e.g., the pontificate of Gregory the Great, the mission of St. Augustine, and the death of St. Columba). We accept the records as relating to 594 and suspect that two records (one in Latin) were available, one perhaps from Bangor, N.E. Ireland, and one from Iona in what is now Scotland. See the important article by A.P. Smyth 'The Earliest Irish Annals', Roy. Irish Acad. Proc., 72C, 1972, 1-48. Certainly this is the first genuine eclipse record from the British Isles (cf. K. Harrison, Studia Celtica, XII-XIII, 1977-8, 25).

596 S.596 Jan.5 ANNULAR BUT UNRECORDED

The track of annularity has noon point in the Atlantic, and passes via Spain and France to an end at sunset in Germany. Ginz.1899 also makes the band of annularity go through the Iberian Peninsula and France (it runs south of Chaves, includes Bilbao and Bordeaux, and passes north of Dijon). Consequently it would not be surprising to find a record of this eclipse, but in fact we have encountered no clear record. The identification in Calv.1620, etc. as the eclipse of the Sun before the death of Guntram, in "scriptores Gallici", is now considered untenable; the eclipse in Fredegar is evidently S.592 March 19.

In the period 200 BC to AD 1000 there are some intervals, usually those of short-lived dynasties, not covered by astronomical treatises; in such periods we find a few references to natural phenomena embedded in the Imperial Annals and in these annals there are only two solar eclipses reported as total. These are the eclipses of AD 516 and 522, both given in Chapter 6 of the Nan Shih (History of the Southern Dynasties) compiled AD 630-650 and assumed to relate to Nanking in S. China.

The date and the day given for these two eclipses, 516 Apr.18 and 522 June 10, are correct and they were almost certainly observed by professional astronomers at the southern capital of Nanking; calculations confirm that 516 was annular but central in the Nanking area. The attribution of totality to the 522 eclipse is confirmed by calculation (Stephenson and Clark 1978, Stephenson 1982, 152; cf. however Newton 1979, 167).

In the Chinese list of eclipses (Hoang 1925) one solar eclipse, that of 579 May 11, is recorded as 'not taking place'. The preceding annular eclipse of 577 Dec.25 and the total eclipse of 574 March 9 would have been noticed by the people, (apart from 594 they are the most important of the half-century), but the Chinese historians did not distinguish them from other (predicted) dates.

# THE SEVENTH CENTURY

## INTRODUCTION

## FIRST RECORDS FROM ENGLAND, ARABIA AND JAPAN

The political map changes completely in this century with the contraction of the Byzantine Empire and the rise of Islam. The cultural map changes very little; Syria remains in the forefront, but Antioch, having been depopulated by plague, is replaced by Damascus, near enough to the desert for the Arab ruling elite to leave during the plague season. The seventh century chronicles - both Syriac and Greek - are known to us mainly through later abridgements (e.g. Theophanes).

Literacy spread geographically in this century to new parts of the world. Arabia, England and Japan now provide us with records of eclipses and comets, and from this century onwards the Chinese and Japanese comet dates combined enable us to correct the dating of European chronicles. Few Islamic sources from this century are extant but astrological calculations of eclipses and conjunctions made later attempted to link the astronomical phenomena with political changes; no doubt Nestorian Christians, Syrians and Persians within the Pax Islamica kept records; predictions of the dates of lunar eclipses must often have been as successful as they were in China and Central America. After our section of Western Eclipses in this century we append brief notes on Chinese eclipses, a Japanese solar eclipse of AD 628 and a Maya lunar eclipse of AD 683.

601 S.601 March 10 (Fri.) NEAR TOTAL IN EGYPT AND SYRIA

An ostracon now in Turin Museum, but found in the village of Djēme (modern Medīnet Habu) near ancient Thebes in Egypt, carries a Coptic inscription first published by L. Stern in 1878 and pointed out to Ginzel by J. Krall. The eclipse it mentions was identified in Ginzel 1883 (655) as S.601 March 10; the identification has been confirmed by E.B. Allen, J. Am. Oriental Soc., 67 (4), 1947, 267-9, and Newton 1979, 388 where further references will be found.

The inscription reads, in Allen's translation: "On the fourteenth of Phamenoth of the fourth indiction, the sun was eclipsed in the fourth hour of the day and in the year in which Peter, son of Palu, was made village official in Djēme".

As the stone was believed to be of the sixth or seventh century AD., and the Egyptian calendar (with Thoth 1 = Aug.29) makes Pham.14 = March 10, and the indiction is correct, and moreover Ginzel found S.601 March 10 to reach a maximum magnitude of 0.92 and 9.54 a.m. at Thebes, Ginzel's identification appears reliable. Allen, searching over a very long period (AD 297-1580), ended up with the same identification, and calculated maximum magnitude 0.88 at 10.02 a.m. at Thebes, in sufficiently good agreement with Ginzel. As the sun rose at 6.08 a.m., totality would come on towards the end of the fourth hour. The track of totality in Schroeter runs by Lower Egypt, Jerusalem, and Caspian Sea. The fact that the record consists of an inscription is notable; most eclipse records are literary, not epigraphic.

There is also literary evidence from several chronicles. The earliest we have encountered is the seventh-century "Chronicle of John, Bishop of Nikiu", translated from Zotenberg's Ethiopic text by R.H. Charles, London 1916. Nikiu lies in the region of the Nile delta (on a route between Alexandria and Memphis; see an article "Nikiu" by H. Kees in Pauly-Wissowa, Real-Encycl.). But the relevant passage on p.163, concerning the reign of Maurice (582-602), mentions Antioch, etc. It is too long to quote in full, but includes:

"And likewise at that time the sun was eclipsed at the fifth hour of the day, and the light of the stars appeared. And there was widespread alarm ..."

No precise date is given, but S.601 March 10 fits well. From Schroeter's tables one may estimate that about 10.35 a.m. there was greatest magnitude almost 0.95 at Antioch (and, it may be added, totality or nearly so at Damascus). The local time of maximum phase, being about half an hour greater in Northern Syria than in Lower Egypt, now falls well within the fifth hour of the day.

The Chronicle attributed to "Denys de Tell-Mahré" (1895), often quoted as "Pseudo-Denys", appears to have been written in the first half of the ninth century. On page 3 we find: "In the year 912 (600-1), there was great darkness in the middle of the day; the stars rose and appeared as at night. They remained about three hours, after which the darkness cleared away and the day shone as before. - This year the emperor Maurice died." The eclipse is almost certainly S.601 March 10. Maurice was actually murdered in 602, about November, but the chronicle misdates most reigns around this time.

The eclipse is also mentioned in the anonymous Chronicon ad annum 846, CSCO, (3) 4, Paris 1903, 174: "In the year 912, there was darkness over the whole earth, and stars appeared in the middle of the day".

The eclipse is clearly mentioned in the Chronography of Elias, metropolitan of Nisibis, who wrote early in the eleventh century. In Delaporte's French translation of 1910, p.77, Elias, quoting as his source the ecclesiastical history of an unknown writer named Aleha-Zeka, says "Year 912. The sun was eclipsed on Friday 10 Adar [March], in the middle of the day; the stars appeared, and there was a violent wind". The year 912 is Seleucid, or "of the Greeks".

The same eclipse is also mentioned by Michael the Syrian, tome II (Michael 1901, repr. 1963), p.373. Although Michael wrote in the twelfth century, his testimony is not to be despised, for he gives correct month and day and reasonable hours, and appears not to be merely copying any of the earlier records mentioned above. "The same day, there was an eclipse of the sun, and darkness, the tenth of ʾAdar [March], from the third to the sixth hour [of the day]." This appears in the section of Michael which relates to the reign of Maurice.

Thus S.601 March 10 forms a clear part of Syrian eclipse tradition. The impression given, of an eclipse occurring wholly or almost wholly in the forenoon, agrees with modern calculations (cf. Newton 1979, 388 and 421).

603 S.603 Aug.12 (Monday) IN EASTERN FRANCE, ETC.

Writing of the eighth year of the reign of Theodoric in Orléans and

Burgundy, usually taken to be AD 603, the Burgundian annals for 584-603 included by the anonymous author commonly dubbed "Fredegar" say "Eo anno sol obscuratus est"; text and translation, 1960, p.15; Latin text, 1888, 130, or in Bouquet, 2, 1869, 421. Although month and day are not given, this solar eclipse can hardly be other than that of 603 Aug.12, which was total in North-East Spain and large throughout France. The identification goes back at least to Petavius 1627 (854).

As a warning against false identifications, we mention two out of a number of much later writers who used Fredegar as a source, directly or indirectly.

Aimoin (d.1008), in the same context, has "Eo anno eclipsis solis facta est" (Bouquet, 3, 1869, 110, or PL 139, 1880, 767). The post-humous editions of Calvisius wrongly identify the eclipse as S.606 June 11, but Bouquet correctly has year 603 in the margin.

The Chronicle of Herimannus Contractus Augiensis (i.e. of Reichenau), who died in 1054, wrongly puts both the death of Gregory the Great and "Sol obscuratus est" as late as AD 605, indiction 8 (SS 5, 1844, 91). Bouquet 3, 1869, 325 corrects editorially to 603. Gregory died in 603 (or 604), and the eclipse is really S.603 August 12. Neither of the two solar eclipses of 605 fits. The wrong chronology seems to be due to the chronicler himself, but it must be remembered that a number of events round AD 600 are nowadays dated two or three years earlier than they were even so late as the seventeenth century.

O'Connor 1952 finds that S.603 Aug.12 had maximum magnitude 0.94 in mid-afternoon in Central Ireland, but there is no Irish record.

604 M.604 July 16-17 (Th.-Fr.) IN MESOPOTAMIA

This deep total lunar eclipse around midnight in the Near East (10.18 p.m. ET or 9.10 p.m. UT in Meeus and Mucke 1981) is mentioned by Elias (1910 78): "Year 915. The Moon was eclipsed on the night of Thursday 16 Tammuz (July)": Elias gives as his source the Chronicle of James of Edessa (cf. CSCO (3), 4, Paris 1903 257), whose report is identical.

606 S.606 June 11

Schroeter shows annularity in Asia Minor, and magnitude about 0.9 at Constantinople, but we have encountered no certain Western record, perhaps because so few original Byzantine records of the time have survived unabridged. For a possible but unlikely identification, see below at AD 610-2 (start of reign of Heraclius). For a false identification, see S.603 August 12 (Aimoin).

610 M.610 March 15 (Sunday) ISLAMIC CALCULATION

This was not, strictly speaking, a Western eclipse, as it was invisible in Europe and the Middle East. Oppolzer gives magnitude 0.42 at 8.31 a.m. UT, his sub-lunar point being in the Pacific. But the list of late seventh-century Islamic horoscopes given in Pingree 1968 (116-7) includes this, as "the third conjunction, in which was made clear the mission of the Prophet of God, and in which a lunar eclipse occurred".

610-2 MISDATED SOLAR ECLIPSE NEAR START OF REIGN OF HERACLIUS

recte 617

The twelfth-century writer, Michael the Syrian, ascribes a solar eclipse to the 'year in which Heraclius commenced to reign over the Romans.' As Phocas was executed on 610 Oct 5, this eclipse has been ascribed to the years 610-612. However, as we explain below, the correct date is 617. (see Proudfoot 1975, 390).

612 S.612 Aug.2 (Wed.). IRISH 'GHOST' (FROM S.SPAIN?)

We know of no certain Western record. Schroeter shows totality in Morocco. The magnitude reached about 0.9 at Seville, where Isidore (d.636) was flourishing, but he does not seem to record it.

There are, however, possible Irish references. Various Irish annals mention a star seen at the seventh (or eighth) hour of the day. E.g. 7th hour in CS (s.a. 614), AC (s.a. 617), and AT (under implied 613?); 8th hour in AU (s.a. 613, probably true 614), and similarly in J. O'Donovan's Introduction to the Annals of the Four Masters (1, 1851, xlviii), where "comata" is editorially supplied to "stella", indicating a comet. No comet is recorded in China.

One may estimate from Schroeter a magnitude of about 0.77 in mid-afternoon in Central Ireland, for which O'Connor calculates 0.77 about 3.04 p.m. Newton 1972 (191) notes the reasonably fitting hour, but considers it doubtful that the record refers to an eclipse, and considers the eclipse to be unidentifiable if it does. It should be noted, however, that eclipse identifications S.615 Jan.5, S.615 June 2, S.616 May 21 and S.617 Nov.4 may be eliminated, as all were distinctly morning eclipses in Ireland, if they were visible there (S.616 looks doubtful from Oppolzer).

Because of the extreme vagueness of the record, we have not thought it worth while to give fuller references to the Irish Annals, which might have derived the information from Spain.

617 S.617 Nov.4, E. MEDITERRANEAN

A solar eclipse in the earlier part of the reign of Heraclius is mentioned by the ninth-century writer George the Monk ("Hamartolos"), Book 9, Chap.22 (Teubner ed., C. de Boor, II, Leipzig 1904, 669, or Migne, PG 110, 1863, 827-8): "There was hard famine and great mortality, and the Sun was darkened, and it rained ashes". The entry occurs in a historical context differently dated by various authorities, but not, as far as we can see, outside the range AD 615-9. We do not know George's authority for the eclipse record, which is not mentioned by Theophanes.

The same source was evidently used by Michael the Syrian, Book II, Chap. 1 (tr. J-B. Chabot, II, Paris 1901 = Brussels 1963, p.401) who misdated it as follows: 'In the year in which Heraclius commenced to reign over the Romans (i.e. 610/611) there was an eclipse of the Sun during four hours'. Both writers mention famine and this must relate to the large solar eclipse of 617 (See Proudfoot 1975, 390). Meeus (personal communication 1982 Aug 23) calculates that 92% of the sun's diameter was eclipsed at Constantinople, sufficient to explain the phrase 'the sun was darkened'.

The annular eclipse of 617 Nov.4 was very large at Constantinople. The track of annularity in Schroeter runs from North-West to South-East from Amsterdam via Vienna, Bucharest, just north of Constantinople, and North-East Asia Minor, to around Babylon. This makes the maximum magnitude at Constantinople about 0.98 around 9.15 a.m.

Newton 1972 (530, 543) says "617 Nov.4?. Valid observation; cannot date", and mentions also S.616 May 21 and S.634 June 1 as other Byzantine eclipses during the reign of Heraclius.

Perhaps, as Newton suggests, the rain of ashes may have been due to a volcanic eruption. This may have darkened the Sun; consequently the record cannot be treated as referring indubitably to a solar eclipse.

This eclipse is clearly mentioned by, ostensibly, Stephanus of Alexandria, who worked at Constantinople under the emperor Heraclius. H. Usener, Kleine Schriften, III, Leipzig and Berlin, 1914 published some whole sections, and extracts from other sections, of a work (Explanation through individual examples of Theon's method for the Handy Tables, Cod.Vat. gr. no. 1059) attributed to Stephanus (and possibly pupils). The year of composition, indicated in several places in the work itself, is equivalent to the Byzantine year AD 618-9 (indiction 7). The examples all relate to AD 617 and 618, and include syzygy calculations associated with S.617 Nov.4 and M.618 Oct.9. We owe the reference to Neugebauer 1975 (2, 1045-51), where much detail may be found.

Information about S.617 Nov.4 occurs in Usener (pp.293-4, 329). "November fourth" is spelled out, the Egyptian month Athyr is correctly given, and so also is indiction 6 (AD 617-8). Any reader who wishes to distinguish text, fragmentary extracts and scholia from one another is referred to Usener and Neugebauer. The identification S.617 Nov.4 is secure, but according to Neugebauer (2, 1049) the solar and lunar eclipse computations (as distinct from mere syzygy computations) are unpublished.

<u>618</u> M.618 Oct.9 (p.m.) E. MEDITERRANEAN LUNAR ECLIPSE

Almost as certain is the mention by Stephanus of this lunar eclipse, see Usener (pp.328-9). Totality appears to be mentioned in the text, October (and Phaophi) and seventh indiction (AD 618-9) in a scholium. Thus at the very least there is written tradition of a total lunar eclipse on 618 Oct.9. This is not surprising, since the eclipse was a deep total one (Oppolzer magnitude 1.72), with mid-eclipse about $18^h$ UT, or say 8 p.m. in the Eastern Mediterranean, an evening time very favourable for observation.

622 M.622 Feb. 1-2. ARMENIA ?

We have encountered no convincing record, but two points may be made.

(a) The eclipse is mentioned in our discussion below of a lunar eclipse during the first winter of the Persian campaign of Heraclius.

(b) One of the seventh-century Islamic horoscopes published in

Pingree 1968 (117) does indeed seem to be calculated for the evening of, as Pingree has it, 622 April 1. The heading (in one version) is "The accession [of the Prophet]. A lunar eclipse occurred before his mission". There was in fact a Full Moon in the evening of 622 April 1 in the Near East, but it was not eclipsed. Oppolzer shows that the three eclipses in the nearest "eclipse season" were:

(1) a partial eclipse of the Sun on 622 January 17,

(2) a deep total eclipse of the Moon on 622 February 1-2,

(3) a partial eclipse of the Sun on 622 February 16.

However, if we understand correctly the variant readings given by Pingree, they suggest that the horoscope was dated (though not calculated for) Feb.2 in one version and Feb.7 in another - enough to suggest some knowledge that it was the Full Moon at the beginning of February, not that at the beginning of April, which was eclipsed. If the details are confused, nevertheless the horoscope seems to provide at least a pleasing little confirmation of the well-known fact that the year of the Hegira was AD 622.

622 M.622 Feb.1-2. LUNAR ECLIPSE IN ARMENIA ?

A lunar eclipse in the first winter (622-3?) of the Persian campaign (622-8?) of Heraclius.

There is a highly problematical account of a lunar eclipse in the "Persian Expedition", a panegyric praising Heraclius, by George of Pisidia, who himself took part in the beginning of the campaign; Greek text and Latin translation, ed. I.Bekker, Bonn 1836; Greek text and Italian translation in Georgio di Pisidia, Poemi I, ed. A. Pertusi, Ettal 1959. The account is much too prolix to be worth quoting; see at least Canto II, lines 249-251, 366-375, Canto III, lines 1-12. Pisides paints a vivid picture of the Persian commander, although needing darkness, delaying his attack on Heraclius on account of an eclipse of the Moon ("the goddess of Persia"). Historians have talked of Heraclius being "saved by an eclipse of the Moon".

Later writers endeavoured to fix the chronology of the events, but may have erred in taking George's eclipse too literally, or even mistaking fiction for fact.

It is known that Heraclius sailed from Constantinople on Easter Monday. The year usually accepted by historians, on the authority of various sources (some contemporary), is AD 622. This certainly agrees with the later Theophanes, who also mentions the eclipse, and places Easter Sunday on April 4; he was, however, quite capable of calculating the date of Easter·for any year he decided on. See Theophanes, I, 1883, 302-5, or 1839, 466-9. If we accept 622 for the start of the campaign, we need a lunar eclipse in the winter of 622-3. But there was none visible in the Near East. Even Stratos (I, 1968, 141) alludes to M.623 Jan. 22 without pointing out that it was invisible to the combatants; Oppolzer gives the maximum magnitude 0.44 of this partial eclipse as occurring at 9.56 a.m. at Greenwich, thus somewhat after noon in the region of Armenia, so that the Sun would evidently be well above the horizon and the Full Moon well below (Oppolzer's sub-lunar point, "Mond im Zenith", is east of Hawaii!). Thus the Persian commander can at most have delayed his attack in consequence of unfavourable astrological calculations. This seems by no means unlikely.

If, however, Heraclius commenced his expedition on Easter Monday of AD 621, astronomers can point to a deep total lunar eclipse entirely visible to the combatants. This was M.622 Feb.1-2, for which Oppolzer gives magnitude 1.81 at 9.48 p.m. at Greenwich (thus in the first hour after midnight in Armenia). Struyck 1740 (154), followed by Struyck-Ferguson, adopted this identification, quoting Hist.Misc., lib.18 (Muratori, 1 (1), 1723, 125), which prints the Latin version of Theophanes by Anastasius; more modern references would be Anastasius, Tripartite Chronography, in Theophanes, II, 1885, 188, or 1841, 145. We do not feel authorized to replace the usual 622 by a very heterodox 621 as the year of the start of the great Persian campaign of Heraclius. If 621 were accepted, there would remain the problem, why the Persian commander, wishing for darkness, should be deterred by a darkness-producing eclipse. The answer would doubtless be (as in the case of the Athenian commander Nicias over a thousand years earlier) that superstition prevailed over reason.

Newton 1972 (666) adds that M.622 July 28 was probably not visible at Constantinople; it was almost certainly not visible in the neighbourhood of Armenia. But that is no great obstacle; it could have been calculated by astrologers. The real objection is that it seems far too much in advance of the winter; the narrative seems to demand an eclipse in the winter, or at the earliest in the autumn.

624 S.624 June 21. ANOTHER IRISH 'GHOST'

The annular eclipse of 624 June 21 and the penumbral eclipse of 625 June 10 have both been invoked to explain Irish and 'Welsh' references to an eclipse in 624 or 625. O'Connor gives magnitudes in mid-Ireland only 47% and 25% respectively, and we do not consider that either eclipse was noticed there. The 624 eclipse may have been considerable (perhaps more than 80%) in Seville, and consequently noticeable there, but it is not mentioned even in Isidore (d.636), e.g. not in Mommsen, Chron. Min. 2 = AA 11, 1894, 481, cf. PL 83, 1862, 1056.

The reference in the original 'Chronicle of Ireland', assembled perhaps at Bangor or Iona in the later seventh century, was presumably what we now find in the Annals of Ulster, namely 'annus tenebrosus' (very dark year), under AU year 624 presumed as usual to have referred to AD 625 (which must be the case if the statement at the beginning of the annal, that the first day of January was a Tuesday, is true). In the AI, (pp.86-7) an eclipse is specified, but this seems to be the rationalization of a later copyist. "The Annals of the Britons" (a compilation of AD 796-801 edited along with Nennius's "History of the Britons" by A.W. Wade-Evans, London 1938, p.89) omit the darkness and state baldly that in 624 "the sun is eclipsed". This is the so-called eclipse of Edwin, but the eclipses of 624 and 625 were not likely to have been noticed anywhere in Ireland or Britain; we therefore consider that the darkness of the year does not refer to an eclipse, unless (as we suspect for 612) the information came originally from Spain.

The term 'pluvialis et tenebrosus annus' in 912 might relate to an eclipse in that year (cf. below), but there too Mrs. M.O. Anderson writes (30 April 1974) that this could mean 'a rainy and literally dark year, the sort of year when the harvest rotted in the ground'.

The darkening of the sun in the 620's is mentioned under c.626-8, and possibly there were volcanic eruptions in this decade which weakened the effective solar heat and light.

625 M.625 Nov.19-20 (Tue.-Wed.). ISLAMIC, BUT NO RECORD ?

We have not ourselves encountered any record, but it may be noted that S.B. Burnaby, Elements of the Jewish and Muhammadan Calendars, London 1901, pp.461, 469, mentions the eclipse as an important event considered in the researches on the early Arabian calendar by Mahmud (Journal

Asiatique, Paris 1858, not seen).

Oppolzer shows the eclipse as a deep total one, with mid-eclipse at 11.51 p.m. at Greenwich; it started at 10 p.m. at Greenwich, thus after midnight in Arabia. Consequently even Western usage places its civil date in an Arabian context as 625 Nov.20; in any case, the Moslem civil day begins at sunset. The eclipse occurred at the Full Moon of the sixth Arabian month (Jamādā II) of AH 4.

626-9 'GHOST' ECLIPSES (VOLCANIC DUST ?)

There are various Syrian records of an "eclipse" of the Sun for nine months, from October to June, in the 17th year of Heraclius, which ought to mean some period within AD 626-8. The earliest full record we have encountered is in Agapius (1912, 452).

Theophanes (9th century) referred briefly to this using a source which Proudfoot terms A III (M.A. Thesis, London University, Personal communication. cf. Proudfoot 1975, 390). Michael the Syrian (Chronicle 11.3 trans. J-B. Chabot. Personal communication from R. Stothers) states: 'In the year AD 626 the light of half the sphere of the sun disappeared, and there was darkness from October to June. As a result people used to say that the sphere of the sun would never be restored to its original state.' The Latin Chronicle of 1234 (CSCO, Louvain, 97, 181) specified 1 year later. Stothers (personal communication) queries whether the unusual 'darkness' not long before 628 referred to in the contemporary work of George of Pisidis (Heraclius 1.81) might refer to this prolonged darkening of the sun.

An acidity layer in a Greenland ice-core reflecting a distant eruption is provisionally dated 622/626 with a possible error of several years (Hammer et al. 1980, 231). Probably therefore the Syrian and Irish reports relate to the effects of the same eruption, evidently a later one than that invoked to explain the ashes that fall in Constantinople about the time of the 617 eclipse.

Further investigations at the Copenhagen laboratory are expected to provide a precise year shortly (Clausen et al. ed. Schove and Fairbridge 1984).

630 M.630 Aug.28. LATER ISLAMIC CALCULATION

An Islamic horoscope for this date is published in Pingree 1968 (118), and is headed "The fourth conjunction. A lunar eclipse occurred before the conjunction". Oppolzer gives a partial eclipse of the Moon, maximum magnitude 0.57 at 12.49 p.m. Greenwich Mean Time (around 3 p.m. in the Near East). Thus the eclipse was quite invisible in the Near East, as indeed the Islamic computations show.

632 S.632 Jan.27 (Mon.). FIRST ISLAMIC SOLAR ECLIPSE

Agapius (1912, 468) says, in a partly illegible passage, "the Sun was obscured", apparently about the time when Mohammed died (632 June 8) and Abū Bakr succeeded. This seems to refer to S.632 Jan.27 rather than S.634 June 1.

The seventh century Islamic horoscopes printed in Pingree 1968 (118) also include one dated 632 January 26 and headed "The solar eclipse indicating the death of the Prophet and the accession of Abū Bakr". The calculations are for the evening before an early morning eclipse.

Oppolzer's track of annularity starts in Southern Libya, passes between Mecca and Aden (nearer the former, where the maximum magnitude would be considerable), has noon-point in the Himalayas, and ends in Siberia.

Agapius also has (1912, 461) a solar eclipse, with stars appearing, in the seventh year of Mohammed (presumably AH 7, which ran from AD 628 May 11 to 629 April 30). However, there was no considerable solar eclipse visible in Arabia or Syria around this time (only several astrologically calculable ones). If the record has genuine substance, there may be confusion with 632. The deep total lunar eclipse in the early morning of 629 March 15 falls in the correct year, but we have not accepted the identification in our table.

We also mention here an Islamic eclipse reference which we owe in the first instance to Richard Bell, 6 Blackit Place, Edinburgh 9. Bukhārī (d.870), in his Sahīh, or collection of Traditions (ed. Krehl, I, p.271; Chap.16, §15), says: "The Sun was eclipsed on the day that Ibrāhīm died, and the people said: 'It is eclipsed for the death of Ibrāhīm'. But the messenger of Allah said: 'The Sun and the Moon are two of the signs of Allah: they are not eclipsed for the life or death of anyone. When you see it, call upon God and pray until it clears'."

"The messenger of Allah" is one of the titles of Mohammed. Ibrāhīm was Mohammed's infant son by Mary the Copt, and is supposed by some to have died in June or July of AD 631. See Sir William Muir's Life of Mahomet, 4, 1861, 160f., or T.H. Weir's one-volume revision, "The Life of Mohammad", Edinburgh 1912, 426-30.

The phrase "on the day that" is not necessarily to be taken seriously in accounts of old eclipses, but we naturally look first for a solar eclipse in AD 631. However, Oppolzer's track makes S.631 Aug.3 invisible in Arabia; no likely track could make it conspicuous. The large Arabian solar eclipses of this decade were S.632 Jan.27 and S.634 June 1. Since the account quoted above seems to refer to an actual spectacle rather than a mere computation, and the eclipse has to occur before the death of Mohammed in 632 June, the eclipse can hardly be other than S.632 Jan.27. If the death of Ibrāhīm and the eclipse thus occurred some seven months apart, and both in the same tenth year of the Hegira, a tradition of their connection could well have arisen. On the other hand, S.B. Burnaby, Elements of the Jewish and Muhammadan Calendars, London 1901, p.461f., mentions (with due caution) the view of the Egyptian astronomer Mahmud, in his researches of 1858 on early Arabian chronology, that 632 Jan.27 was the actual day of Ibrāhīm's death.

634 M.634 June 16-17 (Th.-Fri.) in Arabia. LATER CALCULATION

One of the seventh-century Islamic horoscopes printed in Pingree 1968 (119) is dated 634 June 16 and is headed "The lunar eclipse indicating the death of Abū Bakr and accession of ʿUmar". The eclipse was small; Oppolzer gives maximum magnitude 0.15 only, at 11.30 p.m. Greenwich (hence about 2 a.m. in the Near East). Abū Bakr is usually taken to have died in 634, about August, but this eclipse was evidently calculated at a later date.

639 S.639 Sept.3 NO RECORD EVEN IN THE BRITISH ISLES

Oppolzer shows the track of totality as beginning in the Irish Sea and passing over England and North Germany on its way to a noon-point in Central Asia, then on to China and an end in the Pacific. Schroeter shows most of Wales and Central England as within the band of totality, and the eclipse should have been visible there shortly after sunrise.

Perhaps the weather was unfavourable. We have encountered no Western record at all.

644 M.644 May 27. LATER CALCULATION

One of the seventh-century Islamic horoscopes printed in Pingree 1968 (119) is dated 644 May 27 and headed "The lunar eclipse indicating the death of ʿUmar and the accession of ʿUthmān". Oppolzer gives a total eclipse, magnitude 1.08, at 9.07 UT (thus about noon in the Near East). Consequently the eclipse was not visible in the Near East, the eclipse being, for that region, a purely calculated one. According to Camb. Med.Hist. 2, 1926, 354, ʿUmar was killed on 644 Nov.3. Our next eclipse notice relates to S.644 Nov.5, just two days after the reputed date of ʿUmar's death.

644 S.644 Nov.5 (Fri.). IN SYRIA

This eclipse is clearly mentioned by several authors. The earliest extant reference we have encountered is in Theophanes (died c.817), 1883, 343, or 1839, 524-5): "And an eclipse of the Sun occurred on the fifth of the month Dios [November], on the sixth day of the week [Friday], at the ninth hour [of the day]". For "Dios" there is also an (erroneous) variant reading "December". The passage quoted ends the annal for AM 6136, but a correction +1 is needed for most of the seventh century (see J.B. Bury's ed. of Gibbon, Decline and Fall, 5, 500), and AM 6137 does regularly denote (in the system Theophanes is trying to use) the Byzantine year beginning on AD 644 Sept.1. Cedrenus (ed. I. Bekker, I, Bonn 1838, 754) mentions the eclipse, but without month, day, or hour. The identification goes back at least to the posthumous (1620, etc.) editions of Calvisius.

The broad, curved, band of annularity in Schroeter contains Bergen, Oslo, Gothenburg, Berlin, Warsaw, Budapest, Bucarest, Constantinople, Smyrna, and Bagdad, with annularity occurring at Constantinople around hour angle 18° (1.12 p.m. local apparent time).

The eclipse is discussed in Newton 1972 (543, etc.), where the ninth hour of the day is accepted as looking reasonable for maximum eclipse at Constantinople (cf. Proudfoot 1975, p.419).

The same eclipse is also mentioned by two Syriac writers, Agapius

(1912, 479) and the twelfth-century Michael the Syrian (tr. J-B. Chabot, 2, 1901, 1963, 432). Both writers customarily use Roman months with Syrian names. As their eclipse dates (when given) are almost always impeccable, it is noteworthy that in this case they are both wrong. Agapius has Tishrin II 1 [=Nov.1], though he also says Friday (correct for Nov.5), while Michael has Tishrin I 9 [=Oct.9]; however, the identification of the eclipse is quite clear. Michael also has "at the third hour", presumably of the afternoon. As the local time of maximum phase would be the greater part of an hour later at Antioch than at Constantinople, Michael (if he refers to Syria) agrees more closely than does Theophanes (if he refers to Constantinople) with what one would expect from Schroeter; but Theophanes may also have used a Syrian source.

Agapius places the eclipse in the eleventh year of ᶜOmar (ᶜUmar), usually considered to have reigned from 634 August until his assassination on 644 Nov.3, two days before the eclipse. He reigned from AH 13 until nearly the end of AH 23.

646 S.646 April 21. IN THE MIDDLE EAST

We have encountered no certain Western record (a Chinese mention exists). But the Syrian writer Severus Sebokht, of Ken-neshre near Edessa, is known to have lived until at least AD 665 and to have written a short work on eclipses (see J. Ginsburg, Bull.Amer.Math.Soc. (2) 23, 1917, 366-9, 467). We do not know the date of this work, but wonder whether the author's interest in eclipses may have been aroused by the pair of solar eclipses on 644 Nov.5 and 646 April 21, both large in Syria.

Oppolzer shows the central line of S.646 April 21 as starting off West Africa, and passing via Greece, Constantinople and the Black Sea to the Far East. Schroeter shows the band of totality as crossing Asia Minor from South-West to North-East, with maximum magnitude about 0.9 at Constantinople on one side of the band and also about 0.9 at Alexandria and Edessa on the other side. The maximum phase at Edessa would occur at about 11.05 a.m. local apparent time.

Schroeter makes S.644 Nov.5 almost annular at Edessa, and annular in part of Syria about 2.20 p.m. local apparent time. We have already seen that this eclipse of 644 is mentioned by Syrian writers.

650 S.650 Feb.6. FALSELY ATTRIBUTED TO BEDE

Bede mentions S.664 May 1 (though wrongly as May 3) unmistakably in his Hist. Eccl. of 731 and reasonably clearly in both his small De Temporibus of 703 and his larger De Temporum Ratione of 725. Calvisius (Seth Kalwitz, 1556-1615) and the later editors of his Opus Chronologicum (1605, 1620, 1650, etc.) manage to give both the correct identification and two false ones! Between them they attribute to Bede "tomo secundo" two solar eclipses, S.650 Feb.6 and S.661 July 2, which did occur but appear not to be genuinely mentioned by Bede. As both false identifications were copied in Tycho-Curtius 1666 (XXVIII) and persist in, e.g., Johnson 1874 (33), 1896 (31) and the first also in Chambers 1902 (137), explanations of the errors seem worth while. We consider S.661 in its proper place in our chronological sequence. "Tomus secundus" is the second volume of Bede's Opera, Cologne 1612.

It is the De Temporum Ratione which has sometimes been misinterpreted. The passage occurs in the Sexta Aetas (or AD portion) of the long chronological section Chronicon sive De Sex Aetatibus; e.g. Opera, 2, 1612, 116; ed. Giles, 6, 1843, 326; ed. Mommsen, Chr.Min. 3 = AA 13, 1898, 313.

Bede in three consecutive sentences has

(A) a synod at Rome, in the ninth year of Constantine [should be Constans], October, indiction 8;

(B) Constans sends a gift to Pope Vitalian, and himself visits Rome in indiction 6;

(C) Next year, a solar eclipse, May 3 [really May 1], about the tenth hour of the day.

(A) refers to the Lateran Council, called by Pope Martin (649-53). The synod is usually dated 649 October (correctly in 9. Const. and ind. 8). As for (B), Vitalian was Pope in 657-72. The only sixth indiction in this time was 662-3. It is usually accepted that Constans visited Rome in 663. Calvisius 1605 (681), 1620 (494), etc. wrongly makes (C) occur in the year following (A) instead of in the year following (B).

The above explanation is simpler than that for S.661. Recognition of the S.650 mistake in Calvisius goes back at least to Struyck 1740 (116).

S.650 Feb.6 was annular at sea to the north of Scotland, and of magnitude about 0.75 at London towards sunset; it would be of intermediate magnitude in Northumbria; but we have encountered no genuine Western record of it.

655 S.655 April 12. PROBABLY RECORDED IN SPAIN (cf.666)

Among continuations of Isidore's History of the Goths, Vandals and Sueves is a Continuatio Hispana of about AD 754, edited by Mommsen (Chron.Min.2=MGH AA 11, 1894, 323-369). On page 343 are two eclipse references. The first (§34) is in the reign of the Byzantine emperor Constans II (641-668):

"huius imperio sole medio die obscurato celum stellas prodit".

The second (§36) is in the reign of Reccaswinth, Visigothic king of Spain (sole ruler c. 653-672), and is mentioned immediately after a

Council of Toledo usually assigned, as by Mommsen, to 653:

"Huius temporibus eclipsim solis stellis meridie visentibus omnis Spania territat atque incursationem Vasconum non cum modico exercitus damno prospectat".

Ginzel 1886 (972-4) had already discussed a passage from Roderic of Toledo (died 1248) which is clearly related to the second (§36) passage above. Probably correctly, he had ruled out, as insufficiently striking in Spain, the eclipses S.659 Jan.28 and S.671 Dec.7 (qq.v.). He considered more seriously the two eclipses S.655 April 12 and S.666 Sept.4 (called Nov.4 by oversight). Both were total over much of Spain and large over all Spain, the former more than 4 hours before noon and the latter about 3 hours after noon, so that the reference to "the middle of the day" is apparently to be understood broadly. Ginzel preferred S.666, for reasons which seem weak even in relation to the limited passage from Roderic which he quoted. In relation to the passages from the Continuatio Hispana, and the mention of the 653 Council of Toledo, fresh consideration is needed. It will be noted that both eclipses, S.655 and S.666, fall within both the reigns mentioned; also that it is not clear whether the two passages (§34 and §36) relate to the same eclipse or to two different eclipses. If the eclipse of §34 is derived from some Byzantine source, and "medio die" may be discounted, it could in fact refer to S.644 Nov.5 (q.v.). But the eclipse of §36 seems definitely Spanish, and is probably either S.655 or S.666. Newton 1972 (507, 510-1) discusses both passages under S.655 April 12. We adopt the same course, as perhaps the best that can be done with these weak records.

655 M.655 Oct.20-21. LATER CALCULATION

The Islamic horoscopes in Pingree 1968 (120) include one (IV 12) headed "The lunar eclipse indicating the murder of ʿUthman and the accession of ʿAlī". Oppolzer lists the eclipse as of magnitude 0.78 at 12.35 a.m. at Greenwich, corresponding to around 3 a.m. in the Near East. The murder of ʿUthmān is dated 655 June 17 in Camb. Med. Hist. 2, 1926, 356.

659 S.659 Jan.28. NO RECORD, EVEN AT CONSTANTINOPLE

We have encountered no likely record, but see under S.655 April 12. Probably absence of record is noteworthy, as the central line of this annular-total eclipse passed over Italy and the Balkans, and the magnitude would be greater than 0.75 over most of Europe.

660 M.660 Dec.22. LATER CALCULATION ONLY

The Islamic horoscopes in Pingree 1968 (120) include one (IV 13) headed "The lunar eclipse indicating the death of ᶜAlī and the accession of Muᶜāwiya". In this context, the eclipse is merely a computational result, as it occurred around 2 p.m. in the Near East, where it was consequently invisible. The murder of ᶜAli is dated 661 Jan.24 in Camb. Med. Hist. 2, 1926, 358.

661 S.661 July 2: NO GENUINE WESTERN RECORD

This is the second (S.650 Feb.6 being the first) almost certainly false identification from Bede "tomo secundo" in Calvisius. It differs from the first by its appearing only in the posthumous editions, 1620 (497), 1650 (609), etc., and by its reference to Henry of Huntingdon as well as Bede.

Henry of Huntingdon's History twice mentions an eclipse of the Sun "die tertio Maii" (on May 3), adding on the second occasion "hora decima diei" (at the tenth hour of the day); ii,35 and iii,44, on pp.61 and 100 of T. Arnold's ed. of 1879, and pp.60 and 108 in T. Forester's Engl. tr. of 1853. This so clearly stems from Bede's (HE iii,27) solar eclipse of 664 May 3 (really May 1), "die tertio mensis Maii, hora circiter decima diei", that the eclipse identification is unmistakable. The tenth hour of the day means two-thirds of the way from noon towards sunset (4 p.m. at the equinoxes, later in summer). We might leave the matter there, merely remarking that, to the best of our knowledge, no English (as opposed to Celtic) author of Anglo-Saxon or early Norman times mentions any seventh-century eclipse whatever except S.664 May 1 (normally wrongly as May 3). But in view of the adoption of the S.661 July 2 Calvisian interpretation in, for example, Tycho-Curtius 1666 (XXVIII) and Johnson 1874 (34), 1896 (31), a few words of explanation seem desirable.

The error arises from attempts to relate the eclipse to an ill-dated conversion of the South Saxons, complicated by the fact that their king, named Æthelwalh (but see below), was baptized some time (an interval of very uncertain length) before his people. Thus we have to distinguish

(A) the baptism of Æthelwalh, and his receipt of the Isle of Wight, etc. as a baptismal gift from his godfather Wulfhere, king of Mercia 659-674, who had just conquered the territory.

(B) the conversion of the South Saxons, still under Æthelwalh, by Wilfrid.

Bede refers to (A) and (B) in HE iv, 13, without any precise date, but saying that (A) was "non multo ante", not long before, (B). Bede also refers to (B) incidentally in v,19.

The Anglo-Saxon Chronicle (e.g. tr.D. Whitelock et al., 1961) puts (A) under 661, and the eclipse rightly under 664; it does not mention (B), which, however, is known to have been after Wilfrid's expulsion by Ecgfrith (ASC 678). There is in fact no doubt that (B) occurred, or at any rate began, about AD 681; hence Bede's "not long before" contrasts violently with about two decades in ASC.

The first printed edition of Henry of Huntingdon's History was in H. Savile, Rerum Anglicarum Scriptores post Bedam praecipui (London, 1596), where on folio $182^v$ we find mention of the baptism (A) of Æthelwalh, a three-year interval (doubtless following ASC), and then a solar eclipse on May 3. The same is found on p.318 of the 1601 Frankfurt edition of Savile. "Calvisius" (the posthumous work) probably used one of these editions. For many, Savile's edition of Henry is more accessible in PL, 195, 1855, col.838.

The passage in Savile agrees word for word with that in Arnold (p.61). But whereas Arnold correctly interprets Henry's intentions by referring in the margin to Chronicle (E) as source, and providing editorial years 661 for (A) and 664 for the eclipse, "Calvisius" for some reason puts (A) under 658, and hence takes the eclipse to be S.661 July 2.

In fact, the 3-year interval seems to have no independent value. It arises merely from Henry's having noticed the dates 661 and 664 in ASC. Thus even dating (A) differently does not affect the correct eclipse identification (S.664 May 1).

The date of (A) may doubtless be refined by modern research. F.M. Stenton, Anglo-Saxon England, 3rd ed., Oxf. 1971, p.58, gives "shortly before 675" for the baptism of Æthelwalh. If this refers to (A), we may say that AD 674 makes Bede's error only about half the Chronicle's, but throws no extra light on an eclipse identification which is already quite clear. The dating of (A) is a historical problem, having no direct connection with any eclipse.

It may be added that the name of the king in question consists of Æthel, Athel, Ethel, Adel, Edel, Aedil, etc. followed by wealh, walch, bold, wold, or other variant forms, sometimes Latinized. Savile gives Adelwold, with Edelwalkius as a variant. Modern experts in Anglo-Saxon tend to use Æthelwealh or Æthelwalh.

The eclipse S.661 July 2 certainly occurred, though in the early morning (no "tenth hour of the day"). Schroeter shows it as annular at, e.g., Dublin and Edinburgh. It had magnitude about 0.9 at London, less than an hour (on Schroeter's data) after sunrise, and obviously a greater magnitude in Northumbria. We should be glad to recognize a Western record if we could find the slightest trace of one. But ASC and Henry of Huntingdon do not mention it; nor, after much search, do we find it in the historical or chronological works of Bede. This may appear surprising; but AD 661 is an early date in English scholarship. Bede has done well enough in preserving for us some record of S.664 May 1 (q.v.).

## 664 S.664 May 1. FIRST GENUINE ENGLISH ECLIPSE

This eclipse is famous as the first recorded in England from genuine observation in that country.

The best record, however, comes from Ireland. The Annals of Ulster, known to need year correction +1 over a long period, have under 663 (hence really 664) "Darkness on the Kalends of May, at [or rather in] the ninth hour (in hora nona)" (ed.W.M. Hennessy, 1, 1887, 118-9). The same Kalends of May and "in hora nona" appear, though needing more year interpretation or correction, in the AT (198) and CS (98-9). The hour will be discussed below; ninth is reasonable for Ireland. Thus, apart from some weakness about the year, the original Irish tradition is sound. But the seventeenth century Annals of the Four Masters (ed. J. O'Donovan, 1, 1851, 276-7) have May 3 from the faulty English tradition (see below). The editor does not improve matters by quoting AU as giving the Kalends of May "in $11^{a}$ hora" (for $11^{a}$ read $IX^{a}$).

None of the annals quoted was assembled in its present form before AD 1000, but the Irish sources for this period are known to come from Iona (now in Scotland) and Bangor in N.E. Ireland.

Where Ireland is correct on the day and weak on the year, England (under the influence of Bede) is correct on the year and wrong on the day. Bede is the earliest known author to mention the eclipse, "quam nostra aetas meminit"; the eclipse occurred before he was born, thus doubtless he used verbal or written Northumbrian evidence.

Bede gives day and hour in both his De Temporum Ratione of 725, "quasi decima hora diei, V Nonas Maias [May 3]" (e.g. ed. Giles, 6, 1843, 326; ed. Mommsen, Chron. Min.3 = AA 13, 1898, 313) and in his Hist. Eccles. (iii,27) of 731, "die tertio mensis Maii hora circiter decima diei" (e.g., 1969, 310). There has been much discussion whether or not Bede's error of two days is due to his having dated the New Moon by the ecclesiastical calendar, as suggested by James Ussher (1581-1656). R.R. Newton's discussion of the eclipse, including this point, may be found, as far as Ireland and Britain are concerned, in Newton 1970 (49, 50, 76, 264), 1972 (126-9, 143f, 151, 185-6, 192-3, 606). See also P. Grosjean (1900-64) in Analecta Bollandiana 78, 1960, 239-241, an important reference which we owe to Mr K. Harrison and shall revert to below.

While most manuscripts of the Anglo-Saxon Chronicle are content to mention the year 664 of the eclipse, two give also month and day. Bede's V.non.Mai. (=May 3) is unfortunately mentioned in the influential manuscript E (a copy made at Peterborough of a version which originated in Northern England, probably at York). F's Latin has v.Kal.Mai. (=April 27).

May 3 remained the English tradition. For example, Henry of Huntingdon, who made much use of E, gives (iii,44) May 3, tenth hour of the day, and Florence of Worcester (ed.B.Thorpe, 1, 1848, 25; tr.T.Forester, 1854, 20) gives May 3, about the tenth hour of the day. Recognition of the incorrectness of May 3 goes back, among eclipse chronologists, at least to Bunting 1590 (fol.360).

"Tenth hour of the day" is reasonable for N.E. England. At Armagh and Whitby, whose latitudes are nearly equal, sunset occurred about 7.36 p.m. local solar apparent time, so that if we use seasonal hours (each about $1^h16^m$ of solar time) the ninth hour of the day extends from 2.32 to 3.48 p.m., and the tenth from 3.48 to 5.04 p.m.

There is some confusion about the hour at which the eclipse should (according to astronomical calculation) have occurred. It is certain that the local apparent time of mid-eclipse at Whitby would be about half an hour later than at Armagh (just over 24 minutes being due to difference of longitude, and the remainder to travel time of the Moon's shadow), so that the Irish and English recorded times, given only in whole numbers of hours, might almost indifferently be equal or differ (as they do) by unity. But whereas the data in Schroeter 1923 make the respective local apparent solar times at Armagh and Whitby be about 5.01 and 5.28 p.m., Grosjean (1960), gave 3.26 p.m. and "almost 4 p.m.". However, Richard Stephenson (personal communication) confirms that the Schroeter times are essentially correct, and finds a maximum magnitude at London at around 17.35 local time. Schroeter's times fall late in the tenth seasonal hour for Armagh and Iona, and in the eleventh at the Northumbrian places. But the fairly precise seasonal hours once used by astronomers (e.g. Ptolemy) only very roughly correspond with ecclesiastical canonical hours; see Grosjean, or D. Knowles (e.g. Christian Monasticism, London 1969, 217, 222).

On the chronology of English history around the time of the eclipse (Synod of Whitby, etc.) see F.M. Stenton, Anglo-Saxon England, e.g. 3rd ed. 1971, Grosjean 1960 (255), and recently K. Harrison, Anglo-Saxon England, 2, 1973, 51-70. Stenton, believing that Bede began the Year of Grace, like the Indiction, on Sept. 1, placed the Synod in late September or early October of our AD 663. Grosjean regarded the first half of our AD 664 as most probable. Harrison (p.57) also prefers our AD 664 ; and his Framework of Anglo-Saxon history, Cambridge, 1976, 93.

We may add a few words about European and Syrian records, though neither has anything to add to those of the British Isles. We have amassed about a score of references to continental records, mostly under years between 662 and 670, but they all appear to derive, directly or indirectly, from Bede; for those (the great majority) which give month and day all give Bede's erroneous (but perhaps explicable) May 3, not the correct Irish May 1. Many also give the hour of the day (usually Bede's "quasi decima", but occasionally just "decima"); the cryptic "10. Haracliae" in Ann.Fuld.antiqui (SS 3, 1839, 116*) appears to be a corruption of "10.hora diei", and rather similarly in Ann.breves Fuld. (SS 2, 1829, 237).

Syriac records are silent up to and including Michael (d.1199), but the Chronicle to 1234 mentions, under years corresponding to about 664, a solar eclipse which may be either S.664 May borrowed from Bede, or a genuine Near-Eastern observation of S.666 Sept.4 (q.v.) or S.667 Aug.25.

It is curious that an eclipse so comparatively well dated in Irish and English records has given rise to many ghosts. Several of the principal misdatings were detected in Struyck 1740 (115). See our S.650 Feb.6 and S.661 July 2 above. Paul the Deacon, Hist.Lomb. vi,5 (ed. L.Bethmann & G.Waitz, SrL 1878, 1964, 166) says "His temporibus per indictionem octavam luna eclypsin passa est. Solis quoque eclypsis eodem pene tempore, hora diei quasi decima, quinto Nonas Maias effecta est". The position of the passage is about AD 679-80, but the solar eclipse is almost certainly not the S.679 July 13 of Calvisius 1620 (501), etc., but as Struyck said, S.664 May 1, misplaced by one indiction cycle of 15 years, and exhibiting the English error of May 3 for May 1, as well as Bede's "quasi decima hora diei". The lunar eclipse (probably of 680 June 17-18) will be considered later.

Regino of Prum (d.915) in his Chronicle (SS 1, 1826, 551) mentions an eclipse on May 3, evidently Bede's of 664. But several 16th and 17th century editors manufactured an eclipse in AD 605, through mistaking the peculiar "Annus Dominicae Incarnationis" reckoning used in this part of Regino for the usual AD count, from which in reality it differs by some 64 years.

## 666? S.666 Sept.4? POSSIBLY RECORDED IN SPAIN (cf.655)

We know of no certain record, but of two possible ones. For a possible Spanish record, see under S.655 April 12. The second possible record is Syrian: "In this year there was an eclipse of the Sun and the stars appeared", Chronicle to AD 1234 (CSCO, (3) 14, Louvain, 1937). This is under a year described as

(1) Year 976 of the Greeks, a Seleucid year which, for Syria, ought to indicate AD 664-5, autumn to autumn,

(2) 23rd year of Constans II (641-668), which ought at least to lie within AD 662-5;

(3) 44th year of the Arabs; AH 44 ran from 664 April 4 to 665 March 23. This may be

(A) a record of S.664 May 1 (which did not reach Syria) borrowed from the West. In this case the year is specified tolerably well, but the report is derivative;

(B) a genuine record of S.666 Sept.4, for which, according to Oppolzer, totality ended at sunset in the general area of Upper Egypt and Northern Sudan (West of Wadi Halfa);

(C) a genuine record of S.667 Aug.25, for which, according to Oppolzer, totality began at sunrise near Cairo and occurred not long after sunrise in Northern Arabia.

If the passage refers to (B) or (C), it is rather spoiled by the fact that three moderately consistent attempts to indicate the year have all failed. (B) occurred in AH 46 and (C) in AH 47.

667? S.667 Aug.25. See under S.666 Sept.4.

670 M.670 Jan.11-12. A LATER ISLAMIC CALCULATION

The Islamic horoscopes in Pingree 1968 (120) contain one (IV 14) headed "A lunar eclipse in the sixth conjunction". Oppolzer gives the magnitude as 0.59 at 12.50 a.m. at Greenwich, say between 3 a.m. and 4 a.m. in the Near East, and this eclipse was presumably a calculated one.

<u>671</u> S.671 Dec.7 (Sunday). IN SYRIA AND ARABIA

This eclipse is fairly clearly mentioned by Michael the Syrian (1901, 1963, 456): "In the year 983 there was an eclipse of the Sun on a Sunday in the month of Kanūn I (December)". In Syria, Seleucid 983 normally means AD 671-2, autumn to autumn, which tallies.

The same eclipse appears to be mentioned by S. Ockley (1678-1720) in his History of the Saracens (third ed., Camb.1757, 2, 110; fourth ed., Lond.1847, 367). The eclipse is said to have deterred the caliph Muʿāwiya (AH 41-60, AD 661-680) from removing the pulpit of Mohammed from Medina to Damascus. Ockley says that "the Sun was eclipsed to that degree that the stars appeared". He refers to At-Tabari (839-923) and the later Ibn al-Athir. The year in Ockley's account is rather vague, but apparently in the middle period of Muʿāwiya's reign. Hind's identification, S.671 Dec.7, was accepted in Johnson 1874 (34), 1896 (32), and, in a qualified way, in Chambers 1902 (138). Both Hind and Oppolzer find the eclipse annular; Chambers quotes Hind as finding magnitude only 0.85 at Medina. Hind's identification lies in AH 51, and no other eclipse comes so near to fitting Ockley's account. An annular eclipse with Oppolzer noon-point in Arabia (also with considerably larger magnitude at Mecca than at Medina) seems compatible with some degree of historical truth in the tale. S.667 Aug.25 (see under S.666 Sept. 4) was total, but fits no better on the whole, and seems too early; S.678 July 24, which Oppolzer makes annular near Aden, fits worse and seems much too late.

Ockley begins AH 50 with the death of Al-Mogeirah, who "had lost one of his eyes at the battle of Yermouk [636], though some say that it was with looking upon an eclipse". No eclipse identification appears to be called for.

672? M.672 Nov.10-11? PSEUDO-ECLIPSE IN IRELAND (AURORA OF 674)

recte 674

About 674, a number of Irish annals have a passage which has been taken for a bad eclipse record, but may well be a good record of an aurora. "A thin and tremulous cloud, a kind of rainbow, appeared at the fourth watch of the night on the Friday (not fifth day, as sometimes stated) before Easter Sunday, from east to west, in a clear sky. The moon was turned to blood". It is not clear whether cloud and moon phenomena were simultaneous. See AU, (126-7), under "673" (correctly 674);

CS (1866, 102-3), under "670" (correctly 674);

AC (1896, 108), under "670";

AT (17, 1896, 203);

J. O'Donovan's Intro.to Ann.Four Masters (1, 1851, xlviii), quoting AU. The year-corrections +1 and +4 applied in the first two cases are as given by the editors for this period. There was no lunar eclipse in 674.

A.O. Anderson, (1922, 182) says that, if an eclipse, it seems in 673 or 676. But M.673 May 6 was invisible in Scotland, and M.673 Oct. 31 (p.m.) was barely visible there; M.676 March 5-6 and M.676 Aug.29 were total and well visible (M.675 March 17 was partial and half visible, and M.675 Sept. 9 was small but fully visible). Newton 1972 (654), quoting Chron.Scotorum, mentions M.672 May 17, M.672 Nov.10 and M.673 May 6; but the first and last of these were invisible in Ireland. We list the passage under 672 Nov.10-11, as the date of the only total lunar eclipse visible in Ireland between M.669 July 18-19 (plainly too early) and M.676 March 5-6. But it seems to us easier to suppose that the whole passage refers to an aurora of 674, than to force an eclipse identification of the final sentence. There were maxima of aurorae about 664, 673 and 682 (cf. ed Schove, 1983, Appendix A).

673-6 See under M.672 Nov.10-11.

679 M.679 Dec.22-23. POSSIBLY OBSERVED IN MIDDLE EAST

The last (IV 15) of the seventh-century Islamic horoscopes in Pingree 1968 (121) is headed "The lunar eclipse indicating the death of Muᶜāwiya and the accession of Yazīd". Camb. Med. Hist. 2, 1926, 359 says Muᶜāwiya died on 680 April 18. In this case the eclipse and the death both occurred in AH 60, which began on 679 Oct.13. The eclipse was a deep total one. According to Oppolzer, maximum magnitude 1.64 occurred at 1.38 a.m. at Greenwich, hence around 4 a.m. in the Near East.

680 M.680 June 17-18. LUNAR ECLIPSE AT ROME

The most specific mention is in the Liber Pontificalis (ed. L.Duchesne, second ed., 1, Paris 1955, 350, 356). Under Pope Agatho (c.678-81) we read of a lunar eclipse in indiction 8 (variant, 9) on June 18. Duchesne identifies the eclipse as we do. According to Oppolzer and Schroeter, it was total (magnitude 1.37 about 10.33 p.m. at Greenwich). The day is probably reckoned ecclesiastically (from sunset). Identification goes back at least to Calvisius 1620 (501).

The passage is also printed, as Anastasius (d.897), De vitis Rom. pont., in Migne (PL 128, 1880, cols.807-8), with different variants: indiction 8 (var.3), June 28 (var.18). Ind.8 and June 18 are correct. A note (col.819) gives Paul the Deacon (d.799), Hist. Langob. vi, 5 as the source (see under S.664 May 1). But at this date Lib. Pont. is a contemporary source for Italy; moreover, it is more likely that Paul abbreviated Lib. Pont. (omitting month and day) than that Lib. Pont. filled out Paul's account.

The eclipse is also mentioned, though without month or day, by several later medieval writers. Any aberrant placing in 681 may be rejected, as there was no lunar eclipse in that year.

683 M.683 April 16-17 (Th.-Fri.). LUNAR ECLIPSE AT ROME

As with M.680 June 17-18, there is what is essentially a contemporary Italian record. "At this time, April 16, indiction 11, the Moon was eclipsed after the Lord's Supper (post Cenam Domini); it laboured with blood-red face almost all night and only after cock-crow did it gradually begin to clear up and return to its normal condition". Some MSS. wrongly have April 15, indiction 10. Liber Pontif. (e.g., ed. L. Duchesne, 2nd ed., Paris 1955, pp.360, 362, or Anastasius, (PL 128, 1880, cols.847 ff.), under St. Leo II (pope 682-3). Rather similarly in Herimannus (Contractus) Augiensis and Marianus Scottus (SS 5, 1844, 96 and 544). In 683, April 16 was Maundy Thursday, and April 17 Good Friday. Total eclipse around midnight in Italy. The duration of the eclipse (actually about $3^h40^m$) is somewhat exaggerated. Identification goes back at least to Bunting 1590 (363).

688 S.688 July 3. IN IONA OR IRELAND

An eclipse is mentioned in identical words in three accounts having "The Chronicle of Ireland" as ultimate source. "Obscurata est pars solis", "Part of the Sun was darkened", occurs in Annals of Ulster under AD 688 (AU 1, 1887, 138-9), in Chronicum Scotorum under 685 (CS, 1896, 108-9), and in AT (17, 1896, 210-1). On the Chronicle of Ireland see K. Hughes, Early Christian Ireland: Introduction to the Sources, London 1972, pp. 101 ff; the place was probably Bangor, but possibly Iona.

Newton 1972 (193-4) says rightly that there is no other possible identification within 5 years, and, again rightly, that the years in AU contain random errors as well as the systematic error needing correction + 1. The usual corrections + 1 to AU and (for this period) + 4 to CS both wrongly lead to 689 here, and thus are inapplicable. The + 1 correction to AU over several centuries is merely what statisticians call the "mode" (more frequent than any other single correction). The same phenomenon (AU correction zero, not + 1) occurs also in the case of M.691 Nov.11 (q.v.). Around 690, CS is normally 3 years behind AU, and the correction + 4 given by Hennessy (1866, xlvi) as applying from 652 to 718 merely puts CS approximately on to the "AU + 1" scheme.

O'Connor 1952 (based on Oppolzer) makes maximum magnitude in mid-Ireland 0.65 at 8.42 a.m. Schroeter, whose narrow band of annularity goes through Northern Ireland and just south of Archangel (Arkhangel'sk), leads to a similar result. The magnitude at Iona might approach 0.75, still not large. But perhaps thin cloud and a good observer enabled the eclipse to be recorded.

We have encountered no Western record except from Celtic sources.

690, 691 M.691 Nov.11 (p.m.), etc. LUNAR ECLIPSE IN IONA OR IRELAND

An eclipse of the Moon about AD 691 is mentioned in various Celtic sources. Several, believed largely derived from a lost original which has been named "the Chronicle of Ireland" (see Kathleen Hughes, Early Christian Ireland: Introduction to the Sources, London 1972, pp.101 ff.), mention in similar wording that "the Moon was turned to the colour of blood on the nativity of St. Martin". This is a clear reference to M.691 Nov.11, which according to Oppolzer was partial, with greatest magnitude 0.80 at 5.33 p.m. at Greenwich (beginning 4.01 p.m., ending 7.5 p.m.). Printed sources which include the St. Martin eclipse sentence are:

AU under 691 (1, 1887, 140-1),
CS under 688 (1866, 110-1),
AC under 687 (1896, 110),
AT (17, 1896, 212).

On the chronology of AU and CS see remarks under S.688 July 3.

Several Welsh sources say the same thing, except that they do not mention St. Martin. Over a number of centuries the Annales Cambriae (ed. John Williams ab Ithel, 1860) use a special Welsh era, such that AD - AC = 444 (ab Ithel) or 445 (Newton 1972, 206). On page 8 they have "the Moon was turned to the colour of blood" under Annus 246, meaning AD 690 or 691. Bruts y Tywysogion, or the Chronicle of the Princes (tr. Thomas Jones, 1952, 1, 130) has "the Moon reddened as it were to the colour of blood" under manuscript year 690, editorial year 691. The edition of the same Bruts by John Williams ab Ithel (1860, 4-5) gives the passage under year 692, but the chronology of the Bruts in only approximate, and there was no lunar eclipse during the year which began on 692 January 1. The eclipse appears under manuscript year 690 in John Rhys and J.G. Evans, the Text of the Bruts from the Red Book of Hergest (Welsh only), Oxford 1890, page 257, lines 7-5 up.

There is no lack of other lunar eclipses to which these Welsh sources might refer. For example, in the years 688-93 the only total lunar eclipses, with their greatest magnitudes and UT mid-times from Oppolzer are M.690 May 28 (1.02 at 8.38 p.m.), M.690 Nov.22 (1.64 at 4.48 a.m.), and M.691 May 17-18 (1.23 at 10.45 p.m.); Schroeter gives slightly revised times. Nevertheless, since the Welsh sources contain a certain amount of Irish matter, they may here be borrowing from the Irish, and thus referring to M.691 Nov.11; we notice that Newton 1972 (654) so interprets them.

The Danish Annals of Lund (SS 29, 1892, 192), and the older edition as Annals of Esrom (ed. J.Langebek, 1772), under AD 696 and 695 respectively, have "In this year the Moon was turned to the colour of blood". There was no eclipse of the Moon in 696, but prima facie the reference might be, for example, to any of the following eclipses (all partial), given with their maximum magnitudes and corresponding Greenwich mean times from Oppolzer: M.695 Aug.29 (0.91 at 9.10 p.m.), M.697 Jan.13 (0.25 at 4.01 a.m.), M.697 July 9 (0.77 at 8.46 p.m.), or M.698 Dec.22-23 (0.86 at 10.59 p.m.). However, from similarity of wording and known use of sources from the British Isles, the Danish annals, which are late (begun in the twelfth century) may be referring to M.691 Nov.11, or even any of the total eclipses of 690-1.

693 S.693 Oct.5 (Sunday). FROM S.E. EUROPE TO IRAQ

This important eclipse occurred in the ninth year of the first reign-period (685 Sept. - 695) of the Emperor Justinian II (Rhinotmetus). Before discussing the sources, it will be useful to give some indication of the general locality of the eclipse. The track of totality was essentially Eurasian. Oppolzer shows the central line as starting at sunrise off the West coast of France, proceeding via North Italy, Asia Minor and the Persian Gulf to a noon point on the North coast of the Gulf of Oman (near the seaward end of the Iran-Pakistan boundary), and going on via North India and Vietnam to end at sunset near the Philippines. Here we shall be concerned with the section between North Italy and Basra, and particularly with conditions at Constantinople and Bagdad (founded about AD 762-3, but usable as a representative point in Central Mesopotamia).

We have encountered essentially two good detailed tracks in print, (A) the Schoch-Neugebauer track, (B) the Ginzel 1918-Schroeter track;

for the track in Ginzel 1883 (specified only between the Adriatic and somewhat west of Ankara) may be regarded as superseded by (B). The most extensive tabulation is in Schroeter 1923, which includes the track from Western France to Basra, based on Ginzel's corrections to Oppolzer's elements. Ginzel 1918 has the same basis, so that although it tabulates the track only from about the Dardanelles to the Persian Gulf, the two tracks (Ginzel 1918 and Schroeter 1923) never differ in this stretch by more than 0°.02 (and that very rarely); consequently, to the west of the Dardanelles, the track in Schroeter 1923 may safely be regarded as a western extension of that in Ginzel 1918. But Ginzel's so-called corrections were revised by Schoch in 1926-8, and after Schoch's death in 1929 the track was revised in P.V. Neugebauer 1930 on the basis of Schoch's elements. Track (A) lies to the north of track (B).

The general picture which emerges is that of a total eclipse visible in France during the first seasonal hour after sunrise, in Italy during the second, in Thrace and Constantinople during the third, in North-East Syria during the fourth, and in Lower Iraq (Bagdad, Basra) during the fifth hour.

In particular, the various authors give the following magnitudes and local apparent times:

| | Constantinople | Bagdad |
|---|---|---|
| (A) Schoch-Neugebauer | Total at 8.47 a.m. | 0.986 at 10.09 a.m. |
| (B) Ginzel 1918-Schroeter | (0.984) at 8.50 a.m. | Total at 10.12 a.m. |

The only detail supplied by us is the estimate of magnitude 0.984 at Constantinople for track (B).

At Constantinople, sunrise occurred about 6.18 a.m., so that the third seasonal hour ran from 8.12 to 9.09 a.m., and totality occurred not only during the third seasonal hour, but also during its second half, so that (to the nearest hour) totality may be said to have occurred at the third hour. It was otherwise at Bagdad where sunrise occurred about 6.12 a.m., so that the fifth seasonal hour ran from 10.04 to 11.02 a.m.; totality occurred during this fifth hour, but in its final half, so that totality may be said to have occurred at about the fourth hour.

With regard to visibility of "stars", Neugebauer 1930 (B28) found Mercury 20° West of the Sun, magnitude $0^m$; Venus 43°W., mag. $-4^m$; Mars 20°W, mag. about $1^m.5$; Jupiter and Saturn invisible. We may add that the Sun lay within the triangle Antares-Arcturus-Spica. Totality, or even a little less, would certainly allow some bright stars to be seen.

As far as writers in Greek and Latin are concerned, much the most specific statement is in the Chronographia of Theophanes (died c.817), who wrote at Constantinople from sources said to include city records. Under AM 6186 [AD 693-4] he has: "In this year an eclipse of the Sun occurred in the month Hyperberetaeus [here October], on the fifth day, on the first day [of the week, i.e. Sunday], at the third hour [of the day], so that certain bright stars appeared". (1883, 367; or 1839, 561). As the passage probably refers to Constantinople or at any rate Thrace, "at the third hour" agrees well with modern calculations.

However, the abridged Latin version of the Chronographia by Anastasius (in Theophanes, ed.de Boor, 2, 1885, 233; or in Theophanes, 2, 1841, 187) preserves the hour, but discards the month, the day of the month, and the day of the week. This curious procedure features also in other Latin sources; especially, in exactly the same words, in the Historia Miscella (Muratori 1, 1723, 140), a running together of histories by the fourth-century Eutropius, Paul the Deacon (died c.799) and Anastasius (died 897), revised and augmented by Landulfus "Sagax" about AD 1000. But neither the fact that Paul the Deacon was connected with Pavia nor that Anastasius wrote at Rome makes this a useful Italian record. No justifiable reckoning can associate totality in Italy with any hour of the day later than the second. Mention of the third hour means either that the information is drawn from further East (e.g. Constantinople) or that it is wrong. Similar information, entirely derivative, appears in the Annals of Xanten under 687 (SS 2, 1829, 220), in Sigebert of Gembloux (died 1112) under 695 (SS 6, 1844, 328), and in Alberic of Trois-Fontaines (died 1252 or later) under 695 (SS 23, 1874, 701).

Even the hour is omitted by later Byzantine writers, e.g. George the Monk ("Hamartolos") iv.240 (Migne, PG, 110, 1863, 899-900), Leo Grammaticus (ed. I.Bekker, Bonn 1842, 163-4), and Cedrenus (ed. I. Bekker, 1, Bonn 1838, 773).

References to the Byzantine and associated Latinized accounts go back in chronological literature at least to Calvisius 1605 (690), 1620 (505), etc. and Tycho-Curtius 1666 (xxviii). Recent astronomical discussion is found in Newton 1972 (231, 389, 462, 543-4, 606) and Newton 1979 (376, 421).

The eclipse is also mentioned in several twentieth-century translations (into Latin or modern languages) from Semitic languages. The earliest writer we have encountered in this way is James of Edessa (died c.708), or perhaps rather his immediate continuator, since James himself is said by Elias of Nisibis to have written about 692 (see Brooks, p.197). This particular eclipse occurs in the section of Brooks's Latin translation containing some passages of James of Edessa as preserved in Elias of Nisibis, and we shall see that the Year 75 of the Hegira under which it is placed is wrong. "75$^{th}$ year. In this there was a total eclipse of the Sun on Sunday, the 5$^{th}$ of Teshrīn Prior, at the fifth hour of the day (hora quinta diei)". In Syriac writers, Teshrīn I regularly denotes October. E.W. Brooks, CSCO, Ser.3, Tom.4, Paris 1, Paris 1903, pp.197, 257.

The principal extant Syriac source is Elias of Nisibis (975-c.1049). He is the only Semitic source quoted verbatim in the pioneer paper, Ginzel 1918, which introduced discussion of the Near and Middle Eastern records of this eclipse into astronomical chronology. Besides Ginzel's German, we have used an English translation by Mr. G.C.I. Rawlings, published in duplicated typescript form in Aylesbury Astronomical Society, Bulletin No.37, 1969. Ginzel quotes a German translation of Elias by F. Baethgen. We have used the French translation (Elias 1910, pp.94-5.)

As Ginzel says, "The Chronography of Elias of Nisibis was unknown to the older Orientalists. It was first discovered in the nineteenth century and brought to London. The author has the unusual habit of giving the sources from which he has obtained his information". The last sentence is true in a general way, but we shall find difficulties.

Elias of Nisibis evidently had at least two sources for this eclipse, since he gives it twice over, once correctly under AH 74, and again under AH 75, incorrectly as regards year but otherwise more fully:

(a) "Year 74.- Commenced Tuesday 13 ʾIjar [May] of year 1004 of the Greeks [Seleucid era]. In which there was an eclipse of the Sun on 29 Jamada I, 5 Teshrīn I; the stars were visible."

(b) "Year 75.- Commenced Saturday 2 ʾIjar [May] of year 1005 of the Greeks. In which there was a total eclipse of the Sun on Sunday 5 Teshrīn, at the fifth hour of the day."

Delaporte says that the day should be the 28th of First Jamada (the fifth Islamic month); Ginzel appears content with the 29th. However, in Syriac writers, First Teshrīn regularly means October, and there is no doubt that both passages refer to S.693 Oct.5, which fell in AH 74. (a) is fairly correct on the date; (b) alone gives the hour, which Delaporte describes as written in red ink in his manuscript, which may indicate (see under 350th Olympiad, pp.78-9 in Delaporte) that Elias himself recognized date weakness in (b). "Fifth hour" strongly suggests a Near or Middle Eastern observation, not a borrowing from Constantinople. Where Delaporte has "à la 5e heure du jour", Ginzel quotes "um die 5. Tagesstunde", i.e. about the fifth hour of the day, which tallies well enough with what we have said about the result of modern calculations.

This is where there is difficulty about sources. In general, the annals in Elias contain other items beside eclipses, and when references are given for an annal, it is not always clear which reference applies to which item. Moreover, it is not clear to us, dependent on translations, quite what references Elias gave in this case.

(a) For AH 74, both Delaporte and Ginzel give the references as Khwārizmī and a certain Jésudenah (Delaporte) or Isoᶜdeneh (Ginzel), metropolitan of Basra.

(b) For AH 75, Delaporte gives Khwārizmī and James of Edessa (Brooks), but Ginzel gives Khwārizmī only.

One notes with interest that the involvement of a metropolitan of Basra is possible, and of James of Edessa is probable (in view of his endorsement by Brooks), especially since Edessa and Basra lie only very slightly north and south respectively of the Ginzel-Schroeter band of totality. Nevertheless, the attention of any astronomer may well focus principally on the undoubted mention of al-Khwārizmī who wrote at Baghdad (almost on the central line of the Ginzel-Schroeter track) under the caliphs al-Maʾmūn (813-33) and al-Muᶜtaṣim (833-42). According to Ginzel 1918 (8), the eclipse report is contained in Khwārizmī's "History", on the strength of "local" records available to this author. It would be idle to speculate on the probabilities of Edessa and Basra as places of observation; the eclipse was doubtless widely seen in Mesopotamia.

The eclipse is also mentioned in the Chronicle of Michael the Syrian (1126-1199), Jacobite patriarch of Antioch (Michael, 1901 or 1963, 474). "In the year 1005 of the Greeks, 75 of the Arabs, the Sun was darkened on a Sunday in the month of Teshrīn I, at the third and the fourth hour (à la troisième et la quatrième heure): and there was a profound darkness; the stars appeared." What is interesting here is not so much the wrong year as the mention of the third and fourth hours. "Fourth hour" fits mid-eclipse in both Antioch and North-East Syria, while "Third hour" allows two interpretations: either time of onset of perceptible eclipse in Syria or (perhaps rather more likely) time of mid-eclipse at Constantinople. Chabot gives a reference to Theophanes - whether as a source, or merely as a parallel passage, we are not sure.

698 S.698 Dec.8. IMPORTANT ANNULAR ECLIPSE NOT RECORDED

Some record of this eclipse might be expected to have survived, but we have encountered none. The broad, short, sharply curved band of annularity in Schroeter starts south of Iceland, and includes all the British Isles except the east coast, most of France, most of Italy, most of Greece, all Asia Minor except the north coast, and the northern part of the Caspian Sea. Constantinople, however, lies just to the north, while Jerusalem and Baghdad lie further south.

In England and Ireland the eclipse may have been missed because of the low altitude of the Sun in midwinter, and possibly also because the few minutes of darkness may have occurred in cloudy weather. The absence of any record from the general region of the Mediterranean and Near East is a little more surprising.

A complete list of the seventh-century eclipses (AD 601-700) in Hoang 1925, with their *Oppolzer Greenwich* dates, is:

S.601 Mch. 10
S.616 May 21
S.618 Oct. 24 (Not observable)
S.621 Aug. 22-23
S.623 Dec. 27 (N.O.)
S.626 Oct. 26
S.627 Apr. 21
S.627 Oct. 15
S.628 Apr. 10
S.629 Aug. 24 (N.O.)
S.630 Feb. 18 (N.O.)
S.630 Aug. 13 (N.O.)
S.632 Jan. 27
S.634 June 1 (N.O.)
S.635 May 21-22 (N.O.)
S.637 Apr. 1
S.638 Mch. 21 (N.O.)
S.639 Sept. 3
S.643 June 21-22
S.644 Nov. 5 (N.O.)
S.646 Apr. 21
S.648 Aug. 24
S.660 July 13
S.661 July 2 (N.O.)
S.665 Apr. 21 (N.O.)
S.667 Aug. 25
S.669 July 3
S.670 June 23
S.671 Dec. 7
S.672 Nov. 25 (N.O.)
S.674 Apr. 12
S.675 Sept. 24
S.680 May 4 (Spurious date)
S.680 Nov. 27
S.681 Nov. 16
S.682 May 12-13
S.682 Nov. 5 (N.O.)
S.686 Feb. 28
S.688 July 3
S.691 May 3-4
S.692 Apr. 22
S.693 Oct. 5
S.694 Sept. 24-25
S.695 Feb. 19
S.700 May 23

R. Stephenson points out that the *Sui-shu* in Chapter 21 of the Astronomical Treatise states that on 616 may 21 'the Sun was eclipses and it was complete'. He adds that it was in fact an annular eclipse, the path of annularity lying to the north of the capital, the computed magnitude at Lo-yang being about 0.95.

Hoang's list of lunar eclipses contains none in the seventh century A.D.

N.O. = Not observable

Cohen and Newton (in progress) find no evidence that any seventh-century Chinese 'eclipse' was based on observation.

Japanese eclipses in the early medieval chronicles are, like the Chinese, generally calculated and often invisible. Nevertheless truly annalistic writing begins in 617 and the meteorological details for 628 in the same context suggest that the total solar eclipse recorded for the 2nd day of the 3rd month does indeed correspond to a genuine observation. The reference is W.G. Aston 'The Nihongi' Supplement I of the Trans. and Proceedings of the Japanese Society, London, Vol II, 1896, London, p.155. The eclipse given for 636 Moon 1 Day 1 in the same source (p.167) is nevertheless at an impossible date and conceivably refers to a partial eclipse of 637 (March 31/April 1). The reference under 621 that 'the sun and moon have lost their brightness' suggests a reference (incorrectly dated?) to the volcanic darkening noticed in Europe in this decade.

The Dresden Codex dates near eclipses fall into two main groups, those of AD 447/474 and those of AD 683/872 (Schove 1984a, Table 1). The earliest dates in the second group are those of 683 Oct 28, 709 Jun 13 and 742 Mar 12, the three main bases of the Venus Table. That this Table had eclipse associations was recognized long ago by Spinden, who noted that the intervals between the three dates were eclipse intervals. Adopting our 615,824 correlation, they are all 16 or 17 days before lunar eclipses that were centrally situated in the range 84° - 169°W and hence New 'Eclipse' Moons, which were significant in the Maya world. The expansion of the Venus bases (in the manner which has long been recognized) generates further lunar eclipses in the 8th and 9th centuries of a similar kind. Although we have no evidence of eclipse glyphs attached to any specific seventh century date, it seems probable that lunar eclipses of 683/742 were observed and recorded by Maya astronomers.

# THE EIGHTH CENTURY

## RECORDS FROM WESTERN CHRISTENDOM

Records from Syria are still important, Syriac scholars playing a significant part in the cultural revival which was to take place at the newly-founded capital of Bagdad. However, in this century the growth of chronicle writing in northern England (e.g. Bede), Ireland and, later, on the European Continent, provides us with some good records of eclipses and comets. Byzantine chronicles are known mainly from their abridgment in Theophanes (- 814), which, like so many abridgments, often distorts the chronology of comets and eclipses. A few significant reports of eclipses are available from China (and one possible record in 733 from what is now South Russia) are discussed. Independent records of comets from both Japan and China usually confirm one another and enable us to correct the errors in Islamic and European dates: even Bede's comet of January 729 might perhaps be equated with the Chinese comet of 730, but no Western records of the comets of 707, 767 (?) and 773 have been traced.

706 S.706 July 14 ANNULAR IN S.W. EUROPE BUT NOT RECORDED

Oppolzer gives this annular eclipse a sunrise point in the Bering Sea, a noon point on Baffin Island, and a sunset point in Eastern Algeria. His approximate chart indicated a track through Central Spain, but Schroeter's reliable calculation puts the band rather through the Algarve, the Straits of Gibraltar, Morocco and Algeria. As well as the Iberian Peninsula, Morocco and Algeria, the greater part of the British Isles, France and Italy (including London, Paris, Milan and Rome) have magnitude greater than 0.75 in Schroeter, but we have encountered no Western record.

713 S.713 March 1 N.W. EUROPE BUT NOT RECORDED

The central line of totality has sunrise point in the USA, noon point in the East Atlantic, and afternoon passage via Norway, to a sunset point in the Arctic. In more detail, Schroeter makes the band of totality, missing Scotland, pass between the Faroes and the Shetlands, and on to Vardö in North Norway. His southern line of magnitude 0.75 runs roughly along the French Channel coast and the German Baltic coast, so that the magnitude would exceed 0.75 throughout the British Isles and Scandinavia. But we have encountered no Western record.

716 M.716 Jan 13 (p.m.) RECORDED IN ITALY

During the pontificate of Gregory II "in the 14th indiction there occurred a sign in the Moon, and it appeared like blood up to midnight" (Liber Pontif., e.g., ed. L. Duchesne, 2 ed., Paris 1955, pp.398-411, or Anastasius, De vitis Rom.pont., PL, 128, 1880, cols. 975-6). Although Gregory II was Pope about AD 715-731, so that the 14th indiction could mean either 715-6 or 730-1, the reference is clearly to the deep total eclipse of 716 Jan. 13. This ended soon after 8.30 p.m. at Greenwich, but a little before 9.30 p.m. mean time at Rome. The (later) Annals of Farfa in Italy (SS 11, 1854, 588) put the eclipse in 716. Johnson's date 716 April 12 is wrong.

718 S.718 June 3 TOTAL SOLAR IN SPAIN

S.718 June 3 probable (S.720 Oct. 6 less probable)

Isidore Pacenses, in his Continuatio Hispana (about 754) of the Gothic, etc. history by the more famous Isidorus Hispalensis (Isidore of Seville, died 636), has a passage given with slight variations by various editors and commentators: Roderia of Toledo (died 1248), Calvisius, 620 (512), whence Tycho-Curtius 1666; Ginzel 1886, Mommsen, Chron. Min.2 = AA11, 1894, 356. The gist of the passage is that as Spanish Era SpE 757 (or 758) was beginning, in the 100th year of the Arabs, an eclipse of the Sun occurred in Spain, from the 7th (or 6th) to the 9th hour of the day; some place this eclipse in the time of Hurr, but most in the time of his successor Samh (these being the names of Arab governors in Spain, see Camb. Med. Hist., 2, 373-4).

SpE 757 and 758 are AD 719 and 720. The years of the Hijra, AH 99 to 102, began about 717 Aug.14, 718 Aug.3, 719 July 24 and 720 July 12 respectively. Ginzel asserts that Samh came to Spain in AH 100; Historia de Espana, ed R. Menendez Pidal, 4, 1950, 25, concurs, giving Ramadan 100 (719 March-April) as the date of transfer of office.

Although all the year numbers are a unit or so too large, one must nevertheless agree with the result of the discussion in Ginzel 1886 (974-6): the great Spanish total eclipse of 718 June 3 (Schroeter's band of totality passes north of Lisbon, south of Madrid, and then to Northern Tunisia) fits better than S.720 Oct.6 (total in North Africa) as regards both totality and hour in Spain. If only one eclipse is involved, S.718 must have the preference.

But it is not impossible that the account may have been influenced by both eclipses, e.g., the indication of year may be a compromise by a writer faced with apparently conflicting records and unaware that two different eclipses occurred in the years round 719. The historical chronology about this time seems a little uncertain, but it is probable that 718 June 3 fell in the time of Hurr and 720 Oct. 6 in the time of Samh. If this is true, and one is informed of a division of opinion as to whether "this" eclipse occurred in the time of Hurr or in the time of Samh, one may legitimately suspect conflation.

Newton 1979 (354) refrains from using even the 718 eclipse as one he can use scientifically, but we have allocated this Spanish eclipse 9 and given 4 points also to the North African eclipse of 720.

718 M.718 Nov.12-13 SMALL ECLIPSE IN IRELAND

There is a probable reference to this eclipse in the Annals of Ulster, under year AD 717, needing correction (usually +1): "The son of Cuidin Cutharine , King of the Saxons, dies ... An eclipse of the moon at its full". See AU. The magnitude was only small, Oppolzer gives 0.23 at 11.08 p.m. at Greenwich, and conceivably the entry has been deliberately misplaced from 716 to coincide with the royal death. We have allocated only 6 points for the usual 718 interpretation and allowed 1 point for the possibility that the original record related to the total eclipse of 716.

720 S.720 Oct 6 PRESUMABLY RECORDED IN NW AFRICA

Total in North Africa and probably recorded in a lost Arabic chronicle as 'in the time of Samh' and subsequently amalgamated with the 718 eclipse by a Continuator of Isidore (See 718 June 3). Four points have been allocated on this assumption.

725 M.725 Dec 24 (a.m.) SMALL ECLIPSE PROBABLY UNNOTICED

Probably not recorded in Ireland (See 726 Dec 13-14), although this is the year implied by the annals.

726 S.726 Jan 8 ANNULAR IN ARABIA BUT NO RECORD

Oppolzer's central line of annularity proceeds via the Sahara, to a noon point in Arabia and then via Iran to Siberia. We have not encountered any Western record and Arabian primary sources are mostly lost.

720-760 GAP ON CONTINENT

There is an almost complete dearth of primary records of eclipses from Continental Europe from the early 720s until 760. In compensation, the period has a welcome number of eclipse records from Britain and Ireland, most of them more certainly identifiable than the lunar eclipse which now follows.

726 or 725 M.726 Dec 13-14 (a.m.) TOTAL IN IRELAND

The Annals of Ulster (1887, 176-7) have, under their A D 724, which would normally mean AD 725, "A dark and blood-red moon on the 18th of the Kalends of January", 18th being "xviii" in the Latin. Note that Dec 24 = viiii. Kal.Jan. and Dec 14 = xviiii. Kal.Jan; the style of AU uses iiii rather than iv. Oppolzer gives magnitudes 0.44 (partial) and 1.63 (total) at Greenwich times 6.57 a.m. and 10.32 p.m. respectively. MacCarthy's index to AU (IV, 1901, 140) gives the eclipse of 725; Anderson, 1922, 221-2, tends towards that of 726 and we concur giving 8 points in our table to the 726 hypothesis and only 1 to the 725 possibility.

732 S.732 Mar 1 TOTAL IN S.E. ISLAM BUT NO RECORD

The central line of totality starts in South America to a noon point in the Atlantic, and then via the Sahel and Arabia to the Persian Gulf. The central line is shown by Oppolzer passing rather further from Mecca than in 726, and again we have encountered no Western record. These two 'Arabian' eclipses of 726 and 732 are not in Schroeter but have been mentioned in case an Islamic source might be discovered.

733 S.733 Aug 14. TOTAL SOLAR IN CENTRAL ENGLAND AND CAUCASUS

There is one contemporary account, namely that in the construction of Be 's Ecclesiastical History (1969, 572-3): "In 733 an eclipse of the Sun occurred on xviiii (variant, xviii) Kal. Sept. about the third hour of the day so that its whole orb seemed to be covered by a black and terrifying shield". The dates are Aug.14 (variant, 15), the former being correct, and sometimes degenerating into Aug.19 in later-writers. The most correct of the early copyists from the continuation of Bede (or from another set of Northumbrian annals) date from Norman times; they are Simeon of Durham (e.g., ed. T.Arnold, 2, 1885, 30; EHD, 1, 1955, 239) and Roger of Hoveden (e.g., ed. W.Stubbs, 1, 1868, 4); the hour is also correctly inserted in the latin version of MS.F of the Anglo-Saxon Chronicle, which otherwise gives year only. Newton 1972 (143-7) lists the English sources up to Roger of Hoveden. The hour, when given, is always "about the third".

The central line in Oppolzer has sunrise point in the West Atlantic, proceeds via Southern Ireland and Southern England to a noon point north of the Sea of Azov; and ends at sunset in Indo-China. Ginzel 1883 (660-1) tabulates the band of annularity through Central England.

Modern computations, such as Schroeter make the band of annularity run through Southern Ireland, Central and South Wales, Central England, short of London, East Anglia, Holland, Berlin, the Sea of Azor and the Caspian.

Ginzel, believing the Anglo-Saxon Chronicle to be the most independent source, proposed to find the band ½° south (more would really be needed) to produce centrality at London. But it will be seen that the English account has a strong Northumbrian flavour. Newton 1972 (152-3) corrects the estimated place of observation from the "Canterbury" of Newton 1970 to Jarrow and 1979 (244, 422) estimates its magnitude there as 0.90 ± 0.05.

S.733 Aug.14 (cont.)

Ginzel (Handbuch, III, 1914, 92) took "third hora" at London as 7.20 to 8.29 a.m., and computed maximum eclipse there at about 8.40 a.m. Schroeter would give around 8.48 a.m. The agreement is sufficient to leave the identification unquestioned. The equation of time is negligible (two minutes).

A so-called Scottish record originated in Northumbria; see Lord Cooper, Supra Crepidam, London 1951, 39-40 and Newton 1972 (200-1).

The late Annals of Lund (SS 29, 1892, 192), formerly called the Annals of Esrom in Denmark, follow closely the variant (wrong day, Aug. 15) Northumbrian wording, and say "about the third hour of the day", which points to England. Some half-dozen Frankish annals give year 733 only. This was the year following that (732) in which the Saracens were defeated between Tours and Poitiers by Charles Martel.

A fragmentary primary source of the 730s has been incorporated into the 'History of the Caucasian Albanians' by Movses Dasxuranci (tr. C.J.F. Dowsett, 1961, Oxford University Press). This refers to Cattle Plague, Taxes and Famine in the 720s and is followed by what seems to be a description of a total eclipse at a specific place: 'the borders of Mozu, in the bishopric of Siwnik'. A hermit saw a vision (possibly auroral in origin) and 'announcing the coming of the wrath (of the Lord) throughout the canton, he besought all to pray, and an impenetrable darkness descended over the borders of Mozu, and the earth shook for forty days, and nigh on 10,000 souls were swallowed up; and because of this (the place) was called Vayoc Jor (Valley of Woe)'.

The borders of the bishopric lie in the 50-mile square 45.75°/46.80° East, 38.85°/39.60° North (information from Professor Lang) and the eclipse interpretation, if acceptable, suggested that the track as plotted in Schroeter might have been too far north. However, I consulted Meeus who kindly confirmed (March 1981) that the eclipse was not total (0.9 magnitude) in the region in question; perhaps the rumour of 'impenetrable darkness' came from well to the north of the borders of Mozu.

734 M.734 Jan 24 (a.m.) RECORDED IN N.E. ENGLAND AND IRELAND

Concerning the day, there are two traditions, English and Irish, both wrong, because impossible English sources which specify the day give ii Kal.Feb. (Jan.31), and Irish sources give xi Kal.Feb. (Jan.22). But there was (i) a deep total eclipse on the night of 734 Jan. 23-24, entirely after midnight even in the most westerly Irish longitudes, magnitude 1.62 at 3.10 a.m. at Greenwich (Opp.); (ii) a partial eclipse on the night of 735 Jan. 13-14, entirely before midnight; magnitude 0.87 at 4.41 p.m. at Greenwich (Opp.) Jan.24 is ix. Kal.Feb., Jan.13 is id. Jan., and Jan.14 is xix Kal.Feb.

The English account occurs in almost identical words in the contemporary continuation of Bede (e.g. 1969, 573), copied in Florence of Worcester, Simeon of Durham, Roger of Hoveden and Roger of Wendover: "734. The Moon was suffused with a blood-red for about a whole hour around cockcrow on ii.Kal.Feb. Then blackness followed and finally its own light was restored". Several, including Cont. Bede, add: "735 ... Bede ... died". ASC different, giving no month or day, and putting the lunar eclipse and the death of Bede on the same year 734, though Bede is believed to have died in 735.

The Irish sources are briefer, simply "An eclipse of the Moon on xi. Kal. Feb." AU (under 733, normally meaning 734) and AT.

The English mention of cockcrow favours M.734 Jan.24 (a.m.). The English error of 7 days might result, since Jan. 24 and 31 were Sundays, from incorrect conversion of an ecclesiastically described Sunday into month and day; if not, from corruption of ix into ii. The Irish record would then be 2 days out (ix correspond to xi?). Alternatively, M.735 Jan. 13 (p.m.) may have been reckoned ecclesiastically (from sunset) as Jan. 14 = xix. Kal. Feb., and xix may have degenerated into English ii and Irish xi.

The only record we have encountered from outside the British Isles is in the late Annals of Lund, which, as in the case of S.733 Aug.14, echo Cont. Bede or some similar English source.

744 M.744 January 4 (p.m.); COMET (OR LUNAR ECLIPSE) IN SYRIA

745 "La même année, au mois de kanoun II (janvier), apparut un autre signe sous la forme de la lune; et l' atmosphère fut terne et sombre." Agapius 10 C, PO, 1912, 511. This is about the year (744) of the death of Walid and the contested succession of Yezil III.

Oppolzer gives M.744 Jan.4 a maximum magnitude of 0.42 at 3.48 p.m. at Greenwich (say about 6 p.m. in Syria). Considerations of both language and astronomy might seem to point to this lunar eclipse, if an eclipse were meant at all. However, a comet of 745 January probably explains this reference.

749 S.744 mar 23 SUNRISE AT CONSTANTINOPLE NOT RECORDED

Schroeter shows the band of totality as beginning south-east of Budapest and running via South and East Russia, with Jerusalem and Bagdad within the 0.75 limit. As the magnitude at Constantinople was about 0.9, unfavourable weather may account for the fact that we have encountered no Western record.

752 M.752 July 30-31 TOTAL IN ENGLAND

This was a deep total eclipse of the Moon, maximum magnitude 1.72. Oppolzer makes the whole eclipse lie between 11.42 p.m. and 3.26 a.m. at Greenwich, and totality between 12.43 a.m. and 2.25 a.m. Schroeter agrees within a couple of minutes.

A record of this eclipse occurs in early Northumbrian annals preserved for us in Simeon of Durham: "752. An eclipse of the Moon occurred pridie real. Aug. July 31 ". See Simeon's Historia Regum, Opera, 1, 1868, 19 (Publ. Surtees Soc. 51, for 1867); Opera, ed.T. Arnold, 2, 1885, 40 (Rolls Series); EHD, 1, 1955, 241. The same appears in Roger of Hoveden, Chronica (ed.W.Stubbs, 1, 1868, 6, Rolls Series). Roger of Wendover, Flores Historiarum (ed. H.O. Coxe, 1, 1841, 232; tr. J.A. Giles, 1, 1849, 147) says "752. In the same year there was an eclipse of the Moon, after midnight, on the 31st of July". While Coxe appears to rely, justifiably, on Simeon, Hoveden and Calvisius in recognizing the Moon as the eclipsed body, Giles mistakenly translates "Sun". The additional information, "after midnight", must refer to totality in Northumbria, rather than to the entire course of the eclipse. Any time after sunset on July 30 may count as July 31 ecclesiastically.

753 S.753 Jan 9 ANNULAR (SOLAR) IN ENGLAND AND IRELAND

"753, 15th year of king Eadbert of Northumbria , 5.id. Jan. (Jan.9), a solar eclipse occurred; shortly afterwards, in the same year and month, namely 9. Kal.Feb. (Jan.24), the Moon suffered eclipse, being covered, like the Sun a little earlier, by a fearful black shield". Continuation of Bede's Hist. Eccles., e.g. ed. A. Wheloc, 1643, 491; ed. J.A. Giles, Opera, 3, 1843, 324; ed. B. Colgrave and R.A.B. Mynors, 1969, 574. On some corruption of the passage, see Colgrave and Mynors, and Newton 1972 (153); manuscript dates vary. "15th year" is easily compatible with the year 737 often accepted for Eadbert's accession. Mr K. Harrison refers to Plummers ed. 362 and suggests that the original passage read quarto, quinto ... the quarto being lost sight of (Personal Communication, Feb '84).

Schroeter's broad and strongly curved band of annularity includes Madrid (on the south rim), Lyon, Munich, Vienna, Budapest, Prague (on the north rim), Warsaw and Moscow. Thus the reference to the "fearful black shield" has little appropriateness here; it echoes the account of S.733 Aug.14, for which eclipse (annular in parts of the British Isles) it was quite in order.

The solar eclipse is also mentioned (dark Sun; no month of day) in the AU under 752 (meaning our 753) and in AT O'Connor 1952 makes the greatest magnitude for mid-Ireland be 0.87 at 10.35 a.m.

We have encountered no record except from Britain and Ireland.

753 M.753 Jan. 23-24 NEAR TOTAL

See continuation of Bede, under S.753 Jan.9 above. The lunar eclipse is not mentioned in the AU, but seems to appear in AT and under 749 in AC. (Moon blood-coloured; no month or day). Oppolzer gives maximum magnitude 0.91 at 1.19 a.m. at Greenwich; in both England and Ireland the eclipse would start before midnight and end after midnight.

We have again encountered no record except from Britain and Ireland.

755 M.755 Nov. 23 (p.m.) TOTAL AND JUPITER OCCULTATION IN N.E. ENGLAND

Wrongly under 756, the phenomena are described in almost identical terms by two Northern writers, Simeon of Durham (fl. 1130) and Roger of Hoveden (fl. 1200). "756 ... Moreover, the moon fifteen days old, that is the full moon, was on 8 Kal. Dec. (Nov.24) covered with a blood-red colour; and then the darkness gradually diminished and it returned to its former light. For, most remarkably, a bright star, following and crossing (pertranseunte) the moon, preceded it when it was illuminated at the same distance as it had followed it before it was obscured." Simeon of Durham, Historia Regum, Publ Surtees Soc. 51, 1868, 20; ed. T.

Arnold, 2, 1885, 41; EHD, 1, 1955, 241. Same in Roger de Hoveden, Chronica, ed. W. Stubbs, 1, 1868, 7; tr. H.T. Riley, 1853, 5.

The eclipse occurred in the evening of Nov. 23, which ecclesiastically (reckoning from sunset) may count as Nov. 24. Schroeter 1923 (184), agreeing within two minutes with Oppolzer, makes the beginnings and endings of the partial and total phases be 4.53, 5.57, 7.27, 8.31 p.m. UT (mean Greenwich time); the greatest magnitude 1.43.

With regard to the "bright star", Jupiter was retrograding in opposition (about a degree from the Anti-Sun), and was certainly occulted by the Moon as seen from at least part of England. Struyck 1740 (118) found occultation at London, which still seems likely; modern tables seem also to allow the possibility of occultation at Jarrow (the probable source of Simeon's report). Attempts to identify the observed body with one of the fixed stars, as recently in Newton 1972 (589), have not thrown up any strong candidate, and the Jupiter identification, inherently likely, seems reasonably certain. We shall not attempt to assess the degree of precision of either Simeon's report or any particular set of tables. One might gather from Simeon that mid-eclipse and mid-occultation were simultaneous. That is unlikely; mid-occultation as seen in England probably followed mid-eclipse by an hour or so.

758 S.758 April 12 IMPORTANT SOLAR IN N. EUROPE NOT RECORDED

Schroeter's path of totality runs by the Scilly Isles, Brighton, the Rhine mouths, Berlin and Warsaw, to end North of Astrakhan. The magnitude would exceed 0.75 over much of Europe. In a way, it is surprising that we have encountered no record; but the Anglo-Saxon Chronicle has a large gap in West Saxon entries from 754 to 823, and the 750s are likewise poorly documented on the continent.

750s X.750s GHOST SOLAR

recte 760/4

One solar eclipse ascribed to 755 and a reference to another also in the reign of Pipin (died 768) are given in the Chronicon Luxoviense breve (SS 3, 1839, 221). These are, however, 12th century insertions and the date given, June 4, suggests that the eclipse of 764 June 4 had been borrowed from the Revised Version of the Royal Frankish Annals and misdated. The other eclipse was evidently that of 760, the Byzantine record of which had reached the Latin West through Anastasius. There was an eclipse in 755 in June but it was on the 14th and that was a 'Polar' eclipse invisible on the European Continent.

760 S.760 Aug.15 (Fri) ANNULAR (SOLAR) AT CONSTANTINOPLE

The track of annularity ran in the general neighbourhood of Finland, Russia, the Black Sea and the Caspian Sea, so that the eclipse was larger at Constantinople (almost 0.9) than in Western Europe (about 0.75 in Schroeter at Edinburgh, Paris and Palermo). The chief source seems indeed to be the Greek of Theophanes (1, 1839, 665; 1, Leipzig 1883, 431; Germ. tr., L. Breyer, Graz. etc. 1957, 76), correctly latinized in the abbreviation by Anastasius (Theophanes, 2, 1841, 231; 2, Leipzig 1885, 283), and copied in the Historia Miscella (Muratori, 1, 1723, 158). These all give "August 15, Friday, tenth hour of the day", the date being correct and the hour reasonable (eleventh would be better if mid-eclipse is meant).

At least six other (late) Western European chronicles give date and hour, but usually with either date or hour wrong; the "hora quasi 6" of the Wurzburg Chronicle (Ekkehard; SS 6, 1844, 26) and the "hora 6" of the Flores Temporum of a thirteenth-century Minorite (Eccard's attribution to "Martin" is rejected by the editor, O. Holder-Egger), SS 24, 1879, 226, 233 seem too small if reckoned from dawn and too large if reckoned from noon.

The eclipse is important mainly because it throws light on a problem arising in the chronology of Theophanes. In most of his extensive Chronographia, this writer uses an Annus Mundi and an Indiction, both beginning on Sept.1, such that Ind.= $R_{15}$ + 1, where $R_{15}$ is the remainder when AM is divided by 15. But for a space of nearly half a century (about AD 726-774), Theophanes used Ind.= $R_{15}$ + 2. Whatever may be the explanation of this curious phenomenon, it is clear that one does need to know whether it is AM or Ind. which is out of step (mathematically if not administratively) with the main usage of Theophanes.

This eclipse should fall, by the main usage, in AM 6252 and Ind.13, whereas the usage as deranged over the period mentioned places it in AM 6252 and Ind.14. It is clear that, at least at the time of this eclipse, and so presumably over the whole period of derangement, it is AM which is correct and Ind. which is too large by unity (against the conclusion of many authorities in the 19th and earlier centuries, and indeed the non-astronomical evidence is indecisive).

A fuller account may be found in J.B. Bury's Later Roman Empire, 2, 1889, 425-7, or in his edition of E. Gibbon's Decline and Fall of the Roman Empire, 5, 1898, 524-5. An important result of the chronological revision is that the death of Leo III and the accession of Constantine V as sole ruler fell in 740, not 741. The eclipse had earlier been discussed briefly in Ginzel 1884 (545), and indeed correct identification of the Byzantine eclipse goes back in astronomically oriented works at least to Bunting 1590 (373b). Bury's chronology is further confirmed by the fact that Theophanes mentions a comet in the same year; this was Halley's comet, 760 May-July, see T. Kiang, The past orbit of Halley's Comet, Mem.R.A.S., 76(2), 1972, 53. It may be added that the rival Byzantine year AD 760-1, commencing 760 Sept.1, so often adopted as Theophanes AM 6252 by historians until Bury's rectification, contained no suitable eclipse. S.761 Aug.5, besides falling on a Wednesday, has its Oppolzer totality starting at sunrise in South-East Persia. This eclipse was completely invisible at Constantinople; it would be astonishing to find it mentioned in Theophanes; it has been explicitly rejected by Bury (1889, 426) and by W.T. Lynn (Observatory, 34, 1911, 275).

760 M.760 Aug.31 (a.m.) GHOST LUNAR (LATER CALCULATION)

As a solar eclipse occurred on 760 Aug.15, there were obviously full moons near the beginning and the end of the same month. In fact, they occurred on Aug.1, somewhat after 4 p.m., and on Aug.31, around 3.50 a.m., mean times at Greenwich. There was no umbral eclipse on Aug.1, but for Aug.31 Oppolzer gives a large partial eclipse, maximum magnitude about 0.92, semi-duration 96 minutes.

Unfortunately, we have encountered no record which mentions an eclipse at the correct syzygy. Roger of Wendover (died 1236; unreliable on eclipses in Anglo-Saxon times) has under AD 760 "Eclipsis lunae facta est circa horam noctis mediam kalendis Augusti", "There was an eclipse of the Moon about midnight, on the first of August" (ed. H.O. Coxe, 1, London 1841, 236; tr. J.A. Giles, 1, London 1849, 150). This can scarcely be a misplacing of the laconic Irish "dark moon" of 762 or 763, one "identification" of which is 11.763 June 30 (ii.Kal.Jul.), which occurred early in the morning, considerably nearer midnight.

Struyck 1740 (154) also quotes Florence of Worcester for "Kal.Aug.", but we do not find this eclipse in modern editions (H. Petrie 1848, B. Thorpe 1, 1848, tr. T. Forester 1854), nor in one of Florence's chief sources, the Chronicon of Marianus Scotus, the Irish monk who wrote in Germany (SS5, 1844, 481-568).

The eclipse discrimination between the two Full Moons is not really critical by modern tables, but might have been close for the Toledo Tables (1080) of Arzachel, while the lunar theory of Western Christendom at the time barely reached the modest level of calculating the true day of New or Full Moon.

It was in the extreme west of Europe (above all, Ireland and the Spanish peninsula) that M.760 Aug.31 occurred closest to local midnight. The problem is interesting, and might deserve further research, were not the mention of the wrong Full Moon such strong evidence of a failed calculation (perhaps three centuries or more after the event) rather than a corrupt record.

c 763 M.c763 AMBIGUOUS ECLIPSE IN IRELAND

The Annals of Ulster have a brief "Nix magna et luna tenebrosa", "Great snow and a dark Moon", under AD 761, which with the usual but not invariable correction +1 would mean AD 762 (1, 1887, 226-7; indexed in 4, 1901, 140 as 762 Jan.15). The same sentence appears also in Tigernach, AT, 17, 1896, 260). There are at least five possible "identifications" of this very unspecific report to be found in Oppolzer, with magnitudes and times (mean Greenwich) as follows:

| | | | |
|---|---|---|---|
| (i) | M.762 Jan. 14-15, mag. | 0.41 (partial), mid-eclipse | 12.37 a.m. |
| (ii) | M.762 July 10 | 0.37 (partial) | 10.18 p.m. |
| (iii) | M.763 Jan. 4 | 1.62 (total) | 4.09 p.m. |
| (iv) | M.763 June 29-30 | 1.86 (total) | 1.42 a.m. |
| (v) | M.763 Dec. 25 | 0.81 (partial) | 3.45 a.m. |

Schroeter, who lists only total eclipses, agrees within minutes about (iii) and (iv). No lunar eclipse occurred in 761.

We have seen that MacCarthy (1901) preferred (i). Anderson, 1922, 244) seems to prefer 763 rather than 762. The uncommon magnitude of (iv), occurring in the middle of the night, would have made it very noticeable (granted a clear sky and persons awake to witness the phenomenon). But since the lunar eclipse is mentioned in the same sentence as the great snow, Newton 1972 (194) and 1879 (244) limits consideration to the three winter eclipses (i), (iii) and (v); taking into account the position of the virtually certain mention of S.764 June 4 in the same annals, he prefers (iii). The Moon would rise totally eclipsed. A lunar eclipse of 0.8, unlike a solar eclipse of the same magnitude, is easily noticed and the cold and snow of the 763/764 winter was outstanding in Europe generally. Moreover the summer solar eclipse of 764 occurs in the next annal. We have, therefore, given more points to the Christmas eclipse of 763 in our table. Altogether, the brevity of the account makes it useless to astronomy, and of uncertain value to chronology.

See also under M.760 Aug. 31 (a.m.)

764 S.764 June 4 (Mon) ANNULAR (SOLAR) RECORDED IN IRELAND AND FRANCIA

The band of annularity as shown in Schroeter runs across Europe from Dublin, Liverpool and Hamburg to the north Caspian (near Astrakhan). Apart from Irish records (see below), this eclipse, which occurred about noon (sixth hour of the day) in the Frankish domains, is mentioned with usually correct month, day and hour in the Annals of Flavigny and in Einhard's version of the Frankish Annals and copied in many later sources as follows:

| | | |
|---|---|---|
| Primary: | Annals of Flavigny (France) MGH SS | 3,1839,151 |
| | RFA (revised version only), e.g. in | 1,1826,145 |
| Secondary: | Annals of La Cava (Italy) | 3,1839,87 |
| | Chron. of Herimann of Reichenau (Germany) | 5,1844,99 |
| | Würzburg Chronicle (Germany) | 6,1844,26 |
| Much later: | Annals of Melk (Austria), under 762 | 9,1851,495 |
| | Annals of Admont (Austria), under 762 July! | 9,1851,572 |
| | Flores Temporum, Minorite (Germany) | 24,1879,233 |
| | Chroniques de Saint Denis (France,late) Bouquet | 5,1869,222 |

Several further annals give the year only. The eclipse was discussed, from continental records, in Ginzel 1884 (546). On the Würzburg Chronicle (Ekkehard) see Newton 1972 (359). On Herimann's "Gottwicensis" source, see Newton 1972 (250).

However, only the first two sources can be considered as independent.

The eclipse is also mentioned in Irish annals. The Annals of Ulster (1837,228) have under 762 "Sol tenebrosis in hora tertia diei", "A dark Sun in the third hour of the day". 762 in AU would normally mean AD 763, but here the reference is to S.764 June 4, for which O'Connor 1952 gives maximum magnitude 0.98 at 10.26 a.m. local apparent time at his mid-Ireland point. As he gives sunrise at 3.40 a.m., the third hour would end about 7.50 a.m., more than an hour before his time for the start of the eclipse (9.12 a.m.). As MacCarthy's index (AU 4, 1901, 140) and Anderson 1922 (245) recognized, the eclipse occurred in the fifth hour in Ireland. Tigernach's Annals (AT, 1896, 261) give the same eclipse information as AU.

Johnson 1874 (37), 1896 (34) says "An eclipse of the Sun, about midday, seen in France and England". He must mean that the eclipse was visible (weather permitting) in England, which is very true; but we have encountered no vestige of an English record. The eclipse is not mentioned in the ASC or by Simeon of Durham.

We have encountered no genuine Byzantine record. Bunting 1590 (375-6) uses a Frankish eclipse source. It is only for approximately contemporary events (drought, Turkish invasion) that he refers to Annales Constantinopolitani, lib.22, doubtless meaning, in slightly more modern terms, Book 22 of the Historia Miscella (Muratori 1, 1723, 159). This indeed, like its sources around the date in question (namely the Greek of Theophanes as latinized by Anastasius the Librarian) mentions the drought and the Turkish invasion, but not the 764 eclipse. Consequently we omit further references.

773 M.773 Dec. 3 (a.m.) TOTAL IN IRELAND

There is an undoubted reference in the Annals of Ulster. Under 772 (which with usual correction +1 means AD 773) they have "Luna tenebrosa in ii. non. Dec.", "A dark Moon on Dec.4" (AU 1, 1887, 240-1), rightly indexed by B. MacCarthy (AU 4, 1901, 140) as M.773 Dec.4. The eclipse was total, with maximum magnitude 1.42; Schroeter makes the whole course of the umbral eclipse fall between 12.59 a.m. and 4.37 a.m. at Greenwich, thus entirely after midnight, even in Ireland.

774 M.774 Nov.22-23 NEAR TOTAL IN ITALY (?)

The Annals of La Cava in Italy (SS 3, 1839, 187) have "774 ... Hoc anno facta est eclipsis solis 10. Kal. Sept., die Martis, hora nona, luna quinta". There was no eclipse of any kind on August 23 (Tuesday), but the Moon was eclipsed on the night following Nov. 22 (Tuesday). Consequently Struyck 1740 (154) proposed the emendations "eclipsis lunae 10. Kal. Dec., die Martis, hora nona, luna quinta decima", i.e. a lunar eclipse on Nov. 22, Tuesday, at the ninth hour of the night, on the fifteenth day of the Moon. (These Annals may be based on earlier lost Italian sources which mentioned both this and the August eclipse of 779). As Oppolzer gives an almost total lunar eclipse, of magnitude 0.975, on 774 Nov.23 at mean time 1.51 a.m. Greenwich (hence 2.50 a.m. La Cava), this fits well. Struyck's emendation seems very apt, and is tacitly accepted in the various editions of Struyck-Ferguson (e.g. 1773, p.176), but is not mentioned in the brief treatment in Newton 1972 (463).

777-9 GHOST ECLIPSE OF ROLAND

Roncevaux eclipse: fact or fiction?

The Chanson de Roland, a poetic account of the death of Roland and others in an ambush in the Pyrenean pass of Roncevaux (Roncesvalles, etc.) mentions an eclipse of the Sun which has all the appearance of being fictional. See Newton 1970 (7S ff.), 1972 (313, 320, 324). Indeed, the ambush itself is comparatively thinly documented in reliable annals.

See, for example, Lewis Thorpe, Einhard and Notker the Stammerer, Two Lives of Charlemagne, Penguin Books, 1969, 181-3; also Scholz 1970 (56, 185). There is, however, some evidence for the date 778 Aug. 15 usually assigned to the fight (though pseudo-Turpin, see under S.777 April 12, twice places it on June 16); Europe saw little or (probably) nothing by way of a solar eclipse in 778. But, to stray no further from the historical date, the solar eclipses of 777 April 12, 778 Aug.26 and 779 Aug.16 have all been proposed as having some relation or other to the death of Roland. Probably none of the three eclipses has any real connection with that event, except that the third may have fallen nearly on its first anniversary. There is an undoubted record of the third eclipse, apparently in a non Roland context, but we shall also mention the first two.

777 S.777 Apr 12 CALCULATED ECLIPSE

Calvisius 1605 (710) mentions under AD 777 "An eclipse of the Sun on April 13 a little before the first afternoon hour (paulo ante primam horam pomeridianam), the year before the Spanish disaster (cladem Hispanicam), in which Roland fell". His reference is to turpin of Rheims. We do not see this passage under 777 in the posthumous editions of Calvisius.

This eclipse was penumbral (i.e. nowhere more than partial), with the Moon's shadow missing the Earth's north polar cap. Conjunction was at UT 12 h 58 m (Oppolzer), 12 h 52 m (Goldstine 1973). The eclipse might have been technically "visible", but it would not have been noticed.

Unfortunately, we can point to no evidence that the eclipse was observed. The reference of Calvisius relates to a "De vita Caroli Magni et Rolandi historia" which purports to be the work of John Turpin, archbishop of Rheims (one of the characters in the Song of Roland), but actually composed, like the Song of Roland, some three centuries after Turpin's time. Potthast 1896 (1075-6) gives a long list of manuscripts, editions, translations and discussions. In the English translation, by Thomas Rodd, History of Charles the Great and Orlando, ascribed to Archbishop Turpin..., London 1812, we find only (Ch.26, p.49) that "the Sun stood still for three days" not long after Roland's death. As the latter is twice (Ch.25 and 32) placed on June 16 (no year), this is at most a probable reference to a conventional time of the summer solstice in the year of Roncevaux itself. In other words, the laconic reference of Calvisius to (pseudo) Turpin seems to be for Roncevaux, not for an eclipse. Thus S.777 April 12 appears to be only a computed eclipse; but the fact deserves notice.

778 S.778 Aug 26 CALCULATED ECLIPSE

The Escorial manuscript, which mentions fairly correctly S.779 Aug.16 (q.v.), also mentions a solar eclipse of Spanish Era 816, tertia Kal. Sept. (AD 778 Aug.30). Oppolzer does show an eclipse in Greenland and

Labrador on 778 Aug.26 (vii. Kal. Sept), but it was invisible in Spain and France. The date is evidently the result of calculation (not observation), but the calculated date may originally have been correct, vii having been mis-copied as iii at some stage in the transmission of the result of the computation.

779 S.779 Aug 16 TOTAL SOLAR PROBABLY RECORDED IN SPAIN

Schroeter 1920 shows totality in Southern Spain and North Africa. Ginzel 1886 (980-1) discussed a record of this eclipse contained in a North Spanish manuscript (probably from Oviedo) at the Escorial near Madrid. The Escorial record gives correctly (Spanish) Era 817 (= AD 779) and xvii Kal. Sept. (= Aug.16), but its hour of the day (second) must be assumed to refer to the beginning of the eclipse, as mid-eclipse occurred about the fourth hour of the day in Spain. Ginzel says nothing about Roland or Roncevaux.

Newton 1970 (78 ff.) discusses the same eclipse in relation to the Song of Roland (Roncevaux, 778 Aug. 15) and the possibility that memories of S.634 June 1 and of a military campaign of 636-7 may have influenced the story. The solar eclipse was so striking that the identification has been allocated 6 points but the primary source has not been located.

Newton 1972 (384, 390) also mentions the eclipse as one possible identification of an eclipse defectively recorded in the Annals of Prüfening, near Regensburg (SS 17, 1861, 606). Other identifications which he discusses are S.770 Aug.25 and the well-known S.760 Aug.15 (q.v.), but he concludes that the Prüfening eclipse cannot be identified; no identification fits the whole record.

786 S.786 Apr 3 TOTAL SOLAR IN MIDDLE EAST BUT NO RECORD

Schroeter shows the band of totality as passing north of Alexandria, just north of Jerusalem, and south of Babylon. The northern region of magnitude exceeding 0.75 includes Southern Spain, half Sardinia, Greece, Constantinople and part of the Caspian.

787 S.787 Sept 16 (Sun) ANNULAR SOLAR RECORDED IN THE N.W. AND S.E. EUROPE

The day of the month of this morning eclipse, whose band of annularity extended from Spain via South Italy, Greece, Syria and India to South-East Asia, seems to have suffered to quite an unusual extent at the hands of both Western and Byzantine chroniclers or scribes.

Of some 16 West European authors who give month and day (leaving out of account still more who give year only), almost all are secondary and give Sept.17 (xv. Kal. Oct.). We have found the correct Sept.16 (xvi. Kal. Oct.) only in the Annals of Lorsch (SS 1, 1826, 33) copied perhaps in the Annales Altahenses maiores, of Altaich, modern Niederalteich, in Bavaria (SS 20, 1868, 783). Most sources copy the hour fairly correctly as second or third of the day (for several which say first to fifth hour may be taken to imply mid-eclipse at the third hour). For Lorsch, Ginzel (Handbuch der Chronologie, 3, 1914, 92) finds maximum phase at 7.29 a.m., while from Schroeter 1923 one may estimate about 7.40 a.m.; either falls well inside the second daytime hour, while Newton 1972 (391) also considers the second hour reasonable.

The main Byzantine source is Theophanes. At the end of the annal for AM 6279 (AD 786-7), but with correct eleventh indiction (AD 787-8), he says that on the ninth of September, on the Lord's Day, a very great eclipse occurred at the fifth hour of the day, during divine service. Theophanes, who died about 817, may have relied on memory, or a defective note, for a designation (perhaps ecclesiastical) of the Sunday in question. This erroneous Sept. 9 of Theophanes is copied by Cedrenus.

The references are:

Theophanes: 1839, 716;
1883, 462;
Germ.tr., L.Breyer, Graz. etc. 1957, 119;
trans. Turtledove 1982, 147

Cedrenus: ed. I. Bekker, 2, Bonn 1839, 23.

Since the local apparent times of maximum phase at Lorsch and Constantinople differ by less than two hours on any calculation, the "second hour" at Lorsch and the "fifth hour" at Constantinople are incompatible. For Constantinople, Newton 1972 (545) estimates the third hour from Oppolzer, while from Schroeter 1923 one may estimate early in the fourth hour (say about 9.15 a.m.).

The wording of Theophanes seems to imply a great eclipse at Constantinople, but the maximum magnitudes estimated from Schroeter 1923 are about 0.87 for Constantinople (in Syria it would have been greater) and 0.72 for Lorsch. A probable Italian reference is considered below (S.796).

The eclipse is discussed in Ginzel 1883 (661-3) and in Newton 1972 (especially 391-2 and 544-5). It is sometimes mentioned along with a church council which began at Nicaea on 787 Sept. 24.

788 M.788 Feb.26 (Tues.,a.m.) RECORDED IN FRANCE AND IRELAND

This eclipse is referred to with only slight dating error in the Annales Flaviniacenses (SS 3, 1839, 151): "788 (787). Luna eglypsin pertulit 2. feria, 6. Kal. Martii". The middle of totality (magnitude 1.50) occurred early on Tuesday, somewhat before 4 a.m. at Greenwich (3.55 Oppolzer, 3.58 Schroeter 1923, 3.48 Goldstine 1973); say about 4.10 a.m. at Flavigny-sur-Ozerain (diocese of Autun France). "6. Kal. Martii" would normally mean Feb.24, but in this leap year the annalist must have intended Feb.25, if he really meant "2. feria", i.e. Monday. The eclipse occurred on the night of Monday to Tuesday, but entirely after midnight in all Western Europe.

Slightly less certain, but still probable, is a reference in the Annals of Ulster (1, 1887, 264-5), under 787 (normally meaning 788): "The Moon was red, like blood, on xii. Kal. Mart." This is indexed by B.MacCarthy (AU 4, 1901, 140) as "Feb.18 27 , 788", where his 27 should be 26. J. O'Donovan, in his Introduction to Annals of the Four Masters (1, 1851, x/viii) also quotes "xii. Kal. Mart.", so that the manuscript reading does appear to be wrong.

An alternative, considered in Newton 1972 (654), was M.789 Feb.14 (Sat.). The eclipse was not quite total (magnitude 0.99 in Oppolzer) at UT 6.28 p.m. (Oppolzer), 6.24 p.m. (Goldstine 1973). But this identification requires (a) "xii. Kal. Mart." to be an error for "xvi. Kal. Mart.", (b) Monday to be an error for Saturday, (c) the usual "AU + 1" formula for the year to break down.

791 S.791 July 6 SUNRISE AT CONSTANTINOPLE BUT <u>NOT</u> RECORDED

This annular eclipse is shown in Oppolzer and Schroeter as beginning at sunrise in Eastern Romania, having noon point in Siberia. Although Schroeter's map indicates magnitudes of about 0.9 at Constantinople and 0.75 in North-East Syria, we have encountered no Western record.

795 M.795 April 9 (a.m.)<br>M.795 Oct 3 (a.m.) TWO LUNAR IN N. FRANCE

The Annales Flaviniacenses (SS 3, 1839; 151) say "795. Luna bis osbscurata est", followed by a brief unintelligible phrase. Months and days are not mentioned. Both the above eclipses were total, and both (especially the second) had their later phases invisible at dawn at Flavigny-sur-Ozerain (all under M.788 Feb.26). This seems to be the main primary source for the Continent in this half-century.

M.795 April 9 (a.m.) is also Lynn's unorthodox identification of what we take to be M.796 March 28 (q.v.).

796 M.796 Mar 28 (a.m.) TOTAL RECORDED IN N. ENGLAND

This is the virtually certain identification of an eclipse mentioned in some manuscripts of the Anglo-Saxon Chronicle, around 794-6: "In this year the Moon was eclipsed between cockcrow and dawn" on a day called v. Kal. Apr. (= March 28) in MS.D, and vi.Kal.Apr. (= March 27) in MS.E; F gives no month or day, while A, B, C do not mention the eclipse. See ed.B. Thorpe, 1, 1861, 103; 2, 1861, 49. As usual, some of the annal years are wrong; see, e.g., ed.C.Plummer, 1, 1892, 57; 2, 1899, 64; Whitelock et al. 1961 (37), and Newton 1972 (654). D and E exhibit a northern recension of the Chronicle. The eclipse is dated 796 March 28 by the later northern writers Simeon of Durham (Hist. Regum, 1, Publ.Surtees Soc. 51, 1868, 33-4; ed. T. Arnold, 1, 1885, 57; Eng.tr. in EHD 1, 1955, 248) and Roger of Hoveden (ed. W. Stubbs, 1, 1868, 15; Eng.tr., H.T. Riley, London 1853, 15).

Schroeter 1923, agreeing with Oppolzer, given maximum magnitude 1.06 at 5.59 a.m. (UT), with totality from 5.40 to 6.18 and the whole duration of eclipse from 4.18 to 7.40. Johnson 1874 (37) explains that "between cockcrow and dawn" means between 3 and 6 a.m. ecclesiastical "fourth watch" , while Plummer, 2, 1899, 64, says "Ælfric, following Bede, divides the time from sunset to sunrise into seven parts, of which cockrow is the fifth". Throughout England, the Moon set eclipsed (probably totally), which is consistent with the Chronicle, Simeon and Roger.

W.T. Lynn (Observatory, 15, 1892, 224) would emend the date to v. id. Apr., and identify the eclipse as M.795 April 9 (a.m.), already mentioned above (mid-totality about 4.45 a.m.) The eclipse was similar to M.796 March 28, and cannot be ruled out on purely astronomical grounds, but the suggestion is not very plausible. Lynn seems to have been misled partly by undue regard for a dubious Chronicle year 795 and partly by his own mis-translation of v. Kal. Apr. as March 27; that is a reasonable way of reckoning days, but not how the Romans and their medieval successors reckoned them.

Simeon of Durham puts the eclipse in the same year as the murder of Ethelred, king of Northumbria, and the death of Offa, king of Mercia; these events modern historians agree to have occurred in 796, on April 18 and in late July respectively. To the extent that historians may have relied on the eclipse date, they cannot be quoted as evidence for it. But modern chronology of English history at this period has been enmeshed in places with good chronology of events on the Continent, and consequently may have enough authority, independent of the eclipse, to make the customary eclipse identification virtually certain.

The identification believed correct, namely M.796 March 28, goes back at least to Calvisius 1620 (535), where, however, the reference is to Roger, not ASC or Simeon.

It must surely be by mistake that Struyck-Ferguson, e.g. 1773 (176), gives the place of observation as Constantinople. Struyck 1740 (142) mentions no Byzantine source, and little (probably nothing) of the eclipse would be seen at Constantinople.

796 recte 787 S.796 Sept.6 FALSE DATE FOR SOLAR ECLIPSE

This is one of Oppolzer's p-type eclipses (partial, penumbral, nowhere central), the axis of the shadow cone missing the Earth's north polar cap. The time of conjunction is given as 5.49 a.m. (UT). Such eclipses can be calculated but they were not observed.

A Chinese mention exists, but the eclipse would not be noticeable there either. We have encountered no Western record, only an evident mis-identification by Mommsen, corrected by J.K. Fotheringham; see pages xix-xx of the latter's Latin preface in his Oxford 1923 edition of Jerome's Latin version and continuation of Eusebius, Chronici Canones.

Fotheringham discusses the matter at some length and gives various references. The position may be summarized as follows. The last page of a Carolingian manuscript of Eusebius-Jerome at Lucca in Italy has a marginal note: "a resurrectione domini nostri Iesu Christi usque ad praesens annum Caroli regis in Langubardiam in mense Septembrio, quando sol eglypsin patuit, in indict x anni sunt dcc/xii menses vi". Fotheringham realized that the true identification is not Mommsen's S.796 Sept.6, but S.787 Sept.16, the only suitable September eclipse between Charlemagne's conquest of Lombardy and deposition of King Desiderius in 774 and his being crowned emperor on 800 Dec.25 (notice "regis" in the quotation). Fotheringham finds the eclipse of 796 invisible in Italy, but calculates magnitude about 0.88 for that of 787 at Lucca. While he admits that the year of the resurrection does not tally well. Fotheringham considers the indiction to be more important and in sufficiently good agreement. S.787 Sept.16 fell in Roman indiction 10 and Byzantine indiction 11, whereas the indiction for S.796 Sept.6 is obviously wrong Roman 4, Byzantine 5 . We accept Fotheringham's date.

797-8 S.797-8 March 3 ? VOLCANIC DARKNESS

We have encountered no true record of either of these eclipses, but they need examination to see whether one of them may have any connection with mention (first at Constantinople) of a darkening of the Sun for 17 days. The answer is certainly No.

It is known that in 797 (probably mid-August) Irene blinded her son, Constantine VI, and assumed sole imperial power. Under AM 6289 (AD 786-7) Theophanes, a contemporary, says: "The Sun was darkened for 17 days and did not give off its rays, so that ships went off course, and everyone said that the Sun stopped shining because of the blinding of the emperor; and thus his mother Irene came to power". No eclipse fits. An acidity layer about 797/8 on the Greenland Ice Cap (Hammer 1980 fig.1) indicates that a volcanic eruption must have occurred in the Northern Hemisphere at this time. S.797 March 3 was annular and ended at sunset in the Northern Sudan; it would be invisible, or nearly so, at Constantinople. S.798 Feb.20 was also annular, but ended at sunset between Scotland and Iceland; it would be invisible at Constantinople.

Consequently, it is not worth while to give detailed references to later writers who tell the same tale. Among Byzantines, these include George the Monk (Hamartolos), Leo Grammaticus, and Cedrenus. Among other oriental writers, Hamza al-Isfahani.

Newton 1972 (144, etc.) finds the story in Ralph of Diceto (died 1202 or 1203) and the Annals of Dunstable (c.1297). See also Sigebert, Chron., Ekkehard, Chron. Univ., and Annalista Saxo (all late eleventh or early twelfth century).

800 M.800 Jan 15-16 LUNAR RECORDED IN ENGLAND

Oppolzer gives maximum magnitude 0.84 and mid-eclipse at UT 20 h 38 m, i.e. 8.38 p.m., on 800 Jan.15. Goldstein 1973 gives 8.37 p.m. for Full Moon.

The eclipse is clearly mentioned in the Anglo-Saxon Chronicle; as for M.796 March 28, manuscripts D,E,F give the information (A,B,C do not mention the eclipse). "In this year the Moon was eclipsed at (or in) the second hour of the eve of xvii.Kal.Feb. (=Jan.16)". Thorpe and Plummer have "at", Whitelock et al. "in". At London the eclipse was really from the 3rd to the 5th seasonal hour. As Plummer and Whitelock agree that the Anglo-Saxon means "on the eve of" (i.e. the evening before), the Chronicle does indicate correct month and day. From Thorpe's "hexapla", it seems that D and E rightly give year 800, F wrongly 801.

Johnson 1874 (37) adds "and soon after died King Brihtric...". It is now thought that "soon" is wrong. The death of Brihtric, king of Wessex, and the accession of Egbert, do indeed follow the eclipse in the annal for 800 (D, E), 801(F), in Thorpe's text, which Johnson would rely on, though Thorpe's translation perversely interchanges the order. But the eclipse belongs correctly to 800, whereas the Brihtric-Egbert transition, though dated 800 in A to E and 801 in F, has for some long time now been placed in 802, as it is in Whitelock et al. 1961 (38).

Struyck 1740 (154), followed by Struyck-Ferguson (e.g. 1773), lists Rome as the place but we have not found any continental source.

Chinese 'eclipses' listed in Hoang and elsewhere are again usually calculated and include many that could not have been seen at all. However, in this century a few observations can be detected and they have been studied by Muller and Stephenson (1975), by Newton (1979 Table V 5 and pp. 168, 81 and 87) and by Cohen and Newton (1984. See fig. 3, p. 41). The tracks published by these authors (cf. Newton 1979, 81 and Stephenson and Houlden, in progress See maps in our Introduction) are roughly consistent with those given by Oppolzer (Charts 92 to 96). The term chi is translated as complete by Stephenson and as central by Newton. Stephenson writes (Feb. '84) that the T'ang records (down to AD 879) are to be found in the Astronomical Treatise (Chapter 32) of the Hsin t'ang-shu.

702 Sept 26 NEARLY TOTAL SOLAR

The Annals of the Old Tang History state that their eclipse was like a hook, nearly complete, that it was seen in the capital and in the provinces; the hour is also given (Cohen and Newton, Tables 13 and 14). All modern computations place the track slightly further south-west than given by Oppolzer and are consistent with the Chinese statement. Stephenson writes that the Empress had returned to Ch'ang-an in the winter 'At the capital Ch'ang-an this eclipse of 701 and adds that would be partial - magnitude 0.99 - the track of totality passing a little to the NE.

729 Oct 27 NEARLY TOTAL SOLAR

The same source reports that this eclipse was 'not complete, like a hook' and this confirms that the observation came from the 'North China Rectangle' (31°/35°N, 109°/116°E. See Cohen and Newton, 1984, 26), rather than a little further north-east where it would have been total. Stephenson calculates that at the capital this eclipse would also be partial - magnitude 0.91 - the track of totality again passing to the NE.

754 June 25 NEARLY TOTAL SOLAR

The same source reports that this eclipse was 'like a hook, almost complete'. This was total slightly to the south of the rectangle. Stephenson (Feb. '84) calculates that at Ch'ang-an (the capital from 727 Oct 11 until 731 Oct 21) this eclipse would have a magnitude of only 0.84.

756 Oct 28 TOTAL SOLAR IN EXTREME N.W. CHINA

'First year (of the Chih-te reign period), 10th Moon, day hsin-szu, the first day of the Moon, there was an eclipse of the Sun, and it was complete (chi). It was 10 degrees (in the constellation of) Ti'. The date, day and solar positions are all correct. However, we know that this eclipse was total only in extreme N.W. China and that the emperor Ming Huang had fled from the capital to Szechwan. Newton confirms that the totality ended at sunset at longitude 106°E (Cohen and Newton, 41-42).

761 Aug 5 TOTAL OR NEAR TOTAL SOLAR AT CH'ANG-AN

In the same source we find '2nd year (of the Cheng-yuan reign period), 7th Moon, day kuei-wei, the first day of the Moon, there was an eclipse of the Sun and it was total (chi). All the great stars were visible. It was 4 degrees in (the constellation of) Chang'. Fuller details are given in the CTS (RS). The record appears also in the 'Collected Data' originally AD 805, Extant Version AD 907. See Cohen and Newton 1984, 1 and 43). Newton believes that the eclipse was total a few kilometers north of Ch'ang-an (the capital since 757) and Stephenson writes: 'The remarks of Muller and Stephenson regarding place of observation are probably obsolete and totality was probably witnessed at the capital' and he gives full details of times from the Chiu Tang Shu, T'ien-wen-chih, chapter 36 (see Stephenson and Holden 1985).

780 Feb 10 FAILED PREDICTION ?

The day after a 'blackout of the sun' Schafer tells us, following the account in the New T'ang History, the usual New Year Reception of the Court was severely restricted (E.H. Schafer, 'Pacing the Void', University of California Press, 1977, 169). Calculation confirms that this 'West Asiatic' annular eclipse would not have caused any blackout or even have been noticed by the courtiers and it may have been a failed prediction.

792 Nov 19 PARTIALITY OBSERVED

The Collected Data and the Astronomical Chapter of the Old T'ang History tell us that the magnitude was only 3 instead of the predicted 8. Cohen and Newton confirm that the greatest magnitude in the North China Rectangle was only 0.27, and this seems to be the earliest genuine observation of a solar eclipse with magnitude less than 0.7. Probably the sun's image was watched in its reflection in a bowl of water.

The Venus table in the Dresden Codex has been said to have eclipse 'overtones'. Using the 615 824 correlation (cf. Maya Appendix) this is very clear. The three Venus bases are 16 or 17 days after Visible Lunar Western Hemisphere eclipses: the main base, dated 709 June 13, followed the total lunar eclipse (Central 84°N 22°N) of 709 May 28, the other bases, dated 683 Oct 28 and 742 Mar 12, followed the lunar eclipses of 683 Oct 11 and 742 Feb 24. The Venus bases lead on in the expanded table to further dates such as the Maya date 9.12.16.6.0 which is 16 days after the near total eclipse of 774 Nov 23. Solar eclipse connections exist as well, the first base being 30 days and the last-mentioned date 31 days after a solar eclipse date; these solar eclipses, unlike those of the 5th century, would have been calculated rather than observed.

The first (New Moon) base of what seems to be the final prototype of the Eclipse Table in the same Codex (9.12.10.16.9 13 Muluk on folio 58) is dated 769 Aug 6 and it is 16 days before the lunar eclipse of Aug 22 and 30 days before the solar eclipse of Sep 5.

# THE NINTH CENTURY

## THE CAROLINGIAN RENAISSANCE, BAGDAD SCIENCE, MAYA PREDICTIONS

Scientific observations are now made in the metropolis of Bagdad and there is a general cultural revival in both the Islamic and Byzantine worlds. In Central America, where records of eclipses had been kept over several centuries, and dates were expressed with a place system of numeration, the prediction of lunar eclipses by the Maya was remarkably accurate.

Chronicle writing in the Latin West was usual in the monasteries of the European Continent: the Royal Frankish Annals thus provide us with accurate descriptions of eclipses, comets, sunspots and aurorae. The Carolingian Empire of Charlemagne was soon subdivided in 843, but the chronicle activity continued almost to the end of the century. The eclipse records were borrowed by many later chroniclers.

Ennin's Diary is a daily travel record of a Japanese traveller in China; details of comets and aurorae are included, but such a diary in this millennium is unique.

Islamic and Syriac chronicles from this century are known mostly from later writers, who had access e.g. to Persian records of the period 840/870.

The accuracy of eclipse prediction in Europe as well as Islam improved in this century, so that it is sometimes difficult to determine whether certain eclipses were really observed.

802 M.802 May 21 (a.m.) TOTAL IN ENGLAND

The Anglo-Saxon Chronicle (manuscripts D and E), correctly under AD 802, says that the Moon was eclipsed in the dawn on xiii.Kal.Jan. [Dec.20]. This should be xii.Kal.Jun. [May 21], if the day is taken to start at sunset or midnight. Nevertheless, the identification is certain; there was no other lunar eclipse at dawn in England around this time. This was total at 4.13 a.m. (Oppolzer) at Greenwich.

803 M.803 Nov.2-3 TOTAL IN EASTERN FRANCE

The Annales Flaviniacenses (SS 3, 1839, 151) have "803 ... The Moon suffered an eclipse". The place is Flavigny-sur-Ozerain (Côte-d'Or, Burgundy, Eastern France). Other eclipses mentioned in these Annals appear under their correct years, and, if the year is reliable in this instance also, the record must refer to the total eclipse M.803 Nov.2-3, which had maximum magnitude about 1.21 and a period of totality probably including Flavigny midnight.

806 M.806 Sept.1-2 TOTAL IN FRANCIA AND ENGLAND

RFA has, under 807: "During the previous year there was an eclipse of the Moon on 4.Non.Sept. [Sept.2]. At that time the Sun stood in the sixteenth degree of Virgo, and the Moon in the sixteenth degree of Pisces". The same information (with occasional errors) occurs in many other Frankish annals; none mentions the hour. The eclipse was total (maximum magnitude about 1.37), and occurred around midnight of Sept.1-2. It was dated 806 Sept.1 in the ASC (manuscripts D and E).

On all four Frankish eclipses of 806-7, see Newton 1972, 392-5.

807 S.807 Feb.11 ANNULAR IN FRANCIA, ITALY AND WALES

RFA: "807 ... On 3.Id.Feb. (Feb.11) at midday there was an eclipse of the Sun, during which both Sun and Moon stood in the twenty-fifth degree of Aquarius". Similarly in many other Frankish annals (see the discussion in Ginzel 1884, 546). The southern limit of annularity was rather far north (roughly Belfast, Edinburgh, Trondheim), but Schroeter shows magnitude greater than 0.75 in most of Western Europe. Many sources give the time as "midday", "sixth hour", or "about sixth hour", but some attempt slightly greater precision:

Ann. Juvav. maior. (Salzburg) (SS 1, 1826, 88) (SS 30(2), 1934, 738) 4th to 7th hour

Annals of Farfa (Italy), s.a. 808 SS 11, 1854, 588 3rd to 6th hour

The records give a general impression of mid-eclipse rather before noon, and this is correct. Schroeter's data make the hour of mid-eclipse vary from about 10 a.m. near Dublin and Madrid to about noon near Trondheim and Vienna. For mid-Ireland, O'Connor 1952 finds maximum magnitude 0.92 at 10.00 a.m.

The eclipse is certainly that mentioned, though without month or day, in Welsh records. Brut y Tywysogyon, or The Chronicle of the Princes, ed. T. Jones 1952, 3: "807 was the year of Christ when Arthen, king of Ceredigion [Cardigan], died. And then there was an eclipse of the Sun". Rather similarly in other editions of this Brut (John Rhŷs and J.G. Evans 1890, 258; John Williams ab Ithel, 1860) and in Annales Cambriae (John Williams ab Ithel, 1860).

807 M.807 Feb.26 (a.m.) TOTAL IN FRANCIA, IRELAND (?)

RFA: "807 ... Again on 4.Kal.Mart. (Feb.26) an eclipse of the Moon took place. The Sun stood in the eleventh degree of Pisces and the Moon in the eleventh degree of Virgo". Similarly in a number of other Frankish annals. We have encountered no mention of the hour in the original sources. The eclipse is listed in Oppolzer and Schroeter as total for about 24 minutes only; mid-eclipse shortly before 3 a.m. at Greenwich.

This eclipse is also the most favoured identification of the one mentioned in Irish records: "The Moon was turned into blood". This appears under manuscript year 806 (probably meaning true 807) in AU 1, 1887, 292-3, and under 807 in CS 1866, 126-7. The identification M.807 Feb.26 is accepted in B. MacCarthy's index (AU 4, 1901, 140), but M.806 Sept.1-2 and M.807 Aug.21-22 (qq.v.) cannot be entirely ruled out. An aurora is a further possibility.

807 M.807 Aug.21-22 NEAR TOTAL IN FRANCIA

RFA: "807 ... Again on 11.Kal.Sept. (Aug.22) an eclipse of the Moon occurred at the third hour of the night, the Sun being located in the fifth degree of Virgo and the Moon in the fifth degree of Pisces. Thus from September of last year to September of the present year the Moon was eclipsed three times and the Sun once". Similarly in a number of other Frankish annals; those which give the hour all state "third hour of the night". The eclipse was not quite total; according to Oppolzer, at Greenwich it began at 9.16 p.m. and ended at 12.30 a.m.

On possible Irish mention, see under M.807 Feb.26 (a.m.).

<u>809</u> S.809 July 16 (Mon.) PARTIAL SOLAR IN N.W. EUROPE

This was essentially a northern eclipse, and the original record is probably English. The ASC (manuscript F, Old English - F's Latin is corrupt) has "809. In this year there was an eclipse of the Sun at the beginning of the fifth hour of the day on xvii.Kal.Aug. [July 16], the second day of the week [Monday], the 29th day of the Moon". The narrow band of annularity in Schroeter runs by the Faroes and Trondheim, thus considerably to the north of the British Isles. The magnitude was not more than about 0.7 at either London or the northernmost parts of Charlemagne's domain. The record in various Frankish annals has been thought to be borrowed from the ASC; it may be noted that this is the only eclipse of the Frankish group in the years 806-812 which is not mentioned in the RFA.

Ann. S. Columbae Senon. (SS <u>1</u>, 1826, 103): "809. indictione 2. 17.Kal.Aug. 2.feria, incipiente hora diei 5. eclipsis solis apparuit luna 29". Compared with ASC, this impeccable statement adds the correct indiction and may be the primary source. Similarly in Ann. S. Maximini Trevir. (SS <u>4</u>, 1841, 6); also in Ann. Laudun. et S. Vincentii Mett. breves (SS <u>15</u>(2), 1888, 1294), but the Laon-Metz text is slightly defective and restored, presumably from the Sens and Trier texts.

A similar passage in Bouquet <u>5</u>, 1869, 386-7 is there attributed to four further Frankish sources, but (whatever the manuscripts may say) it needs two corrections: for "Luna XXXIX" read "Luna XXIX", and for "II feria incipiente, hora diei V" read "II feria, incipiente hora diei V". Clearly it was the fifth hour, not the second day of the week, which was beginning. The beginning of the fifth hour (i.e. two-thirds of the way from sunrise to noon) is, as Johnson pointed out, exact for the time of the greatest phase in the London region.

See Ginzel 1884, 548, Newton 1970, 49-52, 259, 1972 (Ch.VI, X, XI).

809 M.809 Dec.25 (p.m.) TOTAL IN FRANCIA AND WALES

RFA: "809 ... An eclipse of the Moon occurred on 7.Kal.Jan. (Dec.26)". This was copied, with the same month and day, and no hour, in a number of other Frankish annals. Since this total eclipse finished before 9 p.m. at Greenwich, and thus before midnight throughout Western and Central

Europe, the recorded date Dec.26 must be attributed to ecclesiastical usage. Nowadays we should say that the eclipse occurred on Christmas day, but we have encountered no Frankish source which says so. However, some Welsh records do mention the fact. Brut y Tywysogyon, or The Chronicle of the Princes, tr. T. Jones, Cardiff 1952, 3: "810 was the year of Christ when the Moon darkened on Christmas day". Similarly in the editions of J. Rhŷs and J.G. Evans, Oxford 1890, 258, and John Williams ab Ithel, London 1860, 8-9. The year 810 is correct for years beginning on Christmas day.

### 810 M.810 June 20 (p.m.) TOTAL IN FRANCIA

RFA: "810 ... In this year both Sun and Moon were eclipsed twice; the Sun on 7.Id.Jun. (June 7) and 2.Kal.Dec. (Nov.30), the Moon on 11.Kal.Jul. (June 21) and 18.Kal.Jan. (Dec.15)". Similarly in numerous other Frankish annals. The first-mentioned eclipse is variously dated, and is discussed below; the other three are clearly genuine.

The deep total lunar eclipse of 810 June 20-21 ended before 10 p.m. at Greenwich, and hence before midnight throughout the Frankish domains, so that the date June 21 invariably recorded must again be attributed to ecclesiastical usage.

### 810 S.810 July 5 CALCULATED SOLAR

Along with three other eclipses of 810 (see under M.810 June 20), a summer eclipse of the Sun is mentioned in RFA, and was very widely copied. For example, RFA (called Einhardi Annales, SS 1, 1826, 198) gives, converting from Ides to modern notation, 8 manuscripts reading June 7, 6 reading June 6, one reading July 9 and one reading July 8. Other sources usually give June 7; but the Annals of Xanten (SS 2, 1829. 224) give June 8, Regino (SS 1, 1826, 565) gives June 9, and the Annales Sithienses (SS 13, 1881, 37) give July 10. The variety of dates is not surprising, as the Franks probably saw nothing whatever (cf. also Newton 1979, 289).

There were, however, solar eclipses on both June 5 and July 5. Neither was more than partial anywhere, and the first was not visible in Europe, or even in the northern hemisphere. The second, with true conjunction at UT $12^h35^m$ (i.e. shortly after Greenwich noon) was visible only in high northern latitudes.

There is little doubt that the record results from mere computation, not observation. Bunting dismissed Regino's record as "hallucination". Calvisius suggested that the eclipse was predicted from imperfect astronomical tables, and not seen. Ginz.1884, 550 regarded the eclipse as that of June 5; as he must have known that this was invisible in Europe, he also must be regarded as favouring the "computation only" interpretation.

The supposed pair of solar eclipses in 810 (summer and November) is the subject of a letter from Dungal to Charlemagne in response to an enquiry from the latter (Bouquet 5, 1869, 635). As the article on Dungal in the Dictionary of National Biography puts it, Dungal gave the Emperor "such an explanation as he could of an event which had not really occurred" (in the Frankish domains). Dungal's letter is discussed in Newton 1972, 595-6.

810 S.810 Nov.30 TOTAL SOLAR IN CENTRAL EUROPE

The strongly curved track of totality in Schroeter goes by Bergen and Berlin, then slightly N.E. of Prague, Vienna, Budapest and Bucarest, and then by Astrakhan. The eclipse is very widely reported, often without hour, as in RFA (see under M.810 June 20), but at least six records do mention the hour. From Salzburg in Austria the Annales Juvavenses maiores (SS 1, 1826, 88; 30(2), 1934, 738) say "fourth to seventh hour". Five other later sources (by no means independent) say "third hour" or "about third hour".

| | | |
|---|---|---|
| Herimann Augien (of Reichenau), Chron. | | SS 5, 1844, 101 |
| Ekkehard, Chron. Wirziburgense | | 6, 1844, 27 |
| Chron. of Melk (Austria) | (Secondary) | 9, 1851, 495 |
| Chron. of Admont (Austria) | ( " ) | 9, 1851, 573 |
| Chron. Suevicum Univ. | ( " ) | 13, 1881, 64 |

Mid-eclipse occurred, in local apparent time, about 11 a.m. in Paris and about noon in Salzburg. Thus the explicit Salzburg record seems correct within an hour. One may note that, a line or two earlier, the Salzburg annalist says exactly the same thing ("fourth to seventh hour") about S.807 Feb.11.

The "third hour" records may also be accepted, if one may interpret them as referring to the beginning of the eclipse, and if one remembers that "third hour", sometimes conventionally translated as 9 a.m., may at this season (end of November) denote about 10 a.m. local apparent time.

Earlier discussions include Bunting 1590, 389, Petavius 1627, 863 and Ginzel 1884, 549; see also Lycosthenes 1557, 341.

810 M.810 Dec.14 (p.m.) TOTAL IN FRANCIA

We have a number of Frankish references to this lunar eclipse, which was briefly total, but we have encountered no medieval mention of the hour. The eclipse ended well before midnight in the whole of Europe (mid-eclipse soon after 6 p.m. at Greenwich). The date is given as 18.Kal.Jan. (= Dec.15) in RFA (see under M.810 June 20) and most other Frankish sources, doubtless reckoning ecclesiastically (from sunset). But the Annales Heremi, of Einsiedeln (SS 3, 1839, 139) give 19.Kal.Jan. (= Dec.14). Dates like 9.Kal.Jan. (Regino, at any rate in SS 1, 1826, 565) doubtless arise from scribal error.

The lunar eclipse mentioned under 810 in Annales Cambriae, ed. John Williams ab Ithel, London 1860, 11, is that of 809 Dec.25 (q.v.), fully dated in Brut y Tywysogyon.

812 S.812 May 14 (Fri.) TOTAL IN SYRIA AND WESTERN ASIA MINOR
PARTIAL PHASE IN FRANCIA

The band of totality in Schroeter goes by Morocco, Palermo, Athens, Smyrna, North Eastern Syria, Northern Iraq and Northern Iran. The curve of magnitude 0.75 to the north goes near Lisbon, Milan, Budapest, Odessa and the Caspian. The noon point is in Morocco. The whole central line in Oppolzer goes from South America to the Himalayas, via a noon point also in Morocco.

As far as original records are concerned, we have found the correct date only in Byzantine and Syrian sources. A contemporary account occurs in Theophanes, Chronographia (Bonn ed., J. Classen, 1, 1839, 771-2; Teubner ed., C. de Boor, 1, Leipzig 1883, 495; German tr., L. Breyer, Graz, etc. 1957, 165). Correctly under his AM 6304 [AD 811-2], Theophanes says: "On the 14th of May, on the sixth day [Friday], a great solar eclipse occurred for 3½ hours, from the eighth hour until the eleventh hour". The Latin translation by Anastasius the Librarian (Bonn Theoph., 2, 1841, 278; Teub. Theoph., 2, 1885, 333) is wrong, giving 5.id.Mai. [May 11], sixth day. The Historia Miscella, Book xxiv (Muratori 1(1), 1723, 176 is even more wrong, giving 5.id.Mai., fifth day [Thursday]. The translation of Theophanes (Turtledove 1982, 175) interprets the sixth day as Saturday.

But a correctly dated record of totality was given in the late 12th century by Michael, xii, 7 (III, 1963, 26), and Muller accepts the Edessa/Harran region as the place (cf. however Newt.1979, 390). "In the [Seleucid] year 1123 [AD 811-2], the 14th of 'Iyar [May], there was a total eclipse of the Sun, from the ninth hour to the eleventh hour, and the darkness was as profound as night; the stars were seen, and people lit torches. Then the Sun reappeared for about an hour". With its starting hour differing in the expected direction, this appears to be more than just a copy of Theophanes, and seems to have come from the Chronicon

Anonymum, ad annum post Christum 812/813 pertinens (CSCO, SS Syri, (3)4, Paris 1903, 196), where the eclipse record is unfortunately spoiled by a lacuna in the version now extant.

The hours in Theophanes and Michael are reasonable for Western Asia Minor and Syria respectively.

Frankish annals have a unanimous error (or peculiarity) in the date. RFA: "812 ... In this year there was an eclipse of the Sun on the Ides of May [May 15] after midday". Similarly in at least eight other Frankish annals, all giving the same erroneous date (apart from a further slip of March 15 instead of May 15).

Several Western sources say simply "after midday". Two are more explicit. The Short Chronicle of Luxeuil in Burgundy (SS 3, 1839, 221) says "hora diei quasi 7", and the Chroniques de Saint Denis (near Paris; Bouquet 5, 1869, 261) say "entre l'eure de midi et de none". As mid-eclipse in these parts fell in about the second hour of the afternoon (eighth of the day), the records seem reasonable, Luxeuil referring perhaps to the beginning of noticeable eclipse.

There is a discussion in Ginzel 1884, 550. See also Newton 1972 (Ch. XI, XV).

812 M.Oct.23-24 POSSIBLE LUNAR IN WESTERN EUROPE

A small partial eclipse, visible throughout Western Europe. See below, under 810-3, "Eclipses before the death of Charlemagne".

813 S.813 May 4 TOTAL SOLAR IN SOUTH-EASTERN EUROPE
PARTIAL PHASE IN GERMANY

The primary account is that of Theophanes in the original Greek (Latin versions wrong). Western accounts are almost worthless, being mainly garbled versions of the Byzantine record.

In Schroeter the whole band of totality starts in the Adriatic, and includes Bucarest, Moscow and Archangel.

Correctly under AM 6305, Theophanes, Chronographia (Bonn ed., J. Classen, 1, 1839, 780; Teub. ed., C. de Boor, 1, Leipzig 1883, 500; German tr., L. Breyer, Graz, etc. 1957, 172; English translation, by Turtledove, 1982, 178) has: "On the 4th of May, an eclipse of the Sun occurred in the 12th degree of Taurus, according to astronomical calculation, as the Sun was rising; and great fear fell on the multitudes".

The Sun was really in longitude 47° (17° of Taurus), but "twelfth degree of Taurus" would have been correct for May 4 about Ptolemy's time. We have thought it sufficient to follow Breyer in translating "κατὰ τὸν ὡροσκόπον" as "according to astronomical calculation". The English translation (Turtledove, 1982, 179) accepts "according to the horoscope".

The Latin translation by Anastasius (Bonn Theoph., 2, 1841, 282; Teubner Theoph., 2, 1885, 337) wrongly has 3.non.Mai. [May 5], whereas the correct date is 4.non.Mai. The Historia Miscella, Book xxiv (Muratori 1(1), 1723, 177) wrongly has 4.id.Mai. [May 12]. The late Chronicles of Alberic (SS 23, 1874, 727) give, apparently from Hugh of Fleury, 3.non.Mar. [March 5]. Lycosthenes 1557, 341-2, has the correct date.

There is a brief discussion in Ginzel 1884, 551. See also Newton 1972 (Ch. XV).

## 810-813 ECLIPSES BEFORE THE DEATH OF CHARLEMAGNE

The above S.813 May 4 is the last of the eclipses recorded anywhere in Europe before Charlemagne's death (814 Jan.28). Eclipses alleged to have portended his death were sometimes mentioned (usually vaguely or inaccurately) by medieval writers. The best known example occurs in Einhard's Life of Charlemagne, §32: "Per tres continuos vitaeque termino proximos annos et solis et lunae creberrima defectio" (SS 2, 1829, 460, and elsewhere). In a note in his English translation (Penguin Books, 1969, 187) Lewis Thorpe enumerates eclipse dates in 806-812, and comments "Einhard tidies this up!" This is charitable, since an abridgement may discard information, but should not introduce error (cf. Schove 1951,135; 1983a, 41).

Nevertheless, one must agree that Einhard compactly gives a correct general impression. In fact, it is quite easy to find a year-beginning which places three solar eclipses (S.810 Nov.30, S.812 May 14 and S.813 May 4) and the death of Charlemagne in four consecutive years. For example, the Byzantine Sept.1 year-beginning does the trick, and is only faintly unnatural in material relating to a Western emperor.

No year-beginning succeeds quite so well with lunar eclipses. But M.809 Dec.25, M.810 June 20 and M.810 Dec.14 were all recorded by the Franks, and M.812 Oct.23-24 (a small partial eclipse, maximum magnitude 0.09 at about 10.26 p.m. at Greenwich) may well have been calculated, though not recorded. Thus the lunar eclipses in the four years or so before Charlemagne's death need only slight allowance for literary licence to count as "very frequent".

814 S.814 Sept.17 CALCULATED SOLAR ECLIPSE

The later Genealogia ducum Brabantiae ampliata (SS 25, 1880, 394) says, with reference to Charlemagne, "quo moriente sol obscuratus est". Other references may exist, as in our reading we have not always noted vague statements of this kind, which frequently accompany accounts of the deaths of famous persons, and smack of superstition. There was no suitable solar eclipse near or after Charlemagne's death (814 Jan.28) until S.814 Sept.17, and this was nowhere more than partial. O'Connor 1952 found magnitude 0.46 just before 1 p.m. at his mid-Ireland point. In Europe the eclipse was probably visible, but of small magnitude. If it had really been seen, we should probably have heard more about it. The report is most likely to have arisen either from a mere computational result, or as an abridgement of the usual allusions to eclipses before Charlemagne's death (see above, under 810-3, "Eclipses before the death of Charlemagne").

817 M.817 Feb.5 (p.m.) TOTAL IN FRANCIA

RFA: "817 ... On the Nones of February (Feb.5) the Moon was eclipsed at the second hour of the night..." This is correct, and was copied by other Frankish sources: a record from Fontanelle (Saint Wandrille) (SS 2, 1829, 293, year wrong, or Bouquet, 6, 1870, 174), and the Life of the Emperor Louis, by "the astronomer" (SS 2, 1829, 621, or Bouquet, 6, 1870, 100). But the Annals of Fulda (SS 1, 1826, 356, or Bouquet, 6, 1870, 206) and the Chronicle of Herimann Augien (Reichenau; SS 5, 1844, 102, or Bouquet, 6, 1870, 224) both wrongly say that the Sun was eclipsed, while the Chroniques de Saint Denis (Bouquet, 6, 1870, 141) wrongly have Kal.Feb. instead of Non.Feb.

The eclipse of the Moon was total and fairly deep, with mid-eclipse about 6 p.m. at Greenwich, so that "second hour of the night" is reasonable.

818 S.818 July 7 PARTIAL SOLAR ECLIPSE (CALCULATED?)

RFA: "818 ... An eclipse of the Sun occurred on 8.Id.Jul. (= July 8)". Similarly, with no hour, in at least ten other Frankish sources (several are mentioned in Newton 1972 and Ginzel 1884, 552). In spite of unanimity about July 8, the true date was July 7, in the early morning.

The band of annularity in Schroeter runs from SW to NE through Reykjavik in Iceland. Even the curve of magnitude 0.75 to the SE runs rather far north (Ulster, Edinburgh, Norway). The eclipse cannot have appeared very striking to the Franks and may have been watched for after a prediction. O'Connor 1952, 70 found maximum magnitude 0.72 at 5.10 a.m. for his mid-Ireland point.

820 M.820 Nov.23 (p.m.) LUNAR IN FRANCIA

RFA: "820 ... The Moon was eclipsed on 8.Kal.Dec. (= Nov.24) at the second hour of the night". (If the date Jan.28 in Scholz 1970, 108 is a correct translation from some manuscript, it is nevertheless wrong astronomically.) Similarly in several other Frankish sources, except that it seems from Bouquet 6, 1870, 180 that one MS. has 9.Kal.Dec. (= Nov.23). The majority date is doubtless due to ecclesiastical reckoning, as the whole eclipse ended about 9 p.m. at Greenwich, and thus before midnight in all Western Europe.

824 M.824 Mar.18 (p.m.) RECORDED IN FRANCIA (CALCULATED?)

RFA: "824 ... The Moon was eclipsed on 3.Non.Mar. (= March 5) at the second hour of the night". This is the only case in which the RFA misdate an eclipse by more than a day or two. Calvisius attributes March 6 to Aimoin, but this date is little better. The total eclipse occurred on the night of March 18-19; as it ended before 10 p.m. at Greenwich, it took place in Europe on March 18 in the usual reckoning. Possibly this was a calculated eclipse.

828 M.828 July 1 (a.m.) TOTAL IN FRANCIA

RFA: "828 ... On the Kalends of July (July 1) the Moon was eclipsed at dawn as it set". Similarly in Vita Hludowici imp. (SS 2, 1829, 632, or Bouquet 6, 1870, 110) and Chroniques de Saint Denis (near Paris; Bouquet 6, 1870, 151). Variant dates (Kal.Jun., 3.Kal.Jul.) occur in all three cases.

A deep total eclipse on the night ofJune 30-July 1 (2.Kal.Jul.-Kal.Jul.), but entirely after local midnight throughout Europe.

828 S.828 July 15 FALSE DATE

We have encountered no Western record of this penumbral eclipse. In Europe it was practically invisible, and can have amounted at most to a sunset eclipse of trifling magnitude. It is true that Newt.1972, 226, in a summary table, gives "828 July 15?" in relation to a passage, "833 ... Sol et luna per eclypsim deficiunt", in the (later) Chronicles of Sigebert of Gembloux in Belgium (SS 6, 1844, 338). But Newt. (226, 233, 397-8) rightly draws attention to similar statements under 832 or 833 (q.v.) in German sources (Annals of Fulda, Xanten, etc.). Admittedly neither 832 nor 833 fits well, the former lacking a solar eclipse visible in Europe and the latter lacking a lunar eclipse credibly noticeable there.

828 M.828 Dec.25 (a.m.) TOTAL IN ENGLAND AND FRANCIA

RFA: "828 ... Luna ... in 8.Kal.Jan. (Dec.25), id est in natale Domini, media nocte obscurate est". Vita Hlud. imp. and Chron. de S. Denis (see M.828 July 1 above) have, respectively, "nocte natalis Domini" and "la nuit de Noël".

The eclipse was briefly total, and occurred on the night of Dec.24-25. It is recorded in England, under 827, in the ASC (various manuscripts), also in Ethelweard and Florence of Worcester. ASC says that the eclipse occurred on "mid-winter's mass night", Florence "Sacrosancta nocte Dominicae Nativitatis". The year 827 is wrong by any reckoning; if a Christmas year-beginning was intended, the year should have read 829.

829 S.829 Nov.30 PARTIAL SOLAR IN BAGDAD

This eclipse is the first of some thirty Islamic eclipses between AD 829 and AD 1004 which were observed by astronomers (as distinct from mentioned by historians) and constitute an important part of our medieval eclipse information. There are 29 Arabian eclipses, observed in Bagdad or Cairo, to be found mentioned in Newcomb 1878 or Newton 1970 or (in 24 cases) both, taken from Caussin 1804 (a French translation, Le Livre de la grande Table Hakémite, of a work by Ibn Yūnus, who died about 1008-9). Caussin 1799 had previously abstracted information about 28 of these eclipses. The records are excellent in general, but weak in a few instances. Newton 1970 (but not Newcomb 1878) also used information about four eclipses between AD 883 and AD 901 reported by Al-Battānī (Albategnius) from Syria (in two cases, Antioch as well as ar-Raqqa), which have been known to European astronomers for many centuries. We shall normally mention only a few salient points about each eclipse, referring the reader to Caussin, Newcomb and Newton for full numerical details.

In the case of S.829 Nov.30, the central line of annularity in Oppolz. starts at sunrise in Egypt, and proceeds via approximately Aden to a noon point in the Indian Ocean. At Bagdad the eclipse was only partial; the astronomical report mentions the beginning (though Newcomb doubts whether it was observed) and the end, the latter at about three seasonal hours after sunrise (half-time between sunrise and local apparent noon).

831 M.831 Apr.30 LATER CALCULATION

Calvisius says that a lunar eclipse on 831 April 30 is mentioned in Frankish annals, and Tycho-Curtius copies this information. We have not encountered mention of this eclipse in Frankish annals, but there was indeed a lunar eclipse on the evening of that day. It reached maximum phase (a little short of totality) shortly after 6 p.m. at Greenwich.

831 S.831 May 15 LATER CALCULATION

Calvisius says that a solar eclipse on "16.Maii fer.2." (May 16, Monday) is mentioned in Frankish annals. Tycho-Curtius copies the day of the month but omits the day of the week. As Calvisius goes on to say, with details, that he finds the eclipse on that day, and as all modern computations show that the date was really May 15, Monday, one wonders whether his 16 is not a mere misprint for 15. Again, we have not encountered mention of the eclipse in the Frankish annals. The eclipse was total very far north (Greenland, Siberia), and consequently not likely to have been noticed in Frankish regions; O'Connor 1952 finds maximum magnitude only 0.44 (at 11.26 a.m.) at his mid-Ireland point, and it is not recorded in Ireland.

<u>831</u> M.831 Oct.24-25 TOTAL IN GERMANY AND WALES

After mentioning the two earlier 831 eclipses just considered, Calvisius says, and Tycho-Curtius copies, that a lunar eclipse on 831 Oct.24 is mentioned in Frankish annals. We have not encountered the complete date in Frankish sources; but the Annals of Xanten and their Appendix (SS <u>2</u>, 1829, 225, 236) do both mention a lunar eclipse in 831 October. The eclipse was total, with magnitude about 1.28, and mid-eclipse occurred about 11 p.m. at Greenwich; the eclipse began in the late evening and ended in the early morning.

In Welsh annals, the eclipse appears under 831 with correct date viii.Kal.Nov. (Oct.25) in the Brenhinedd y Saesson, or Kings of the Saxons, version. There are other versions in which it appears sometimes without month or day, sometimes with wrong month. For details, see Brut y Tywysogyon, or The Chronicle of the Princes [tr. Thomas Jones, Cardiff, Univ. of Wales Press, 1952, xi-xiii, 4, 130].

<u>832</u> M.832 Apr.18 (p.m.) TOTAL IN N.E. FRANCE AND W. GERMANY

The important and frequently edited Annals of St. Bertin are our chief authority; see the Bibliography for several editions. These annals state that in 832 an eclipse of the Moon occurred on 13.Kal.Mai. (April 19) or 14.Kal.Mai. (April 18), according to which manuscript is followed. The primary Annals of Xanten and their Appendix (MGH SS <u>2</u>, 1829, 225, 236) both record an eclipse of the Moon in 832 April, but do not specify the day.

All other records which we have encountered are linked with an alleged solar eclipse in the same year (probably a failed prediction, see "non-eclipse of 832 May 4") and moreover are faulty on the date of the lunar eclipse. Peter the Librarian (SS 1, 1826, 417 or Bouquet 6, 1870, 205) gives 13.Kal.Mai., but under 833; however, his year numbers are normally one unit too large at this point. The lunar eclipse is wrongly dated as 2.Non.Jun. (June 4) or 13.Kal.Jun. (May 20) in the Annals of Fulda (SS 1. 1826, 360 or Bouquet 6, 1870, 210), and as 13.Kal.Jun. in Herimann (SS 5, 1844, 103 or Bouquet 6, 1870, 226).

The eclipse was total, and mid-eclipse occurred soon after 8 p.m. at Greenwich. The eclipse took place on the night of April 18-19; since it ended before midnight throughout Western Europe, we use April 18 as the date, but medieval ecclesiastical usage preferred April 19. The correct identification goes back at least to Calvisius, whose crudely calculated magnitude (1.28) agrees well with modern computations (1.31).

832 X May 4 PREDICTED INVISIBLE SOLAR ECLIPSE

For 832, some of the sources specified under M.832 April 18 mention also an eclipse of the Sun. The Annals of Fulda give the date as 5.Non.Mai. (May 3), while Peter the Librarian (under his 833, the usual 832) and Herimann give 2.Non.Mai. (May 6). There was in fact only a near miss on May 4; the genuine southern-hemisphere eclipse of 832 April 4 was invisible in Europe. Probably the authors quoted are reporting a failed prediction. Discrepant 'evidence' about the date of an imagined or invisible eclipse is not uncommon; compare "S.810 June 5 or July 5" above.

833 CONFUSED PREDICTIONS

"833. Sol et luna defecerunt per eclipsin", Annals of Xanten, Appendix (reference as under M.832 April 18). Similarly in at least three other sources: a Rheims chronicle (Bouquet 6, 1870, 241) Sigebert (SS 6, 1844, 338 or Bouquet 6, 1870, 234), and Alberic (SS 23, 1874, 731). There was an eclipse of the Sun (total in Africa) on 833 Sept.17, but the only eclipse of the Moon in 833 (on April 8) was so minute (magnitude 0.01) as to be practically imperceptible. Probably all the passages are abbreviations of some original such as Peter the Librarian's "Anno 833 ... eclypsis solis 2.Non.Maii (May 6), lunae vero 13.Kal.Maii (April 19)", which really refers to the solar non-eclipse (near miss) of 832 May 4 and the genuine lunar eclipse of 832 April 18, both already discussed. In other words, the only eclipse of the years 832 and 833 unquestionably recorded is M.832 April 18. See also the end of our account of S.828 July 15.

835 M.835 Feb.17 (a.m.) TOTAL IN WESTERN GERMANY

Both the Annals of Xanten and their Appendix (SS 2, 1829, 226, 236) state, under 835, that the Moon was eclipsed in February. The eclipse was total (magnitude 1.52), with mid-eclipse about 2.30 a.m. at Greenwich. The whole eclipse took place after local midnight both there and in the Rhineland.

<u>838</u> M.838 Dec.5 (a.m.) TOTAL IN NORTH-EASTERN FRANCE

The Annals of Saint-Bertin at Saint-Omer state, under 838, that the Moon, 15 days old, was eclipsed in the middle of the night on the Nones of December (Dec.5). The eclipse was total (magnitude 1.23), with mid-eclipse about 3.50 a.m. at Greenwich. The whole eclipse took place after local midnight both there and in Northern France.

<u>840</u> S.840 May 5 (Wed.) TOTAL SOLAR IN W. GERMANY AND ITALY

The documentation of the total eclipses of the Sun in 840 and 878 is more copious than that of any other eclipse of the century. Moreover, the report of the eclipse on 840 May 5 (the day before Ascension Day) is so frequently followed by that of the death of Louis the Pious on June 20 that 840 June 20 as the date of his death is one of the most certain pieces of chronological information. The eclipse is known astronomically to have occurred on 840 May 5, and the small number of 839's and 841's found in the sources are to be written off as wrong.

The track of totality on Schroeter's map has Bordeaux, Lyon, Milan, Venice and Bucarest near its southern edge, and Tours, Basle, Munich and Vienna not far north of its northern edge. Mid-eclipse in local apparent time varied from about 12.30 p.m. on the Atlantic coast of France to about 2.15 p.m. at Vienna. On average, and mostly even individually, the reports agree with this within an hour. There are over two dozen reports of the hour, and no reports mention an hour of the day less than 6 or more than 9.

The eclipse is discussed in Ginzel 1883, 663f.; see also Ginzel, Handbuch der Chronologie, <u>3</u>, 1914, 92. In view of the extensive bibliography (42 items) in Ginzel 1883, it seems unnecessary to go into further detail here. We have accumulated over a dozen further references, but the only ones which mention a time of day (sometimes with wrong date!) are:

| | MGH SS | Bouquet |
|---|---|---|
| Chronicon Brixiense | <u>3</u>, 1839, 240 | |
| Heriman.Aug.Chron. | <u>5</u>, 1844, 104 | <u>6</u>, 1870, 227 |
| Mariani Scot.Chron. | <u>5</u>, 1844, 550 | <u>6</u>, 1870, 228 |
| Agnellus, Pont. Ravenna | | <u>6</u>. 1870, 307 |

These four give respectively quasi hora nona,
post sextam diei horam,
inter 8. et 9. horam,
meridie ... usque ad horam nonam.

There is no independent record from England, the mention by Florence of Worcester being a later borrowing from the Continent.

Most of our additions to Ginzel's list are now given in Newton 1972, partly corrected in 1979, 289-290. Newton 1979 (189, 319, 354, 462-3) assumes, partly because stars were seen, that its magnitude at Xanten was $0.88 \pm 0.05$. This record and that of Fulda are independent records and are scientifically useful; most other records are obviously borrowed.

841 S.841 Oct.18 (Tues.) PREDICTED AND WATCHED FOR IN FRANCE

This annular eclipse, partial in France, is clearly referred to in Nithard's History, Book II, Ch.10: "Dum haec (or hanc) super Ligerim juxta sanctum Fludualdum consistens scriberem, eclipsis solis hora prima, feria tertia, 15.Kal.Nov. [Oct.18] in Scorpione contigit". "Feria tertia" is from SS 2, 1829, 661, "tertia" being correctly supplied by the editor, G.H. Pertz; the frequent "prima feria" (e.g. Bouquet, 7, 1870, 22) is wrong, whatever the manuscripts say.

Scholz 1970, 154 translates: "While I was writing this at the Loire[35] near St. Cloud, an eclipse of the Sun occurred in Scorpio in the first hour[36] of October 18, a Tuesday". His note 35 says "St. Cloud on the Seine is meant". His note 36 says "Six o'clock in the morning", and presumably implies from sunrise onwards. Editors appear to agree that S. Fludualdus means S. Clodoaldus. Pertz suggested "Saint Claude supra Blois?" as the place of observation. Newton 1972, 328 recognizes some uncertainty, but approximately follows Pertz. In fact the place of observation is so far from the central line that its precise position is immaterial; anywhere within rather a wide region containing both Blois and Paris would fit the record. Ligerim may or may not be a slip for Sequanam.

The central line of annularity starts in Morocco and runs ESE. Both the general run of the track, and the fact that O'Connor 1952 finds magnitude only 0.38 at 7 a.m. sunrise in mid-Ireland, show that the eclipse might have been noticed in France - but not with magnitude more than about 0.5 anywhere near Blois or Paris. Probably this had been predicted and was watched for by Nithard.

The Sun's longitude was about 209°.3, so that strictly the Sun was in the last degree of Libra rather than in Scorpio.

<u>840-841</u> THE SAME SOLAR ECLIPSES IN S.E. EUROPE

The last two solar eclipses seem to have been recorded in Constantinople, from observations in the northern and southern parts of the Empire respectively.

A passage in 'Theophanes Continuatus', Book IV (Bonn Theophanes, Vol.III, 1838, 203), referring apparently to the beginning of the reign of the child Michael III and his regent-mother Theodora, mentions that, "two solar eclipses having occurred", Theoctistus, chief minister during the regency, (i) campaigned unsuccessfully against the Abasgians (in West Georgia), (ii) made another unsuccessful campaign, (iii) unsuccessfully attacked the Arabs in Crete. The passage is clearly hostile to Theoctistus.

Since Michael's father Theophilus died on 842 Jan.20, and the temporary capture of Crete is dated 843 (by H. Grégoire, Camb.Med.Hist., <u>4</u>(1), 1966, 106, 115), the two eclipses mentioned in our heading may serve as at least provisional identifications, to give the passage a place in our eclipse accounts. From Schroeter one may estimate that S.840 May 5 (q.v.) had a fairly large magnitude, say about 0.87, at Constantinople (greater further north), and that S.841 Oct.18 (q.v.) had a modest magnitude, say about 0.45, at Constantinople (greater further south). One would expect two eclipses mentioned in the same phrase to have occurred reasonably close together; our identifications are suggested as having some <u>prima facie</u> suitability, but the second eclipse would have been weak even in southern Greece.

<u>842</u> M.842 Mar.30 (Thurs. a.m.) TOTAL IN GERMANY AND FRANCE

The Annals of Fulda (SS <u>1</u>, 1826, 363 or Bouquet, <u>7</u>, 1870, 160) have "842 ... in the same year an eclipse of the Moon occurred iii.Kal.Apr., on the Thursday (quinta feria) before Easter, at the tenth hour of the night (decima hora noctis)". The correct iii.Kal.Apr. (= March 30) is a minority reading; most manuscripts have "in Kal.Apr.", perhaps by an easy scribal error ("in" for "iii"). Peter the Librarian (SS <u>1</u>, 1826, 417 or Bouquet, <u>7</u>, 1870, 158), abbreviating Ann.Fuld., has 3.Kal.Apr. but no hour. "Decima hora noctis" may refer to the beginning of totality, which lasted approximately from 3.40 a.m. to 4.50 a.m., Fulda apparent time. The magnitude of the eclipse was 1.27.

A contemporary but fragmentary chronicle from Fontanelle (Saint-Wandrille, near Caudebec, Seine-Inférieure) puts the fact that the eclipse occurred on Maundy Thursday otherwise (SS 2, 1829, 301-2 or Bouquet, 7, 1870, 41): "On the third of the Kalends of April, on the very day of the Lord's Supper (ipso die coenae Domini), before dawn brought an end, the Moon was eclipsed, starting from the top (a summo incipiens)". Only the very end of the eclipse would be lost in the dawn at Saint-Wandrille (the final phase would be more affected at Fulda). The last words of the report are correct; in the latitude of northern France, when the Moon (as in this case) has a slight southern latitude during an eclipse occurring towards moonset, consideration of a diagram of the celestial sphere quickly shows that the first contact of the Moon with the Earth's shadow occurs not too far from the point of the Moon's rim furthest from the horizon (nearest to the zenith).

The eclipse was discussed by Calvisius, and rather more fully by Petavius 1627, 867.

843 M.843 Mar.19 (p.m.) TOTAL IN FRANCIA

We have found this eclipse recorded only in Nithard, Hist., Book 4, Ch.7: "At this time, on 13.Kal.Apr. (= March 20) an eclipse of the Moon occurred". The date is that given, for example, in SS 2, 1829, 672 and Scholz 1970, 174; Bouquet 7, 1870, 33 does give 14.Kal.Apr. (= March 19).

The eclipse was total (magnitude 1.17), and ended about 11 p.m. at Greenwich, consequently also before Nithard's midnight. Thus Bouquet's date is in modern style, but March 20 can be justified, as usual, by ecclesiastical usage. The eclipse is omitted from Scholz's index.

The Maya seem to have observed this eclipse and to have used the date as the Full Moon base of their Eclipse Table (see below).

848 NON-ECLIPSE

Under the year AD 848, in which he places the slaying of Ethelred [the second], king of the Northumbrians, and the beginning of the 18-year reign of Osbert, Roger of Wendover (d.1237), Flores Historiarum, says: "In the same year there was an eclipse of the Sun, at the sixth hour of the day [noon], on the Kalends of October [October 1]." (Ed. H.O. Coxe, 1, 1841, 285; tr. J.A. Giles, 1, 1849, 180, and D. Whitelock, Engl. Hist. Docum., 1, 1955, 256).

However, Roger's eclipse information is completely spurious; there was no eclipse anywhere on the day mentioned. Neither of the genuine solar eclipses of 848 (June 5 and Nov.29) was visible in England, and changing only the year gives no help. While gross misplacement of some eclipse may be considered, the explanation may be simpler. New Moon occurred on 848 Oct.1, about 4 a.m. at Greenwich. It is easy to imagine that sometime between the ninth century and the thirteenth, probably later rather than earlier, some astrologer-astronomer, computing this conjunction, found the right day and the wrong hour, and misguidedly inferred a solar eclipse.

849 S.849 May 25 LARGE SOLAR ECLIPSE IN SCOTLAND NOT RECORDED

Schroeter's final stretch of the band of totality includes Eastern Iceland, the Faroes, Shetlands and Orkneys, and ends at sunset in the North Sea off Southern Norway. The magnitude appears to be not less than 0.9 throughout Scotland, and O'Connor 1952 found 0.83 for mid-Ireland. But we have encountered no Western record.

852 S.852 Mar.24 LARGE SOLAR ECLIPSE IN N. EUROPE NOT RECORDED

Schroeter's final stretch of the band of annularity includes Cornwall, Devon, London, Northern Belgium, Holland, Berlin and Northern Poland, and ends south-east of Moscow. Rather surprisingly, we have encountered no Western record.

854 M.854 Feb.16-17 LARGE PARTIAL IN BAGDAD

This large partial lunar eclipse, given by Ibn Yūnus (tr. Caussin) as observed in the late evening of 854 Feb.16 at Bagdad, is omitted, apparently because considered weak, by Newcomb 1878, 45, but is discussed in Newton 1970 (146, 148, 215, 229); the details seem sufficiently clear to guarantee at least the identification.

854 M.854 Aug.11-12 TOTAL IN BAGDAD

The beginning of this total lunar eclipse was observed at Bagdad, according to Ibn Yūnus as translated by Caussin. The beginning occurred around 3 a.m. Bagdad mean time, or Greenwich mean midnight. See Newcomb 1878 (§5) and Newton 1970, 146, 229.

856 S.856 Jan.11 MEDIUM SOLAR ECLIPSE IN SPAIN NOT RECORDED

Schroeter 1923 shows the final stretch of the band of totality as passing over Morocco and Algeria, and ending to the south-west of Sardinia. It shows magnitude more than 0.75 in most of Spain and Portugal and in South-Western France, while O'Connor 1952 finds magnitude 0.52 in Central Ireland. But we have encountered no Western record.

856 M.856 June 21-22 PARTIAL LUNAR IN BAGDAD

This partial lunar eclipse (magnitude about 0.6) was observed at Bagdad, according to Ibn Yūnus. The beginning occurred about $3^h20^m$ a.m. Bagdad mean time ($0^h20^m$ UT). See Newcomb 1878 (§5) and Newton 1970 (146, 148, 215, 229).

861 M.861 Mar.30 (a.m.) TOTAL IN N.W. EUROPE

This is the last of the four eclipses mentioned in the Annals of Saint-Bertin at Saint-Omer: "861 ... 4.Kal.Apr. (= March 29), after the eighth hour of the night, the Moon, 14 days old, turns entirely dark (tota in nigredinem vertitur)". 4.Kal.Apr. is the date given in the chief manuscript, but the edition of Grat et al. (1964, 84), though it gives this in a footnote, uses 3.Kal.Apr. (= March 30) in the text.

The record appears also, in the same words, in Bouquet 7, 1870, 274, from the (later) Chronicon Elnonense (of the monastery of Saint-Amand, near Tournai).

The eclipse was total (magnitude 1.27), and occurred entirely after midnight of March 29-30 in England and France, so that March 30 is the true date. Totality at Greenwich lasted from about 2.40 a.m. to 3.55 a.m., so that "after the eighth hour of the night" is intelligible.

863 S.863 Aug.18 SOLAR IN S. EUROPE NOT RECORDED

This annular eclipse provides one of the more striking cases of apparent lack of Western record; such an eclipse half a century earlier would almost certainly have been recorded by the Franks. Schroeter 1923 shows the band of annularity starting at sunrise in Northern Spain, and proceeding via Southern France (Nice), Northern Italy (Milan), Northern Yugoslavia, between Budapest and Bucarest, and then via Odessa and Astrakhan to the Northern Caspian. The eclipse was of magnitude greater than 0.75 over most of Europe; O'Connor 1952 found 0.71 for Central Ireland.

865 S.865 Jan.1 TOTAL SOLAR IN IRELAND

This total eclipse of the Sun is clearly mentioned in Irish records. Under manuscript year 864 (true 865) the Annals of Ulster, 1, 374-5, have "An eclipse of the Sun on the Kalends of January, and an eclipse of the Moon in the same month". The solar eclipse is correctly indexed as S.865 Jan.1 by B. MacCarthy in AU 4, 140. No Irish annals give more detail, and some give less. The manuscript year is always wrong, but long since rectified editorially (together with one or two slips). See

AI 1951, 132-3; CS 1866, 158-9; J. O'Donovan, introduction to Ann. Four Masters, 1, Dub. 1851, xlix; Duald MacFirbis, Ann. of Ireland, Three Fragments, ed. J. O'Donovan, Dub. 1860, Fragm. III, 162-3; AC 1896, 141. The hour of the day is nowhere stated, and we have encountered no record except from Ireland.

Oppolzer's approximate central line is sharply curved and short (as eclipse tracks go). It starts off Labrador, has noon point in the East Atlantic, passes in the early afternoon over Northern Ireland and Scotland, and ends in Norway. For his mid-Ireland point, O'Connor 1952 finds magnitude 1.00 (totality or nearly so) at 1.16 p.m., local apparent time. Schroeter 1923 shows a narrow band of totality passing from SW to NE over Northern Ireland, north of Edinburgh, north of Bergen, and ending short of Trondheim.

This is a striking example of the value of a certainly identifiable eclipse in establishing chronology. Those manuscripts which give month and day are unanimous on "Kalends of January", and there was no other solar eclipse on January 1 during the ninth century AD.

865 M.865 Jan.15 (p.m.) TOTAL IN IRELAND AND GERMANY

On the record in the Annals of Ulster, see under S.865 Jan.1. The lunar eclipse also is correctly indexed, as M.865 Jan.15, by B. MacCarthy (loc. cit.). Several of the Irish sources mentioned under S.865 Jan.1 include the lunar eclipse, but none adds any further information. The day of January is not stated, but a total lunar eclipse did occur in the early evening of Jan.15. Magnitude about 1.06; mid-eclipse $18^h25^m$ UT, according to Oppolzer, and Schroeter; full moon $18^h33^m$ UT, according to Goldstine 1973.

This lunar eclipse is also almost certainly that referred to in the primary Annals of Xanten (SS 2, 1829, 231): "866. In January an eclipse of the Moon occurred". This can only be M.864 Jan.27 or M.865 Jan.15; both were total during the evening, with magnitudes about 1.32 and 1.06 respectively. There is no reason to think that a Xanten year (AX) at this period can be two units out, but it can be one unit too large; see, for example, Weiss 1914, 70-71, where AX 868 = AD 867 and AX 864 = AD 863.

866 S.866 June 16 PARTIAL SOLAR IN BAGDAD

A total eclipse, observed as partial at Bagdad in the early afternoon (mid-eclipse around 1.30 p.m. local time). Oppolzer shows the central line of totality as starting in the Atlantic (off Eastern Brazil), continuing via West Africa to a noon point in Upper Egypt (near the Sudan border), and proceeding via Arabia and India to an end near Indonesia. Newton 1970 finds the magnitude stated for Bagdad (between 7 and 8 digits, i.e. about 0.625) reasonable. Newcomb 1878 did not use this eclipse. See Caussin 1799, 1804 and Newton 1970 (146, 149, 241-2, 247).

866 M.866 Nov.26 (a.m.) CALCULATED BY A BAGDAD ASTRONOMER

A lunar eclipse of small magnitude (0.125 Ibn Yūnus, 0.06 Oppolzer). Not included in Caussin 1799 nor used in Newcomb 1878, but included in Caussin 1804 and discussed in Newton 1970 (146, 149, 229). Newton, 149 tends to think, probably rightly, that Newcomb did not use this Islamic record because of doubt whether the data were observed or calculated.

We shall discuss here only the time of mid-eclipse. The $9^h31^m$ seasonal time of night at Bagdad, quoted by Newton, 146, 149 from Caussin 1804, is approximately equivalent (neglecting refraction and solar semi-diameter) to $4^h7^m$ local apparent time, or $3^h57\frac{1}{2}^m$ local mean time, or $1^h0^m$ UT. Newton, 229 appears to imply about $0^h55^m$ UT, a value doubtless more finely computed, and to be preferred. Such reductions may be compared with mid-eclipse at $0^h51^m$ UT (Oppolzer) and full moon at $0^h44^m$ UT (Goldstine 1973). It seems that the Islamic value was not observed, but calculated by a competent astronomer, well in touch with the real motions in the heavens. The identification of the eclipse appears beyond doubt, and a slight umbral eclipse appears probable.

<u>867</u> M.867 Nov.15 (p.m.) TOTAL IN BAGDAD

This is the first in time of several Persian and Arabian eclipses pointed out to Ginzel by the orientalist A. Müller, of the University of Königsberg, and described in Ginz.1887. Ginz., 713, quotes at-Tabarī as stating that "the Moon was eclipsed and was completely invisible or [at any rate] disappeared for the most part" on the night of AH 253 Dhū 'l-Qaʿda [XI] 14, equivalent to AD 867 Nov.15. This date is correct. The eclipse was total, of magnitude about 1.23 (Ginz.) or 1.25 (Oppolz., Schroet.), with mid-eclipse about 7.12 p.m. Bagdad mean time (Ginz.), 7.20 (Oppolz.), 7.26 (Schroet.), full moon approximately 7.20 (Goldstine 1973). Ginz. correctly says that the whole course of the eclipse was visible at Bagdad [assuming clear sky].

871 M.871 Sept.2-3 (Sun.-Mon.) POSSIBLY OBSERVED IN IRAQ

This is another Tabarī record; the eclipse appears from Ginz.1887, 714 to have been imperfectly predicted, but it would be fully visible in Mesopotamia in good weather, and anyway is chronologically significant, marking a disastrous month for the inhabitants of the city of Baṣra. Ginzel quotes at-Tabarī to the effect that in AH 257, in the month of Shawwāl, a rebel known as Al-Khabīth [the Reprobate], claiming ʿAlid descent, decided to invade al-Baṣra, and noticed that a lunar eclipse was due on the night of Tuesday, Shawwāl 14; this date converts to AD 871 Sept.4 (Tuesday), and is wrong by either one or two nights, according to what is meant by "the night of Tuesday". All calculators find a total eclipse (magnitude about 1.125), with mid-eclipse shortly before Bagdad mean midnight of Sept.2-3 (Sunday-Monday). The error did not spoil Al-Khabīth's action; his Zanj (African) forces surprised Baṣra during the Friday services of 871 Sept.7, according to C. Brockelmann, Hist. of the Islamic Peoples, Engl. tr. Lond. 1949, 134. A massacre of the inhabitants followed.

On the Zanj rebellion in the Lower Euphrates region, see also G.E. von Grunebaum, Classical Islam, Engl. ed. Lond. 1970, 105; W. Muir, The Caliphate, rev. T.H. Weir, Edinb. 1915, 544-5; Camb. Med. Hist., 4(1), 1966, 682 (Grunebaum), 702 (M. Canard). On the ʿAlids, see Encyl. of Islam, 1, Leiden & London 1960, 400-3 (article by B. Lewis); on al-Baṣra, ibid. 1085-7, esp. 1086 (article of S.H. Longrigg).

873 S.873 July 28 ANNULAR ECLIPSE (MOON IN SUN'S BODY) IN IRAN

A description, unique in this period, of an eclipse in Nishapur, Iran (36°13'N, 58°49'E) by Abu al ʿAbbās al-Irānshahrī, has been published by B.R. Goldstein in Archiv. Internat. d'Hist. des Sci., 1979, 29 (No.104), 101-2. According to al-Biruni this astronomer "recorded that the body of the moon was in the middle of the body of the sun such that the light of the remaining portion of the sun surrounded it uneclipsed; thus it is clear that the apparent diameter of the sun exceeds that of the moon". Stephenson 1982, 162 points out "The last clause, an interpolation by al-Biruni, is of course wrong, but Tycho Brahe later arrived at the same mistaken conclusion quite independently".

878 M.878 Oct.15 (Wed. a.m.) TOTAL IN IRELAND AND GERMANY

Apart from Irish and English records (see below), there are a number from Europe, including several with correct dates, but only the Annals of Fulda and (copying them) Peter the Librarian mention the hour, both giving "last hour of the night". In relation to 878 Oct., the Annals of Fulda appear to give "III Idibus"[Oct.13] or "in Idibus" [Oct.15], according to manuscript (see e.g. SS 1, 1826, 392, Bouquet 8, 1871, 38 and Ginz.1883, 671, item 22). Peter gives the correct date (SS 1, 1826, 418 or Bouquet 8, 1871, 98).

The eclipse was total, with magnitude about 1.07. At Greenwich, mid-eclipse was about 4.25 a.m., and the eclipse ended soon after 6 a.m. The Fulda mean times would be about 39 minutes later. Thus "last hour of the night" is reasonable for Fulda.

The Annales Heremi (of Einsiedeln) and Annalista Saxo are wrong in saying "on the 16th day of October", as the eclipse occurred after midnight on the night of Oct.14-15. We have encountered no version of the Annales Vedastini (of Arras) which is not seriously corrupt in relation to this eclipse.

There is a clear record in the Annals of Ulster, 1887, 392-4. Under 877, but with epact clearly indicating normal AD 878, we find "An eclipse of the Moon on the Ides [15th] of October, the 14th of the Moon, about the third vigil, on the fourth day of the week [Wednesday]". In his index, AU 4, 1901, 140, MacCarthy says "vigil 3 [4.30 a.m.]". This is an intelligible conventional interpretation, but mid-eclipse would be around 4 a.m. in Ireland, and this time also lies within the third vigil, if three watches per night are meant.

A late English record is unsatisfactory. Roger of Wendover (d. circa 1237), Flores Historiarum (ed. H.O. Coxe, 1, Lond. 1841, 335; Engl. tr., J.A. Giles, 1, Lond. 1849, 213-4), has, wrongly under 880, "In the same year there was an eclipse of the Moon after midnight on the third of the Ides of October [Oct.13]". Roger may have used a manuscript which reproduced the erroneous Fulda variant reading.

The eclipse was mentioned in Calvisius 1605, etc.

878 S.878 Oct.29 (Wed.) TOTAL SOLAR IN IRELAND, ENGLAND AND GERMANY

All solar eclipses recorded in the British Isles and Europe under years 874 to 880 really refer to the total solar eclipse of 878 Oct.29. We explain below why the 879 eclipses were not visible.

Discussion of S.878 Oct.29 goes back at least to Gemma Frisius, Kepler, Calvisius, Riccioli and others. Ginzel 1883, 669-73 and Newton 1972 (Ch. VI-XI, XIII) both list over 30 medieval sources, containing various amounts of detail; the eclipse is discussed by these authors, and also in P.V. Neugebauer 1930 and Newton 1969, 1970 (49, 52, 264).

Schroeter 1923 gives geocentric conjunction a little before $13^h30^m$UT; Goldstine 1973 gives about $13^h32^m$. The track shown in Schroeter is strongly curved (concave towards the north), and his band of totality includes Reykjavik, Belfast, Liverpool, London (just), Amsterdam, Brussels, Berlin (almost), Prague (just), Warsaw, and ends short of Moscow. His calculated apparent times of mid-eclipse are about 1 p.m. in Ulster and about 2.45 p.m. at Vienna and Berlin. Since a vast literature exists, we can mention only a few of the main medieval sources and even of modern discussions.

The Annals of Ulster (1, 1887) give under annal 877 (normal AD 878) the correct month and day, and are also correct on the hour, since "about the seventh hour of the day" here means somewhat before 1 p.m. (O'Connor 1952 finds magnitude 0.97 at $12^h48^m$ local apparent time in central Ireland).

Other English sources more or less contemporary with the eclipse are the Anglo-Saxon Chronicle and Asser's Life of Alfred the Great, neither of which gives month or day. The Chronicle gives, usually under 879 (but MS. C under 880) a solar eclipse

(i) for one hour of the day (ane tid daeges, with minor variants),

e.g. ed. C. Plummer, 1 (text) 1892, 2 (notes) 1899; Engl. tr. D. Whitelock et al. 1961.

Asser gives, under 879, a solar eclipse

(ii) between None and Vespers, but nearer None (inter nonam et vesperam, sed propius ad nonam),

e.g. ed. F.W. Stevenson, 1904, re-issued with addition by D. Whitelock 1959; various English translations.

The annal year 879 is no longer a problem. In the course of a long note, 280-6, Stevenson 1904 virtually rejected S.879 March 26 (q.v.), which special calculations made by A.C.D. Crommelin confirmed as only slight in England. This part of Stevenson's note may now be by-passed (see Whitelock's 1959 addition, p.cxxxvi), as it has become widely accepted (since M.L.R. Beaven's paper, The Beginning of the Year in the Alfredian Chronicle, Engl. Hist. Rev. 33, 1918, 328-42, and now Harrison 1976, 117) that at the period in question the ASC (and, following it, Asser) used a year running from autumn to autumn (though probably not beginning with the Bedan indiction on Sept.24). Thus an October eclipse in our 878 fell in 879 of ASC and Asser (a conclusion to which Stevenson came uncommonly close in his footnote on p.282). Year-beginnings Sept.1, Sept.24 and Oct.1 are all compatible with this, so that no discrimination between them need be attempted here. The Bedan indiction seems never to have been used.

The hour does, however, present a difficulty. Statement (i) in the ASC may pass, as a rough approximation to the duration of substantial eclipse; e.g., it is about the length of time during which more than half the Sun's diameter is usually eclipsed at places near the central line. But Asser's (ii) is uncomfortably late for England, where mid-eclipse almost certainly occurred before None. In Schroeter 1923, mid-eclipse at London seems to be totality about 1.42 p.m. local apparent time. In other calculations it is usually 5 or 10 minutes earlier; e.g., one may roughly estimate totality at 1.35 p.m. from J. Maguire, MN 45, 1885, 400, see also 46, 1885, 25, and near-totality at 1.33 p.m. from P.V. Neugebauer 1930 (B28). On the other hand, None at London, if reckoned half-way between apparent noon and sunset (as Stevenson intended for Fulda, see below), cannot have been before 2.20 p.m. apparent time. It is true that mid-eclipse might be considered as at the beginning of the ninth seasonal hour of the day, but this seems a strained interpretation; it is unusually clear in this case that the office of None is referred to.

Since several later English sources (Annals of St. Neots, Florence of Worcester, Simeon of Durham, Roger of Hoveden) followed Asser's (ii) rather than the Chronicle's (i), the discrepancy is of some importance. Plummer 2, 95 thought that Asser "altered the time given by the Chronicle to suit" S.879 March 26. Stevenson 1904, noting possible observation in Flanders or West Germany (p.lxxviii), thought perhaps Asser was then near Fulda (286); apart from a necessary correction to be mentioned presently, this solves the problem, except that Asser is not known to have been there. Newton 1972, 209 takes the place of observation as St. David's, in Wales; if true, this increases the difficulty. Altogether, Asser's hour remains as mysterious as his career is elusive.

The qualification needed by Stevenson's 'solution' is that his calculations for Fulda need corrections. On 285 he applies a 15-minute equation of time with the wrong sign, so finding apparent times half an hour too early. At the season of the year in question, apparent times have always (throughout recorded history) been later, not earlier, than mean times (cf. his line 10). Ignoring seconds, but preserving Stevenson's evident intention to place None and Vespers at the third and fifth seasonal (variable) hours after noon (i.e. half and five-sixths of the way from apparent noon to sunset), his Fulda apparent times of 1.52 p.m. and 3.27 p.m. (lines 22-3) need changing to 2.22 p.m. and 3.57 p.m.; sunset was about 4.44 p.m. Fulda apparent time. Allowance for refraction, etc. would increase these times by only a few minutes, a negligible quantity in the present context.

We estimate the local apparent time of total eclipse at Fulda as about 2.33 p.m. (using Schroeter) or about 2.25 p.m. (using P.V. Neugebauer 1930). These eclipse times are indeed slightly later than None (2.22 p.m.), but the margin is half an hour less than Stevenson imagined. In fact the plain "post nonam horam" of the Primary Annals of Fulda (e.g. SS 1, 1826, 392), which report the Sun obscured for half an hour, and stars visible, is at least as accurate as Asser's more elaborate statement.

This brings us to the Continental records, largely from France and Germany. Besides the Annals of Fulda, there is a contemporary record by Regino of Prüm (e.g. SS 1, 1826, 590), who says "circa horam nonam". One can make a large dossier of European sources from Ginzel 1883 and Newton 1972 and 1979, 280, but most are not contemporary. However, the tradition is fairly sound; the year is frequently wrong (or "wrong"), but month and day, when given, are almost always correct, and the hour, when given (overwhelmingly from "about eighth" to "after ninth"), is almost always broadly correct, not differing by more than one hour from what one would expect for the locality. It is usually unclear whether "hora nona" refers primarily to the ninth hour or the office of None, but the point is unimportant if the two times were meant to coincide. Peter the Librarian (e.g. SS 1, 1826, 418, or Bouquet 8, 1871, 98) puts the time of this eclipse as "hora nona officii". The general sense is close to "mid-afternoon".

An interesting English charter referring to a grant of land near Taunton, on the river Tone, in Somerset, SW England, was written a few days after this eclipse, but only a copy survives (information from Dr. Eric Poole) and its date was for a long while obscure. It ran:

"Anno ab incarnatione domini nostri Jhesu Christi DCCCCLXXVIIII. Indictione XIIII kal' November in hunc annum sol obscuratum fuit". This mixed Latin, if translated as it stands, would mean "In AD 979 (sic). Indiction 13. November 1 in this year the sun was obscured". This was included as the Saxon-Latin Charter No.549 in 'Cartularium Saxonicum' by W. de G. Birch (London, 3 vols., 1885-1893, 169-170) who realized that it was 9th, not 10th, century, but who played the 'identification game' badly by listing, in his footnote, only the eclipses from 879 to 881, and who supposed that 880 March 14 was the eclipse in question. Historians have suspicions about authenticity, but the place is very near Schroeter's belt of totality.

This eclipse was total in western Iceland, and, as it is recorded in the Icelandic Annals (AD 880), it has been argued that the original observation could have come from Iceland. The relevant quotations are thus included by Thorkelsson as the first Icelandic eclipse. T. Thorkelsson 'Sonnen und Mondfinsternisse nach Islandischen Quellen bis zum Jahre 1734', Visindafelag Islandica, XV, 1933, 1-39 (Reykjavik). Certainly, the Scandinavians from 874 onwards were migrating (probably via W. Scotland, Dr. A. Smyth tells us) to S.W. Iceland, as illustrated in the maps based on the Landnama-Boc (see A.C. O'Dell 'The Scandinavian World', fig.125, p.339, 1957, London).

The Icelandic versions as they stand are mostly translated: "About three hours after midday the sun was so much darkened that the stars could be seen", but these are presumably derived from the Flatyjarannall of Flateyarbok III version "The sun was darkened so much about the ninth hour of the day that the stars could be seen in the sky". The Royal Annals (Konungsannall or Annales Regii) give this in Latin: "Sol hora diei IX[2] ita obscurata est, ut stelle in caelo apparuerunt". The chronology of these various versions is not known, but the reference to the ninth hour proves that the observation was not made in Iceland. It probably came from Fulda (cf. Newton 1972, 233-4, 495 ff.).

879 S.879 Mar.26 (Thurs.) FALSE IDENTIFICATION

Numerous English and Continental writers mention under 879 a solar eclipse now known to be S.878 Oct.29 (q.v.). The penumbral solar eclipse of 879 March 26 could not have been noticed.

879 M.879 Oct.4 (Sun., a.m.) FALSE IDENTIFICATION

A false identification of a lunar eclipse mentioned for 880 Oct.13 (an impossible date) by Roger of Wendover (13th C), see under M.878 Oct.15. It is true that M.879 Oct.4 was total and entirely visible in England (greatest magnitude 1.28 at $3^h 54^m$ UT), but no earlier English source mentions this lunar eclipse, and Roger's information seems to come from a faulty Continental record of M.878 Oct.15.

880 S.880 Mar.14 (Mon.) FALSE IDENTIFICATION

There is no genuine record of this eclipse. Oppolzer makes centrality end at sunset in longitude 22°W., latitude 40°N., a little to the N.E. of the Azores. For England, the identification of a few records as referring to S.880 Mar.14 is certainly a red herring; several mid-19th century editors and translators of Florence of Worcester and Roger of Hoveden gave this identification, where a modern commentator would give S.878 Oct.29. We are equally unable to suggest any genuine record from Europe or Iceland - only a few '880' annals, of late composition (Sigebert, Alberic, etc.), really referring to S.878 Oct.29. In short, we mention S.880 Mar.14 only as an identification now discarded by astronomers, but stated as a fact by chronicle-commentators in works which are still standard.

881 S.881 Aug.28 FALSE IDENTIFICATION
recte
891 Posthumous editions of Calvisius (1620, 1650, etc.), and hence Tycho-Curtius 1666, mention an eclipse of the Sun on 881 Aug.28 as noted in Frankish annals. We have found a solar eclipse of 881 mentioned only in the second continuation of the Annales Juvavenses maximi, of Salzburg (SS 30(2), 1934, 742), and the description does not fit:

"DCCCLXXXI. The Sun was obscured from the third to the sixth hour." This indicates a morning eclipse.

The eclipse of 881 Aug.28 was annular. Oppolzer gives the track of centrality as starting at sunrise in the United States, having noon point in the Atlantic, and passing via approximately Morocco to end at sunset in East Africa. The small eclipse (magnitude perhaps 0.3) visible in Austria would occur there well on in the afternoon, but it would not be likely to be noticed.

The Salzburg record seems to refer to S.891 Aug.8 (q.v.). This eclipse was large at Salzburg around 10 a.m. local apparent time. If this conjecture is correct, DCCCLXXXI is a mistake for DCCCLXXXXI.

882 M.882 Feb.7 (a.m.) TOTAL IN SWITZERLAND (AMBIGUOUS DATE)

Oppolzer and Schroeter give this total eclipse a maximum magnitude of 1.27 a few minutes before $2^h 30^m$UT.

The Annales Heremi, of Einsiedeln (SS 3, 1839, 140) give "Eclipse of the Moon" under annal year 883, but along with events normally dated 882 (deaths of Louis III, king of France, and of Pope John VIII). They therefore probably refer to this eclipse or to M.882 Aug.3 (q.v.), rather than to M.883 Jan.27 or M.883 July 23 (qq.v.).

The Annales Alamannicorum, Continuatio Sangallensis Tertia (SS 1, 1826, 52) give "882. Eclipse of the Moon". This apparently refers to one of the two 882 eclipses mentioned above.

Any interval of one year, whenever beginning, taken within 882-3 contains at least two lunar eclipses visible in Western Europe. Thus the evidence of annals which, like the two Swiss ones quoted, give only the year (no month or day) is essentially incomplete, and the identification ambiguous. They may derive from the St. Gall record considered under 893/4. We shall see that the two summer lunar eclipses are recorded with complete dates in Islamic sources.

882 M.882 Aug.3 (a.m.) TOTAL LUNAR IN BAGDAD

Oppolzer and Schroeter give this total eclipse a maximum magnitude of 1.44 at $2^h 3^m$ and $2^h 9^m$UT respectively.

The contemporary record by the Bagdad historian at-Tabarī (AD 839-923) is discussed in Ginz.1887, 712. Tabari describes the Moon as setting eclipsed on AH 269 Muharram 14, an Islamic date which converts to AD 882 Aug.3. Ginzel finds maximum magnitude 1.44, with time-table, in Bagdad mean time (we omit decimals of a minute)

$3^h 27^m$, $4^h 27^m$, $5^h 11^m$, $5^h 54^m$, $6^h 57^m$,

the last two phases (end of totality and end of eclipse) being invisible at Bagdad, through occurring after moonset. As Bagdad is 44½° east of Greenwich, Ginzel's mid-eclipse at $5^h 11^m$ corresponds to $2^h 13^m$UT.

For a possible European record, see under M.882 Feb.7.

882 S.882 Aug.17 (Fri.) BAGDAD SOLAR ECLIPSE PROBABLY CALCULATED

Ginzel 1887, 712-3 discusses a solar eclipse recorded by Tabarī as having occurred in the same month of Muharram. Perhaps because it was a sunset eclipse at Bagdad, and the Moslem day begins at sunset, Tabarī, while mentioning Friday, gives the date as AH 269 Muharram 29, which the usual tables convert to the Moslem day falling within AD 882 Aug.17-18 (Fri.-Sat.). The eclipse was penumbral, i.e. nowhere more than partial. Geocentric conjunction was on Aug.17 at $15^h 3^m$UT (Oppolz.), $14^h 54^m$UT (Goldstine 1973). Sunset at Bagdad occurred about 6.33 p.m. local apparent time, and for this moment Ginz. finds an eclipse of magnitude about 0.15 (using Oppolzer's elements) or 0.17 (using Ginzel's elements). Ginzel also remarks that this is the smallest eclipse known to him as having been observed before the telescope came into use; the reader will notice that only about 5' of the Sun's diameter was obscured, and we feel that it was predicted rather than observed. We have nevertheless allowed 1 point in our table on the assumption that an eclipse had been predicted and the particular sunset was watched by astronomers.

883 M.883 Jan.27 (a.m.) FALSE IDENTIFICATION

Oppolzer and Schroeter give this total eclipse a maximum magnitude of 1.10 at about $2^h 24^m$UT. For a barely possible record from Einsiedeln, see under M.882 Feb.7.

883 M.883 July 23 (p.m.) LARGE IN SYRIA

Oppolzer gives this eclipse a maximum magnitude of 0.93 at $17^h6^m$UT. Besides the barely possible record from Einsiedeln, see under M.882 Feb.7, there is in this case a definite record by Albategnius (al-Battānī, c.850-929). It is the first in time of four eclipses (two solar and two lunar) between 883 and 901 mentioned in Ch.30 of his Astronomy and used by European astronomers since at least Regiomontanus. Albategnius gives information about this eclipse for ar-Raqqa (latinized Arracta), whose latitute and longitude as given in Newton 1970, 25, namely 36°.0N. and 39°.0E., will suffice for most purposes. The date is given as Tammūz (July) 23 of the year 1194 of the Era of Alexander the Great (Tārīk Dhū'l-qarnaini), by non-Arabs usually called the Seleucid Era (julianized Syro-Macedonian form, running from autumn to autumn in Syria, beginning as from the autumn of 312 BC). Dates of the eclipses of Albategnius in the Era of Nabonassar and in the Era from the death of Alexander in 323 BC abound in chronological literature. The modern edition and Latin translation of Albategnius is that of Nallino (1903, etc.).

Albategnius gives the magnitude as a little more than $\frac{5}{6}$ and says that mid-eclipse occurred a little later than 8 equal hours after noon. Oppolzer's time yields 7.42 p.m. mean time, or 7.36 p.m. apparent time, at ar-Raqqa. The difference from the time given by Albategnius is just tolerable. Writers in the pioneering half-century (1583-1632) of eclipse chronology noticed no discrepancy, because inaccurate astronomy (European full moon calculated up to about an hour too early) was compensated by inaccurate geography (Arracta imagined much too far East). The two errors are related, since times of lunar eclipses then provided a good way of finding longitudes. See Bunting 1590, 402, Calvisius 1605, 743; 1620, 568, Petavius 1627, 868, and Lansberg 1632 (eclipse M14; Opera Omnia, 1663, 54-5), reckoning European time on the meridians of Königsberg, Leipzig, Paris, and Goes (Holland) respectively. More modern discussions occur in Halley 1693, 918, Struyck 1740, 86, 144, Nallino 1903, 56f, 226f, and especially in Newton 1970 (146, 149, 215, 229).

885 S.885 June 16 NEAR TOTAL SOLAR IN SCOTLAND AND N. IRELAND

The only records known to us come from Scotland and Ireland. This is not surprising. Schroeter shows the band of totality, which runs from SW to NE near the British Isles, as skirting the North-West coast of Ireland, i.e. the extreme North-West parts of County Mayo and County

Donegal (soon after 9 a.m. local apparent time) and continuing via Northern Scotland and the Shetlands (before 10 a.m.), Bergen, and just north of Oslo, to Archangel, etc.). The band of totality given by J. Maguire, MN 45, 1885, 400 (see also 46, 1885, 25) includes a sizable part of North-West Ireland, perhaps a third of the whole island (between about 9.0 and 9.20 a.m. local apparent time), and most of Scotland (between about 9.20 and 9.45). O'Connor 1952 finds magnitude 0.95 at 9.06 a.m. local apparent time in Central Ireland. The equation of time is negligible. The eclipse is discussed in Newton 1972 (Ch.VII).

The only record which mentions the day comes from Scotland. The Chronicle of the Kings of Scotland, version A (W.F. Skene, Chronicles of the Picts and Scots, Edin. 1867, 9) says that in the ninth year of Eochaid [Eocha], king of the Britons [of Strathclyde], "on the very day of Cyricus [St. Cyr], an eclipse of the Sun occurred". The day mentioned is June 16, and it fixes the year (885) which might otherwise be uncertain; see Anderson 1922, 363-4.

The eclipse is also mentioned, without month or day, in the Annals of Ulster (under AU 884 = AD 885), 1, 1887, 402: "An eclipse of the Sun, and stars were seen in the sky". In B. MacCarthy's index (AU 4, 1901, 140) the date 885 June 16 is correct, but the time 10 a.m. perhaps refers to geocentric conjunction, given as 10.20 UT in Oppolzer and in Goldstine 1973, and as 10.26 UT in Schroeter 1923. In the CS (1866, 168-9) we find under 885 the same eclipse entry as that of AU. In his Introduction to the Annals of the Four Masters, 1, Dublin 1851, xlix, J. O'Donovan correctly recognizes 885, not 884, as the true AD year of the eclipse.

887 S.887 Oct.20 FALSE IDENTIFICATION

A false identification of Byzantine records of a solar eclipse fairly early in the reign of Leo VI (886-912). The identification was doubted in relation to Cedrenus by the revisers of Calvisius; they found magnitude only about 0.4 at Constantinople, see Calv.1620, 569; 1650, 676, etc. Not in Calv.1605, 744, since Calvisius in his lifetime referred only to Albategnius and S.891 Aug.8 (q.v.). S.887 Oct.20 is the first of six alternative identifications of the Byzantine records rejected by Ginz.1883, 673-5 in favour of S.891 Aug.8. J.B. Chabot accepted S.887 [Oct.20] without demur, in relation to Leo Grammaticus; see Chabot's edition of Michael the Syrian, III, 1963, 119. It is true that several of the Byzantine eclipse accounts occur after events in the second year of Leo's sole reign, but S.887 Oct.20 was of much too small a magnitude in the Byzantine domains to account for the appearance of stars. See under S.891 Aug.8 for references to Symeon the Logothete, Theophanes Continuatus and Leo Grammaticus, as well as the later Cedrenus and Glycas. The eclipse was total in parts of West and East Africa. Other false identifications are noted in our check-list.

891 S.891 Aug.8 (Sun.) ANNULAR SOLAR DESCRIBED IN EUROPE AND ASIA

An annular eclipse. The band of annularity in Schroeter 1923 includes Brest (about 8.30 a.m.), Tours and Orleans (about 9 a.m.), Venice (about 10 a.m.), Gallipoli (about 11.30 a.m.), and Damascus (about 12.40 p.m.), the times being local apparent times; the noon point is in southern Asia Minor (Pisidia) and the eclipse ended at sunset in the Indian Ocean. We shall discuss in turn records from (i) Syria, (ii) Byzantium, (iii) Western Europe (Newton 1979, 372 accepts Salzburg as the place).

(i) al-Battānī (Albategnius) observed a partial eclipse at ar-Raqqa (Arracta, about 39°.0E., 36°.0N.) on Sunday, SE 1202 Ab 8 = AD 891 Aug.8 (see under M.883 July 23). He observed magnitude more than two thirds (probably area, not diameter, according to Newton 1970, 149) at one seasonal hour (about 1⅛ equinoctial hours, i.e. $1^h 7\frac{1}{2}^m$) after (apparent) noon. From Schroeter one may estimate for ar-Raqqa a magnitude about 0.9 (of diameter) somewhat less than one equinoctial hour after apparent noon. The agreement is reasonable.

This observation is discussed in Newton 1970 (25, 146, 241-2, 247) with reference to the modern Latin translation in Nallino 1903. Discussion goes back at least to Regiomontanus. We ourselves have noted mentions in Scal.1598, 402; 1629, 428, Bunting 1590, No.59, Calv.1605, 745, Petav.1627, 869, Ricc.1653, 370, Tycho-Curt.1666, xxxiii, Halley 1693, 917, Struyck 1740, 85, 144, and Nallino 1903, 56f, 226f.

A further Syrian observation is probably the source for Michael the Syrian's (fl. c.1195) report discussed by Newton 1979, 390. Michael says (ed. J-B. Chabot, III, 119): "... there was an eclipse of the sun at midday, so that the stars could be seen in all the sphere of the heavens". Newton points out that Michael specified the year incorrectly as 888/889 and that, for an eclipse that could not have exceeded 0.95, the statement about the stars is exaggerated.

(ii) The Byzantine records are at least six in number, and have been discussed in Ginz.1883, 673-5 and Newt.1972, Ch.XIV. The identification S.891 Aug.8 is certain, but the records are conflicting. Ginz.1883 rightly rejects six alternative identifications (S.887 Oct.20, S.888 Apr.15, S.889 Apr.4, S.889 Sept.28, S.890 Aug.19, S.892 Feb.2), the first of which we have included in our list. S.887 Oct.20 had already been proposed in Calv.1620, 569; 1650, 676, etc., and S.889 Apr.4 in Struyck 1740, 144.

The Byzantine post-Theophanic writings (AD 813 onwards) are complex, see Camb. Med. Hist. 4(2), 1967, 229-30, 235-6 (F. Dölger). It would lead us too far to set out what each author said; some idea may be gained from Ginz.1883 and Newt.1972, and we shall give some Byzantine references below. The broad Byzantine story (there are variants) is that a considerable eclipse (always "stars appeared", sometimes "night") occurred, usually at, or in, the sixth hour of the day. Symeon's Annals (10th C) have "in the third hour"; Leo Grammaticus appears to substitute a duration, "for six hours", which is fanciful and contributes nothing to identification. The eclipse usually comes after a revolt which several authors put in the second year of Leo VI (886-912), who is known to have become sole ruler on the death of Basil I (886 March 1); Symeon especially puts both revolt and solar eclipse in the brief annal for the second year of Leo. The revolt was that of Hagio (Egio, etc.) in the Longobardic army in Southern Italy; Hagio defeated Leo's commander, Constantine, Controller of the Table (ho tēs trapezēs), who was slain.

The trouble is that while S.891 Aug.8 was large at Constantinople (magnitude not less than 0.9) and occurred at and in the sixth hour (compare about 11.45 a.m. local apparent time there in Schroeter), it fell in about the sixth year of Leo. S.887 Oct.20 (q.v.) occurred in about the second year of Leo, but, besides being an afternoon eclipse at Constantinople, was not nearly large enough there to account for "stars", etc. It is possible that one writer transferred a large eclipse in Leo's sixth year to Leo's second year for literary effect, and so misled other writers. It should be added that the late author Glycas puts the eclipse in a different context (Leo's sending of Abbot Santabaren into exile). The possibility that this considerable eclipse has been associated with events maybe several years removed destroys its value for purposes of Byzantine chronology.

We give references to four 10th century sources (not necessarily in chronological order) and two later ones. In each case we give a reference to the Bonn Corpus, and to Migne, Patrol. Gr., which in these cases provides post-Bonn reprints of pre-Bonn editions. BTC denotes the Bonn volume containing Theophanes Continuatus ... Symeon ... George, ed. I. Bekker, Bonn 1838.

(a) Symeon Magister et Logotheta, Chronographia (Annales), BTC 701; PG 109, Paris 1863, col.763-4. (Early 10th century.)

(b) Georgius Monachus (cont. by Symeon?), Bivi tōn neōn basileōn (Vitae imp. recent.), BTC 852; PG 109, 1863, col.913-4. Also in Georgius Monachus Hamartolus (cont. by Symeon), Chronicon syntomon

(Chronicon breve), PG 110, 1863, col.1093-4; not in C. de Boor's critical edition of George the Monk's Chronicle, which ends about AD 842 and omits continuations.

(c) Theophanes Continuatus, Book VI (by Theodore Daphnopates, see Camb. Med. Hist. 4(2), 1967, 230), BTC 356. PG 109, 1863, col.373-4. In the portion on the reign of Leo, the 'Leonis Imperatoris Imperium' of Newton 1972.

(d) Leo Grammaticus, Chronographia, ed. I. Bekker, Bonn 1842, 265-6. PG 108, 1861, col.1097-1100. (11th century.)

(e) Georgius Cedrenus, Synopsis Historiōn (Compendium Historiarum), ed. I. Bekker, 2, Bonn 1839, 252-3. PG 121, 1894, col.1139-40. (11th century.)

(f) Michael Glycas, Biblos Chronikē (Annales), ed. I. Bekker, Bonn 1836, 553. PG 158, 1866, col.553-4. (12th century.)

(iii) The only detailed West European reference we know is spoiled by an unfortunate error. In the Gesta abbatum S. Bertini Sithiensium (Acts of the abbots of Saint-Bertin at Saint-Omer), by Folcwin the Deacon (SS 13, 1881, 623), we find "an eclipse of the Sun occurred on 18.Kal.Sept. (= Aug.15), at the second hour; and great drought in May, June and July; and a comet appeared". Thus the eclipse is dated one week too late; perhaps at some stage in transmission an ecclesiastical description of the Sunday in question has been converted erroneously to day of the month. Mid-eclipse at Saint-Omer seems from Schroeter to have been about 9.05 a.m. local apparent time, so that "second hour" is acceptable only if it refers to an early stage of the onset of the eclipse. As the magnitude at Saint-Omer was about 0.93, it is not surprising to find this record. There are now very few primary West European chronicles. "891. Comet. Eclipse of the Sun." occurs in some manuscripts of the third St. Gall continuation of the Alamannic Annals (SS 1, 1826, 52); "891. Comets and eclipse of the Sun." occurs in the Annales Corbeienses (of Corvey, near Minden, Germany; SS 3, 1839, 3); while "891. Comet. Eclipse." occurs in the third part of the Annales Laubacenses (of Lobbes, near Charleroi, Belgium; SS 1, 1826, 52). The West European records are discussed in Newt.1972, 331 and passim, who now accepts 1979, 354, 372, a record from Ann. Juvavenses "The sun was obscured from the third to the sixth hour" as a Salzburg record of reliability 0.5 in his scale.

891 M.891 Aug.23 SPURIOUS RECORD

Calvisius 1605, 745 mentioned a lunar eclipse as observed by Albategnius at the first morning hour of AD 891 Aug.23. This allusion was rightly omitted in the posthumous editions of Calvisius. The small eclipse (Oppolzer magnitude only 0.07) occurred around $10^h$UT, thus in full daylight in Syria, where it was consequently invisible (Sun well above the horizon, Moon well below the horizon). If Albategnius really did mention the eclipse, it can only have been calculated, not observed.

c.893 NON-ECLIPSE IN IRAN

The Sun failed for three hours in 894 (recte 893: as it was before the earthquake) according to the Persian Chronicles of Hamza of Ispahan. This has been ascribed to a solar eclipse of 893 Dec.3 (e.g. in Muralt). Although there had been a total eclipse in N.W. India in 892 Feb.2, this would seem to be due to haze, as "In the evening a black wind began to blow". These extracts sent by Miss C.M. Botley are from the Latin edition of the Annals ed. J.M.E. Gottwaldt, Leipzig, 1844 and 1848 (cf. D.M. Dunlop, Arab Civilizations to AD 1500, 1971, 114ff.).

893 M.893 Dec.26 (p.m.) TOTAL RECORDED IN ARMENIA

The Islamic-dated records of (c), M.893 Dec.26 (p.m.), are discussed in Ginzel 1887, 713-4. There is a contemporary source, namely at-Tabarī (AD 839-923). According to Ginzel, Tabarī (iii/4, 2129) mentions the observation as made at Dabīl, and gives the date as AH 281 Shawwāl 14, which transforms to AD 894 Dec.17; but on 894 Dec.16 (q.v.) there was only a very small lunar eclipse, invisible in Iran. The location of Dabīl is not Ardabil in Azerbaijan (38.2°N., 48.3°E.) as Ginzel and others have assumed, Dabīl being the Arabic for Dvin (Duvin) in Armenia, S.W. of Lake Sevan (40°N., 44°E.). Others have ascribed it to 'Outer India'. Misinterpretations of the date and place are discussed by N.N. Ambraseys and C.P. Melville in A History of Persian Earthquakes (Cambridge University Press, 1982, 175). This eclipse and that of 901 were remembered because one or both were followed by a severe earthquake.

Elias 1910, 117 gives AH 280 Shawwāl 14, which transforms to AD 893 Dec.27, the day starting at sunset on 893 Dec.26.

Ginzel makes the middle of the total eclipse be at $10^h54^m.5$ p.m., Ardabil mean time. Tabarī apparently describes the report of the Shawwāl (tenth month) lunar eclipse as arriving [at Bagdad?] from Dabīl in the twelfth month.

## 893-4 AMBIGUOUS ECLIPSE OF ST. GALL

Lunar eclipses of 893 and 894. Let us consider the following four eclipses of the Moon, the mid-eclipse times and magnitudes being taken from Oppolzer:

| | | UT | Mag. |
|---|---|---|---|
| (a) | M.893 Jan.6 (a.m.) | 5.45 | 1.24 |
| (b) | M.893 July 2-3 | 22.58 | 1.67 |
| (c) | M.893 Dec.26 (p.m.) | 19.30 | 1.26 |
| (d) | M.894 June 21-22 | 0.11 | 0.60 |

(c) is certainly mentioned in Islamic records (see below), but the European records we have encountered are weak.

Some manuscripts of the third St. Gall continuation of the Alamannic Annals (SS 1, 1826, 53) have, as two consecutive annals:

"893. Eclipse of the Moon. Arnulf in Moravia.
William killed. Engilscalch blinded. Robert killed.
Alamanni into Italy.

894. Eclipse of the Moon. Arnulf into Italy with army."

The year-numbers at this point appear reasonable (maximum error one unit), and at 891 are confirmed with certainty by the solar eclipse in that year. Assuming them to be correct, probably the 893 annal refers to (a) or (b), and the 894 annal to (c) or (d), since (c) falls in 894 in a year starting on Dec.25. The probability that the annalist's year-pair (893 and 894) is correct is greatly strengthened by the fact that the four eclipses (a) to (d) are the only lunar eclipses visible in Europe during the quadrennium 892-895. Thus even rather rough records provide a little confirmation of chronology.

## 894 S.894 June 7 TOTAL SOLAR IN EXTREME N. EUROPE NOT RECORDED

Schroeter's band of totality passes south of Iceland and includes the Faroes, Central Norway, Central Sweden, and then runs just south of Archangel to the Urals. O'Connor 1952 finds magnitude 0.80 at 9.32 a.m. in Central Ireland. However, we have encountered no Western record.

Some scientifically useful observations, available for the solar eclipses of 808 and 888, are used by Newton 1979, 81 and 168 ff., who plots the track of the latter eclipses.

808 S.808 July 27 PREDICTED AND OBSERVED AT CH'ANG-AN

The 'Collected Data' (Cohen and Newton 1984, 1 and 36) report that "His Highness" told the Prime Minister that an eclipse had been predicted for that date and "when it came to the date in question, it was just as he said - everything was verified!" The Emperor was apparently in the capital at the time, but the magnitude was only 0.4! It could have been observed in reflection.

879 S.879 Apr.25 FALSE REPORTS AND DATE

This 'eclipse' was described as _chi_ or _complete_ (_jih_ in Cohen and Newton), but this is not the date of an eclipse at all. The solar position given is correct for the time of year and Dr. Stephenson suggests an abortive prediction.

The annular eclipse of 873 July 28 had been central in extreme NE China, but that eclipse is not mentioned in the Chinese records.

888 S.888 Apr.15 FALSELY TERMED COMPLETE

Perhaps the term _chi_ (_jih_) was now being used loosely. This eclipse was termed complete, but Cohen and Newton (Table 13, p.36) point out that it was an annular eclipse in SE China and that its greatest magnitude even there was only about 0.91. A full discussion of these two eclipses is given by Cohen and Newton.

AD 819

Stories of Indian eclipses associated with the life of the Buddha are believed to be later pious invention (cf. discussion in D.J. Schove, JBAA 58, 1948, 178-190 and 202-204). However, an early solar eclipse is recorded in a year 500 at Valabhi Samvat (c.22°N, c.72°E) in Savrashtra, western India. The era is believed to have begun in AD 318/9 so that AD 819 June 26 is a plausible date. However, Dr. K.J. Virji, in his Ancient History of Sausashtra (Bombay, 1955, 112-114), points out that other dates are favoured by other authors.

The Maya were remarkably successful in predicting lunar eclipses in this century, the Full Moon bases of the Eclipse Table, using the new correlation, being the date 842 Mar.30. The New Moon bases of the same table (cf. Bricker & Bricker 1982) enabled them to predict solar eclipse dates as well, but most of these would never have been visible in Mesoamerica. They numbered the Moons of each eclipse half-year on a scale 1 to 5 or 1 to 6 as required, and paid special attention to the New Moon at the commencement of an eclipse moon. They were interested also in the connections of the expanded Venus table with visible lunar eclipses. They used in addition the 4 x 819-day soli-lunar cycle, some of their 'astronomical' dates yielding remainders of 816 when the Mayan Day Numbers were divided by 819. With our present limited ability at decipherment we cannot prove that the eclipses below were actually observed; but we here suggest that the Maya were wise after the event and that, after observing a lunar eclipse, they commemorated the date of the preceding eclipse New Moon.

837 July 6 SUNRISE SOLAR ECLIPSE

This sunrise eclipse had a magnitude of 0.8 at Tikal, and at Yaxchilan a Maya astronomical date (9.15.19.14.14 or MDN 1411 134) connected with the 819-day cycle (noted by Thompson 1950, 213) converts with the new correlation to 837 July 5. This date is connected by the same cycle to the 846 dates discussed below.

842 Mar.15 BEGINNING OF ECLIPSE MOON (BASE A)

This is the date 9.16.4.10.8, 12 Lamat or the first New Moon base of the Eclipse Table in the Dresden Codex. At this date there was no solar eclipse anywhere on Earth (although there was a 'near miss'), but the additions specified by the Maya to the two New Moon bases (A and C) allow prediction of all solar eclipses, anywhere on Earth, through the next 33 years (letter from Professor H.M. Bricker 21 March 1983). This is equally true if the conventional date (AD 755) is used.

842 Mar.30 LUNAR ECLIPSE BASE (BASE B)

This date 9.16.4.11.3, 1 Akbal is the Full Moon base of the Eclipse Table and was visible on both sides of the Atlantic. The additions specified by the Maya lead to successful predictions of 46 out of 50 lunar eclipses in the 33-year period.

842 Apr.14 END OF ECLIPSE MOON (BASE C)

This date 9.16.4.11.18, 3 Etznab is the second New Moon base and corresponds to an overall partial solar eclipse which would not have been visible to the Maya.

846 June/July PREDICTED ECLIPSES

The Quirigua (F) date 9.16.8.16.10 (846 June 25) was connected by Thompson 1950, 213 with the 837 solar eclipse date. Although the solar eclipse two days later (June 27) was invisible, the lunar eclipse of July 12 would have been total and (weather permitting) visible.

847 June/July ECLIPSE MOON

A Maya date 9.16.8.16.10 (846 June 24) is one of a set that Teeple, 1931, 92 supposed would mark the beginning of eclipse half-years. In each case his dates now prove to be about 15 days before visible lunar eclipses (July 2), but the date is two days off the (invisible) solar eclipse date of June 16. At Quirigua an eclipse glyph confirms the association.

847 S Dec.11 IMPORTANT SUNRISE SOLAR ECLIPSE

This eclipse was nearly total (0.96) but it occurred at sunrise at Tikal on a date that the Maya would have termed 9.16.10.7.9. No record of this date is known to us, but the name of the ruler inaugurated at the so-called Astronomical Congress of 849 at Copan has been translated New-Sun-at-Horizon (cf. J.B. Carlson, Copan Altar Q: The Maya Astronomical Congress of AD 763 ? in ed. A.F. Aveni, Native American Astronomy, 1977, University of Texas Press, Austin and London, 101-109). A Moon sign is associated with the date.

849 M May 11 and Nov.4 LARGE VISIBLE ECLIPSES

These two visible eclipses are 15 days after two further Eclipse New Moons commemorated by dates (9.16.11.14.7 and 9.16.12.5.4 discussed by Teeple, op.cit. p.59). Furthermore the Astronomical Congress pictured on Altar Q at Copan is dated two days earlier than the lunar eclipse.

850 M Oct.24 VISIBLE ECLIPSE

The New Moon dates that so regularly precede visible eclipses by 15 days were presumably engraved <u>after</u> the observation of the eclipses as reliable indicators of Eclipse New Moons. In this case Stela D at Quirigua contains the date (with a Moon sign) 9.16.13.4.17 which converts to 16 days before the lunar eclipse and thus 1 day before the Oppolzer solar eclipse (a partial one visible to the Maya). This date is correctly predicted by the expansion of the Eclipse Table using the second base (cf. Kelley 1977, Table 5.1).

857 M June 11 IMPORTANT VISIBLE ECLIPSE

Fifteen days before this eclipse brings us to the date 9.17.0.0.0 (the date of an invisible solar eclipse), a round-figure date at Piedras Negras and elsewhere but one with glyphs that suggest comparison with the Eclipse Table date of 842 Mar.15 (Kelley 1977, 63). This date is correctly predicted by the Eclipse Table as No.32 in the expansion of the 1 Akbal base.
(cf. Kelley 1977, Table 5.1.)

876 Nov./Dec. INVISIBLE ECLIPSES

The date 9.17.19.13.16 which converts to 876 Nov.19 has glyphs which suggest an eclipse and it is indeed the date of a solar eclipse. That eclipse was total in South America; 16 days later on Dec.5 there was a lunar eclipse, but it was equally invisible to the Maya.

## THE TENTH CENTURY

### EGYPTIAN OBSERVATIONS, CLIMATIC CHANGES AND EUROPEAN DOCUMENTATION

Islamic science continued to develop not only at Bagdad and in Syria (e.g. ar-Raqqa or Ray on the Euphrates) where observations were made from AD 877 to 919 or 933), but also in Spain and finally in Egypt at Cairo (from AD 977 to 1007).

The cold climate of the early tenth century (cf. Schove and Fairbridge 1985) coincided with a Dark Age as far as European chronicles are concerned, but in the warm second half of the century chronicles in the Holy Roman Empire became significant, this period being described as the Ottonian Renaissance. However, most of Europe now participated, the Greek chronicles of Constantinople and the annals of the British Isles and France being especially important.

901 S.901 Jan.23 PARTIAL SOLAR IN SYRIA AND ARMENIA

This is the third of the four eclipses long known to have been observed by Albategnius in Syria (M.883 July 23 and S.891 Aug.8 have already been treated). Totality occurred in Russia and Siberia, but the magnitude in Syria estimated from Schroeter 1923 is 0.7. We have encountered no record except from Syria and Armenia. The civil date was Kānūn II (January) 23 in the year 1212 of, effectively, the Seleucid Era (julianized Syro-Macedonian form, running from autumn to autumn in Syria). The Sun, according to Albategnius, was a little more than half obscured at 8.20 a.m. at Antioch (36°.2E, 36°.2N), and "therefore" less than two-thirds obscured at 8.30 a.m. at ar-Raqqa (38°.9E, 36°.0N).

A recent important discussion appears in Newton 1970 (146, 149, 242, 247); see also Nallino 1903 (56f, 228f). The former, p.149, considers that the fractions of the Sun said to have been obscured probably relate to area.

Earlier discussions occur in Halley 1693 and Struyck 1740, 85, and, less importantly, in Riccioli 1653, 370 and Tycho-Curtius 1666, xxxiii.

The solar eclipse appears to have been associated, like the lunar eclipse of 893, with an earthquake in Dabil (= Dvin in Armenia). However, our earliest source, Ibn al-Jauzi (V1, 27) numbers the casualties as 150,000 in both cases, and subsequent sources are confused about the location as well (see N.N. Ambraseys and C.P. Melville, A History of Persian Earthquakes, 1982, Cambridge, 175 n.33).

901 S.901 Aug.2-3 TOTAL IN SYRIA

The last of the four Syrian eclipses of Albategnius occurred in SE 1212 Ab (August). He gave it as very little short of total about 3.20 a.m. at Antioch and about 3.35 a.m. at ar-Raqqa. These times imply mid-totality about 1 a.m. at Greenwich, in only moderate agreement with UT $0^h 28^m$ (Oppolz.), $0^h 31^m$ (Schroe.). The equation of time was small (about $5^m$, mean time minus apparent time).

For modern discussion see Newton 1970 (146, 149, 215, 220f, 229, 235) and Nallino 1903, 57f, 231f. Earlier discussions appear in Bunting 1590, 406, Scal.1598, 402; 1629, 428a, Calv.1605, 748; 1620, 573; 1650, 680, Petav.1627, 871, Lansberg 1632 (eclipse M15; Opera Omnia 1663, 55), Ricc.1653, 370, T-C 1666, xxxiii, Halley 1693 and Struyck 1740, 144. Riccioli also refers to Regiomontanus, Epit. Almag., lib.5, prop.21.

903 S.903 June 27 (Mon.) TOTAL SOLAR IN S. SPAIN

A total eclipse whose central line (according to both Oppolzer and Schroeter) ended at sunset in Algeria after passing over Algarve and the Straits of Gibraltar. Stephenson and Muller mention a report (defective in detail) of visibility at Córdoba; see Muller & Stephenson 1975, 483 and Muller 1975, 8.48. In Guráieb's Spanish translation of Ibn Hayyān (Cuadernos de Hist. de España, 29-30, 1959, 351; see under S.912 June 17), correctly under AH 290: "In this year occurred a solar eclipse seen at and near Córdoba, on Wednesday, a day before the end of the month D'ul Huŷat". The eclipse really happened before the sunset which separated Islamic Monday and Tuesday, and near the end of the month Rajab

904 TOTAL IN CENTRAL ENGLAND

M.904 May 31-June 1 or M.904 Nov.25-26

One or other of these total eclipses is undoubtedly referred to in the Mercian Register, a separate component (for AD 902-24) of the ASC, best seen in MSS. B and C, where it is inserted as a solid block (after the annal for 915), rather than in MS. D, where some attempt was made to weave the information into the ASC annalistic pattern. See F.T. Wainwright, The Chronology of the Mercian Register, EHR 60, 1945, 385-92. The record says "904. In this year the Moon was eclipsed." The chronology of the Mercian Register is generally good.

C. Plummer wrote (Two of the Saxon Chronicles Parallel, 2, 1899, 117): "There was an eclipse of the Moon in 904, but as there were lunar eclipses also in 901, 902, 903, 905 and 907, this is not much help in fixing the chronology". In fact, the lunar eclipses in the years around 904 were all either invisible or insignificant in England, whereas both total lunar eclipses of 904 itself would (assuming good weather) be striking. As the Mercian Register uses a midwinter year-beginning throughout its length, either eclipse fits the record. See Wainwright for fuller discussion.

The eclipse is also mentioned in later English writers, e.g. Florence of Worcester (under 905) and Henry of Huntingdon. That the 905 of Florence needs rectifying to 904 was recognized in Struyck 1740, 144 and in B. Thorpe's edition of Florence, 1848, 119.

906 S.906 April 26 NO GENUINE RECORD EVEN IN MIDDLE EAST

An annular eclipse. The central line starts in the Atlantic, runs via North Africa to a noon point in Libya, and proceeds via Asia Minor to an end in China. Schroeter shows a band of annularity passing over Lower Egypt (west of Alexandria), Cyprus, East Turkey, Tabriz (N.W. Iran), Baku and the Caspian; magnitude greater than 0.75 at Constantinople, and one may estimate about 0.82 at Jerusalem and about 0.85 at Bagdad.

Ginzel 1886, 979-980 gave this as the identification of a solar eclipse mentioned by Glycas, Bonn 1836, p.556; PG 158, 1866, col.555-558. This is a mis-identification. For Glycas mentions a solar eclipse in much the same context (Pantaleo, Samonas, Leo VI, Alexander) in which other authors mention a lunar eclipse, probably 908 Mar.20 (q.v.). It seems most likely that Glycas made a mistake.

908 M.908 Mar.20 ECLIPSE OF SAMONAS IN CONSTANTINOPLE

There is ample mention of a lunar eclipse, apparently this one, in Byzantine writers, e.g. those quoted under S.891 Aug.8, except that Glycas says solar eclipse, and Zonaras (lunar) may be added.

What the Byzantine writers mention is an eclipse, lunar except in Glycas, towards the end of the reign of Leo VI (886-912). Pantaleo, metropolitan of Synnada, interpreted the eclipse, in different terms (both hinting at misfortune to the second in rank) to his friend Samonas (Leo's chief minister) and to Leo VI himself. The eclipse was in fact followed by the fall of Samonas, the death of Leo VI (912 May 11) and that of his brother and successor Alexander (913 June 6). Using the extant life of Samonas, his deposition was dated by Jenkins 13 June 908 (see R.J.H. Jenkins, The Flight of Samonas, Speculum, 23, 1948, 235); thus the eclipse cannot be either 912 Jan.7 or 911 Jan.17 and Jenkins' suggestion of 908 Mar.20 is evidently correct.

The Byzantine references are (see AD 891 for fuller details of (a) to (f) editions):

(a) Symeon (writing c.912); PG 109, 1863, col.773-4 (Primary source).

(b) George Monachus (continuation of early 10th C.), Vitae imp. recent., BTC 1838, 869-70; 1863, col.931-2, Chron. breve, 1863, col.1121-2.

(c) Theoph. Cont. (late 10th C.), 1838, 376; 1863, col.393-4.

(d) Leo Gramm. (early 11th C.), 1842, 284; 1861, col.1117-8.

(e) Cedrenus (late 11th C.), 2, 1839, 272; 1894, col.1159-62.

(f) Glycas (late 12th C.) (solar eclipse, sic) 1836, 556; PG 158, 1866, col.555-8.

(g) Zonaras (early 12th C.), Epit. Hist., 3, 1897, 453-4; Teubner ed., 4, 1871, 47; 5, 1874, cxv; 6, 1875, 150; PG 135, 1887, col.75-6.

912 M.912 Jan.7 (a.m.) DARK YEAR IN IRELAND

A total eclipse, magnitude 1.26 according to Oppolz. and Schroe., with mid-eclipse at UT 4.19 a.m. (Oppolz.), 4.25 a.m. (Schroe.). It was perhaps referred to in Ireland.

Calv.1620, 576; 1650, 682, etc., i.e. the posthumous editions, claimed that this eclipse, in the 13th year of Edward [the Elder, who succeeded Alfred the Great, probably in 899 Oct.], was recorded by "English writers". The information is more or less copied in T-C 1666, xxxiii, Struyck 1740, 144, and Johnson 1874, 42. We have not found any English source.

The Annals of Ulster, 1, 424-7, appear to give "a rainy and dark year" under both AU 911 (AD 912) and AU 912 (AD 913). "Dark year" has a tendency to indicate a solar eclipse, but Anderson 1922, 403 comments "in 912 there were eclipses of both moon and sun". The relevant ones were M.912 Jan.7 and S.912 June 17 (q.v.). No known volcanic eruption fits (cf. AD 620s). Moreover the magnitude of the solar eclipse (below) was only 0.9 in Holland and could not have caused darkness in Ireland (J. Meeus, letter 1981).

912 S.912 June 17 (Wed.) TOTAL SOLAR AT SUNSET IN SPAIN
ANOTHER 'DARK YEAR' IN IRELAND?

A total eclipse. The final stretch of the band of totality shown in Schroe.1923 passes over Portugal and South-West Spain, and likewise ends in the Mediterranean.

There are records from Hispano-Arabic sources, for information on the first of which we are indebted to Professor R.L. Harvey, of King's College, University of London. We have encountered no convincing record from Hispano-Christian sources.

The normal Islamic date corresponding with AD 912 June 17 is AH 299 Shawwāl 28 (Wednesday). Ibn Adarī has (see, for example, Fagnan 2, 1904, 247) "In this year, on Wednesday 28 Shawwāl (18 juin 912) the Sun was totally eclipsed before sunset; the stars appeared, and most of the criers attached to the mosques hastened to call the Sunset Prayer, which was in fact said. But the Sun reappeared and brought back light, then really set".

A very similar account of the same eclipse is one of the select few Western solar eclipse reports found useful for the exacting purposes of Muller & Stephenson 1975 and Muller 1975 (on page A3.2, for 916 read 912) and cf. Newton 1979, 354, 372. Their source is Ibn Hayyān (988-1076),

Al Muqtabis, Vol.III, in J.E. Guráieb's Spanish translation of M.M. Antuña's Arabic text (Chronique du règne du calife umaiyade Abd Allah à Cordoue; Paris, Geuthner, 1937). Guráieb's translation appeared, in many instalments, in Cuadernos de Historia de España, 13-32, Buenos Aires, 1950-60, a sequence which includes several double volumes with single pagination. The report of S.912 June 17 (Wednesday) thus occurs in 31-32 (1960), 321, under AH 299; Stephenson and Muller take the inference of totality to relate to Córdoba. ʿAbdallah ruled at Córdoba from 888 until his death on 912 Oct.15 (Camb. Med. Hist. 3, 1922, 418-9). See also under S.903 June 27.

A similar account passed into modern histories, for example, the History of the Domination of the Arabs in Spain, by the nineteenth century Spanish historian J.A. Conde. This work, which is largely founded on Arabic sources, says Shawwāl 29, but also Wednesday, and is referring to the same eclipse. See Conde 1840, 177; 1854, 362.

Further details of editions and translations of Ibn Adarī and Conde (sometimes given incorrectly as Condé) will be found in our Bibliography.

There are possible poetic allusions to the same eclipse, see J.T. Monroe, 'The Historical Arjuza of Ibn ʿAbd Rabbihi, a tenth-century Hispano-Arabic epic poem', J. Amer. Oriental Soc., 91(1), 1971, 67-95 (verses 14 and 22 on p.81).

Greatest magnitude in Central Ireland is given as 0.92 in O'Connor 1952 and appears about 0.9 from the map in Schroe.1923. The eclipse may be referred to in AU, see under M.912 Jan.7.

913 S.913 June 7 TOTAL SOLAR IN EGYPT BUT NO RECORD

A total eclipse, with central line running from the equatorial Atlantic via Egypt/Sudan, a noon point in the Red Sea, and Arabia, to an end in the region of Indonesia. We have encountered no Western record.

916 S.916 Apr.5 ANNULAR IN NEAR EAST BUT NO RECORD

Schroe.1923 shows the band of annularity passing over Lower Egypt, Jerusalem, and the Southern Caspian. No genuine record has been found (see 939).

917 S.917 Sept.19 ANNULAR IN BAGDAD BUT NO RECORD

Both Oppolz. and Schroe. make annularity to have started at sunrise in Northern Iraq, with large magnitude at Bagdad. Surprisingly, we have encountered no Islamic reference.

921 M.921 Dec.17 (Mon. p.m.) TOTAL IN IRELAND

A total eclipse, with magnitude about 1.22, and mid-eclipse at UT $18^h51^m$ (Oppolz.), $18^h57^m$ (Schroe.). The Annals of Ulster, 1, 1887, 440-3, have, under AU 920 = AD 921, "An eclipse of the Moon on 15.Kal.Jan. (Dec.18), a Tuesday, in the first hour of the night". Ecclesiastical usage fully allows the association of the Monday evening eclipse with Tuesday. Identification has never presented any problem, and is given correctly by, e.g., B. Mac Carthy in AU 4, Dublin 1901, 140.

923 M.923 June 1 (p.m.) PARTIAL IN BAGDAD

Oppolz. gives a partial eclipse at Bagdad, magnitude about 0.66, with middle and end of the eclipse at UT $17^h26^m$ and $18^h51^m$ respectively. This is one of the eclipses in the Ibn Yūnus manuscript, see Caussin 1799, 6; 1804, 112. The end was timed at Bagdad by noting that the altitude of $\alpha$ Cygni was 29°30'. Newcomb 1878 (§5, p.46) quoted Caussin 1804; Newcomb's reduction of the UT of the end of the eclipse gave $18^h56^m33^s$ observed, $18^h49^m49^s$ tabular. The eclipse has recently been discussed, again using Caussin 1804, in Newton 1970 (146, 149, 215, 229).

Schroeter's initial stretch of the band of totality runs roughly South-East from a start on the East coast of the Black Sea, and shows magnitude at Bagdad as greater than 0.75 (one may estimate about 0.78).

This is one of the Ibn Yūnus eclipses, see Caussin 1799, 71; 1804, 114, Newcomb 1878 (§5, p.46), Newton 1970 (146, 242, 247). The middle and end were observed at Bagdad by several astronomers in collaboration. It is reported that at mid-eclipse (greatest obscuration) the Sun was 8° high and the magnitude was 9 digits (i.e. 0.75); also that at the end the Sun was 20° high and the time was $2\frac{1}{5}$ seasonal hours after sunrise. As Newcomb 1878 unaccountably rejects the time of mid-eclipse, we may give a little detail.

Disregarding seconds, the reduction of the observed times gives

| | | |
|---|---|---|
| mid-eclipse | UT $4^h22^m$ (Newcomb), | $4^h19^m$ (Newton) |
| end | " $5^h33^m$ " , | $5^h29^m$ " |

If Newton's calculated semi-duration of $1^h1^m$ is correct, the observed semi-duration of about $1^h10^m$ is about $9^m$ too long. If one looks at the other Islamic eclipse between AD 866 and AD 977, the impression one gets from Newton is that in 923 mid-eclipse was observed slightly too early and the end slightly too late; in that case mid-eclipse was observed less than $10^m$ too early. On the other hand, Newcomb found the end observed $17^m$ too late; in which case mid-eclipse was observed about $8^m$ too late. Newcomb may have had reason for saying (p.46) "Of course, the first observation can have no astronomical value", but the numerical value is quite likely less than $10^m$ in error. Newton does welcome justice to the Islamic astronomers by retaining their time of mid-eclipse.

925 M.925 Apr. 11 (p.m.) TOTAL IN BAGDAD

A total eclipse, maximum magnitude about 1.09 (Schroe.). This is one of the Ibn Yūnus eclipses; the beginning and end were observed at Bagdad in the early evening, see Caussin 1799, 7; 1804, 116, Newcomb 1878 (§5, p.46f), Newton 1970 (146, 150, 229). A wildly erroneous stellar altitude at the beginning has to bc discarded, see Newcomb, 46, Newton, 150, but enough sound information remains to provide good observed times of both beginning and end of the eclipse. Disregarding seconds, all reductions give observed beginning and end at about UT $16^h23^m$ and $19^h49^m$ respectively, both in good agreement with tabular values, such as

$16^h27^m$ and $19^h46^m$ (Newcomb),
$16^h25^m$ and $19^h49^m$ (Oppolzer),
$16^h30^m$ and $19^h54^m$ (Schroeter).

926 M.926 Apr. 1 (Sat. a.m.) TOTAL IN FRANCE

This eclipse of the Moon early on the kalends of April is clearly referred to in the Annals of Flodoard (Frodoard) of Rheims (e.g. SS 3, 1839, 376; various editions and translations). One MS. also says that the eclipse occurred on Easter Saturday (sabbato sancto paschae), which is correct, as Easter Sunday in 926 fell on April 2. One might think from Flodoard's description that the eclipse was only partial, but it was in fact total. The eclipse is also mentioned, though only vaguely, by the later writer Richer (e.g. SS 3, 1839, 583, or ed. and tr. R. Latouche, Tome I, Grenoble 1930, 102-3).

Mentions (sometimes with discussion) of the Flodoard account occur in Kepler, Astronomiae Pars Optica, 1604, 274 (Ges. Werke, 2, 1938, 239, 452), Calv. 1620, 580; 1650, 686, etc., Ricc.1653, 370, T-C 1666, xxxiii, Struyck 1740, 144, Johnson 1874, 42.

927 M.927 Sept.13-14 (Th.-Fri.) SMALL BUT OBSERVED IN BAGDAD

A small partial eclipse, observed during the later part of "Friday night" at Bagdad, according to the Ibn Yūnus list, see Caussin 1799, 7; 1804, 118, Newcomb 1878 (§5, p.47f), Newton 1970 (147, 150, 215, 229). Maximum magnitude about 0.3 was observed, compared with calculated 0.21 (Oppolz.), 0.22 (Newton). Times of beginning, middle and end were given in Caussin 1804, but only that of the beginning was confirmed by a stellar altitude (Sirius 31° high, east of the meridian). Disregarding seconds, reduction of this gives beginning at

UT $0^h51^m$ (Newcomb), $0^h55^m$ (Newton),

in fairly good agreement with calculated

UT $1^h4^m$ (Newcomb), $0^h59^m$ (Oppolzer).

Newcomb attributes the slight earliness to the keen eye of the young observer ("mon fils Aboulhassan") detecting the penumbra before the arrival of the umbra.

928 S.928 Aug.18 SMALL SOLAR OBSERVED IN BAGDAD

An annular eclipse, whose path starts at sunrise in East Africa and has noon point near Indonesia. According to the Ibn Yūnus account, at Bagdad the Sun rose with a little less than a quarter of its surface eclipsed, and its altitude at the end of the eclipse was 11°53'20". See Caussin 1799, 7; 1804, 120, Newcomb 1878 (§5, p.47f), Chambers 1902, 140-1, Newton 1970 (146, 150, 242, 247). Ignoring seconds, the observed time of the end reduces to UT $3^h29^m$ (Newcomb), $3^h34^m$ (Newton), compared with Newcomb's calculated tabular UT $3^h19^m$. The Sun was observed distinctly by reflection from a water surface. Fuller quotation from the records of the eclipses of 927/8 are available in Stephenson and Clark, 1978, 33.

929 M.929 Jan.27-28 TOTAL IN BAGDAD

A total lunar eclipse, magnitude 1.20 (Oppolz., Schroe.). The beginning was observed at Bagdad, but the time as it has reached us via Ibn Yūnus and Caussin is defective. See Caussin 1799, 7; 1804, 122, Newcomb 1878 (§5, p.47f), E.B. Knobel, MN 39, 1879, 339, and Newton 1970 (147, 229, 232).

The time of the beginning is given in two ways

(i) Arcturus was east of the meridian at altitude 18°. Ignoring seconds, this gives UT $20^h6^m$ (Newcomb), $20^h10^m$ (Newton, 229). But these are about an hour too early; the beginning was about $21^h4^m$ (Newcomb,

Oppolzer), $21^h8^m$ (Schroeter). Knobel, noting that 18, 33 and 38 are easily confused in the Arabic script used, said "there can be little doubt that the real observed altitude was 33° or 38°, which has been erroneously rendered 18 by a transcriber or translator". Since rough calculation shows that changing 18° to 33° or 38° would make the observed time later by about $1^h.2$ or $1^h.6$ respectively, the first of these emendations (i.e. for 18° read 33°) would certainly reduce the error to a tolerable amount.

(ii) Unfortunately, the other way in which the time is specified supports the 'wrong' time. The beginning is said to have been observed about "5 unequal hours after sunset". Newton 1970, 232 rightly finds agreement "almost to the minute" with "altitude of Arcturus 18°". Our rough calculations give about UT $20^h9^m$.

Thus unless the observed time of the beginning was at some stage wrongly given in one mode of statement and then correctly converted to the other mode (see Newton 1970, 232), the Bagdad time has to be discarded for astronomical purposes, as it was by Newcomb and Newton. But the identification of the eclipse seems certain.

<u>932</u> M.932 Nov.16 (a.m.) NEAR TOTAL IN SYRIA AND BAGDAD

According to R.R. Newton (Mem.Roy.Astr.Soc. <u>76</u>(4), 1972, 114-5), al-Birūni in a work of AD 1025 described how al-Hashimi timed this eclipse at ar-Raqqa (= Ray) and calculated it for Bagdad, in order to try to find the longitude difference between the two places. The attempt was unsuccessful, as ar-Raqqa came out East instead of West of Bagdad! Oppolzer gives the eclipse as not quite total (magnitude 0.975), with mid-eclipse at UT $4^h3^m$ (about 7 a.m. Bagdad mean time).

932 S.932 Nov.30 SMALL SOLAR IN IRELAND NOT RECORDED

A total eclipse whose central line ended between Iceland and Ireland. O'Connor 1952 found greatest magnitude 0.63 at sunset at his central Ireland point, but we have encountered no Western record.

<u>933</u> M.933 Nov.4-5 TOTAL IN IRAQ

A total lunar eclipse, magnitude 1.425 (Oppolz., Schroe.). It is the last of the Bagdad series of eclipses listed by Ibn Yūnus; his later eclipses (AD 977-1004) were observed at Cairo. See Caussin 1799, 8;

1804, 124, Newcomb 1878 (§5, 47f), Newton 1970, 147, 229. The information concerns the time of the beginning. Arcturus was East of the meridian at an altitude of 15°. Ignoring seconds, this reduces to UT $1^h18^m$ (Newcomb), $1^h22^m$ (Newton, 229). The beginning was also $9^h56^m$ unequal (night) hours after sunset, which rough calculation reduces to UT $1^h14^m$. These times are in reasonable agreement with tabular values such as UT $1^h16^m$ (Newcomb), $1^h31^m$ (Oppolzer), $1^h32^m$ (Schroeter).

934 S.934 Apr.16? DARKNESS IN ENGLAND

The ASC has no entries for the year 936 but there seems to be a veiled reference associating the solar eclipse with the battle of Brunanburh of 937 (probably after 24 September). The poem describing the battle was in the portion 924-955 written up by the Third Scribe at Winchester after 955. The quotation that follows may merely mean 'from dawn to dusk' but the use of the word daennade, translated 'grew dark', is suggestive. The relevant quotation (ASC MS.A) is:

.... The field
Grew dark with the bold warriors after the sun,
That glorious luminary, God's bright candle,
Rose high in the morning above the horizon
Until the noble being of the Lord Eternal
Sank to its rest.

In the B, C and D versions the phrase bold warriors of the Parker Chronicle is changed to blood of men, perhaps in an attempt to make sense of the darkness.

The three-year discrepancy between eclipse and battle would possibly have been within the limits of poetic licence about 950. However, Dr. Dumville, Mr. Harrison and Professor H. Loyn are all doubtful about my suggestion (personal communications to DJS) and I have thus allotted only 1 point for identification. This is the only important British solar eclipse after AD 664 without documentation.

c.936 AURORA IN GERMANY

The annular solar eclipse of 934 April 16 has also been the usual identification of a phenomenon mentioned by Widukind of Corvey (d. circa 1004), Res gestae Saxonicae, Book II, Ch.32 (SS 3, 1839, 446, many other editions and some translations). Referring to portents (including the comet of Autumn 941) in the reign of Otto I, he says people were terrified since "indeed before the death of king Henry many omens occurred, such as that the brightness of the Sun out of doors in an unclouded sky appeared almost nil, but it streamed indoors, red as blood, through the windows of houses". This reads rather like an eclipse of a low Sun. Widukind of Corvey is a primary source for most of the reign of Otto I.

The accepted date for the death of Henry I, the Fowler, is 936 July 2. Oppolzer lists four other solar eclipses, all total or annular, between 934 April 16 and 936 July 2, but all are more or less irrelevant to Central Europe. S.934 April 16, on the other hand, was a late afternoon eclipse whose track of annularity in Schroeter runs right across Germany. On Schroeter's data, the Sun's altitude at greatest phase was some 15° over the Belgian coast, 9° in Northern Bavaria, and 6° in Moravia.

The fact that the phenomenon, whatever it was, is mentioned in a reference back from the reign of Otto I to the reign of Henry I has caused various later writers to have placed it, or to have been read as placing it, in the wrong year. See Sigebert of Gembloux in Belgium, under 937 (e.g. SS 6, 1844, 347, or Bouquet 8, 1871, 313); from him, the Annals of Waverley (ed. H.R. Luard, London 1864); Annalista Saxo (e.g. SS 6, 1844, 604); John Tritheim (Trithemius), Chronicon insigne monasterii Hirsaugensis, Basle 1559, 36. The last is a history of the Benedictine abbey at Hirsau, near Kepler's birthplace at Weil der Stadt (near Stuttgart).

Newton 1972 (97, 157, 226, 234-5) considers the solar phenomenon probably not an eclipse. It is impossible to prove or disprove that it was, and it is true that the eclipse we mention preceded Henry's death by over two years. But it seems to us that by about 941, and still more by about 968, when Widukind was writing, "before the death of king Henry" may have seemed a sufficient reminder of the approximate time of a striking event.

The identification S.934 April 16 was adopted in Calv.1620, 583; 1650, 688-9, T-C 1666, xxxiii, and Johnson 1874, 42, but rejected in Struyck 1740, 144. The identification is in any case too weak to be of any special use to astronomers and chronologists.

A sunspot maximum occurred about 937 (Schove 1983a, 156 and 371) and an aurora in 935/6 is a more likely explanation than the eclipse of 934.

936 M.936 Sept.3-4 BLOOD-MOON IN FRANCE

"936. King Henry dying at the same time, contention about the kingship arises between his sons; in the end, the sovereignty fell to his eldest son Otto. On 2.Non.Sept. [Sept.4] the Moon, 14 days old, was covered by the colour of blood, and was seen to illuminate the night very little." It is then stated that Pope John [XI] died, and Leo [VII] succeeded. (Annals of Flodcard (d.966), e.g. SS 3, 1839, 382, or Bouquet 8, 1871, 191.) The kings mentioned are Henry I (the Fowler) and Otto I (the Great). Henry died on 936 July 2, and the papal change occurred in 936 January.

The total eclipse of the Moon occurred entirely after midnight in Germany and North-Eastern France, and even at Greenwich the partial phase began only some 5 or 10 minutes before midnight. Thus the date given by Flodoard is correct.

939 S.939 July 19 (Fri.) TOTAL SOLAR IN THE MEDITERRANEAN

This eclipse is correctly dated in several sources, and is often mentioned as happening about the time of a great victory of Ramiro II (Ranimirus, R(h)adamirus, even Rameses), the Christian king of León, etc., over ʿAbd-ar-Rahmān III (Abderahamen, Abdaram, Addaram, etc.), the famous

Saracen leader in Spain, at Simancas in AD 939, probably between Monday, July 22 and Monday, Aug.5 inclusive (Aug.1 is the date given by E. Lévi-Provençal in Historia de España, dir. R. Menéndez Pidal, 4, Madrid 1950, 292). As the eclipse was twice discussed by Ginz. (1883, 1886), who listed many sources in his 1883 discussion, and moreover is much mentioned in Newt.1972 and 1979, 320, we may limit our account to salient points.

The band of totality in Schroe. runs through, or very near, Lisbon, Madrid, Barcelona, Nice, Rome, Bucarest, the Sea of Azov, and the North Caspian. Most sources which mention an hour say third hour of the day; the actual hour was the second in Spain and the third in Northern Italy.

Accurate sources mentioned in Ginzel 1883 include

Ann. Sangall. maiores (dicti Hepidanni) SS 1, 1826, 78

Chronicle of La Cava (Italy) Muratori 7, 1725, 961-2

which give about third hour and third to almost fifth. The latter reads like an eye-witness account, and the same description is given in the Annals of Monte Cassino, Italy (SS 3, 1839, 172), except under wrong year 938, though correct indiction 12. We would add to Ginzel's references a good account in Vatican MS. 1340, quoted in Liber Pontificalis, ed. L. Duchesne, 2nd ed., II, Paris 1955, 244, under Stephan VIII (939-42). Converting from Roman numerals, it reads: "In his time, in the year of the incarnation 939, the Sun was obscured from the second to the third hour, July 19, twelfth indiction". A secondary record from Pisa is noted by Newton 1979, 320.

Several later sources are more than a unit out in the year. Sigebert (SS 6, 1844, 348) and Annalista Saxo (SS 6, 1844, 605) have 944. See also under S.912 June 17 and S.916 April 5 in our list. The late Annales Blandinienses (of Blandigny near Ghent); e.g. SS 5, 1844, 25) have a solar eclipse at the third hour under 951 July 19 (an impossible eclipse date); according to Newt.1972, 237f this refers not to S.939 July 19, but to S.418 July 19. Such sources are clearly not primary ones.

Although its best records come from about the longitude of Italy, S.939 July 19 (Friday) is important in Spanish chronology, on account of its undoubted though elusive proximity in time to the battle(s?) of Simancas (etcetera?). The eclipse figures prominently in one of the investigations included in R. Dozy, Recherches sur l'histoire et la littérature de l'Espagne pendant le moyen âge. Ginz.1886 quotes the second (1860) edition; the edition available to us has been the third (Paris and Leiden, 1881), in which the discussion of the campaign of 939 occupies pages 156-70 of Tome I. We may notice two points.

(i) The expected Hispano-Christian source for this period is Sampiro of Astorga (d. 1041). He gives the date of a great victory as "II.feria, imminente festo Sanctorum Iusti et Pastoris"; as the feast of Saints Justus and Pastor is Aug.6, and as its eve, Aug.5, fell on a Monday in 939, the date seems clear. Sampiro does not mention the interval between eclipse and victory; indeed, it appears that he does not mention the eclipse at all. Ginz. quotes the relevant passage from Sampiro, including as usual, before the victory, "Tunc ostendit deus signum magnum in caelo et conversus est sol in tenebras in universo mundo per unam horam". But according to Dozy (Recherches 1, 1881, 158) this eclipse sentence is an interpolation; thus it seems safer to regard it as outside the Sampiro canon.

(ii) It is a known irony that the contemporary Arabic record of the Spanish eclipse and Christian victory occurs in the 'Golden Meadows' of Masʿudi (c.900-956), a much-travelled native of Bagdad who never saw Spain. Dozy quotes extensively from the 19th century edition and French translation of Barbier de Meynard; it is now possible to refer to Les Prairies d'Or, traduction française de Barbier de Meynard et Pavet de Courteille, revue et corrigée par Charles Pellat, Tome II, Paris 1965. On p.346 we find: "'Abd ar-Rahmān led ... The battle he fought with Ramiro ... took place in the month of Shawwāl in the year 327 ..., three days after the eclipse which was seen in that month". This clearly places a battle on AD 939 July 22 (AH 327 Shawwāl 1). The eclipse was actually on AH 327 Ramadān 28, but the new moon nearest to the beginning of Shawwāl is easily thought of as the new moon of Shawwāl.

The same 3-day interval also appears in modern histories. We have noted it in Conde 1840, 209, 211; 1854, 424. But this author, whose work first appeared in 1820-21, stated his sources only generally in his Prologue, and he carries little authority after the devastating criticism of Dozy.

Closer inspection makes (i) and (ii) seem at odds, in that the Aug.5 of Sampiro appears to belong to an earlier stage of the campaign than does the July 22 of Masʿudi. Presumably ordinary non-astronomical records clear up the paradox, since Lévi-Provençal (loc.cit.) states categorically that the battle of Simancas occurred on exactly 939 Aug.1 (AH 327 Shawwāl 11).

940 M.940 Dec.17-18 (Th.-Fri.) TOTAL IN SYRIA

This total eclipse is clearly mentioned by Elias 1910, 129 under the year AH 329: "The Moon was eclipsed on the night of Friday, the 15th of Rabīa I [940 Dec.18], and during Saturday night the caliph Ar-Rādī died". Elias gives as general reference for his 329 annal Tābit, son of Sinān. Ar-Rādī reigned c.934-40.

Oppolz. and Schroe. give maximum magnitude 1.225 about UT $14^h 17^m$ (say 5.15 p.m. Bagdad mean time). As the Islamic day begins at sunset, even the early evening belonged to Friday rather than (as we should say) to Thursday.

946 S.946 March 6 TOTAL IN ARABIA AND S. EGYPT (NO RECORD)

The track of totality begins in the Central Atlantic, proceeds via equatorial Africa to a noon point in the Red Sea, and continues via Arabia and the Persian Gulf to a sunset point in Mongolia. We have encountered no record, although it would have been noticed in Egypt.

c.949 NON-ECLIPSE IN ITALY

Solar non-eclipse about 949?

We have four (late?) references to the Sun's becoming like blood, or ruddy and blood-red, during the reign of Lothar II (948-50) in Italy. We have concluded that this was not an eclipse. See

| | |
|---|---|
| Annales Polonorum | SS 19, 1866, 614 |
| Martin of Oppau, Chron. pont. et. imp. | 22, 1872, 464 |
| Chron. Univ. Mettens. | 24, 1879, 510 |
| Instituta regalia ... Longobardorum | 30(2), 1934, 1459 |

(951) See under S.418 July 19 and S.939 July 19 for 951 July 19 error in Annals of Blandigny (Ghent).

955 M.955 Sept.4-5? TOTAL IN FRANCE

Bouquet 8, 1871, 299 gives, in a "Fragmentum historiae Francorum, ex antiqua membrana Floriacensis coenobii", a passage which mentions both a lunar and a solar eclipse. We can tentatively identify the former, and perhaps the latter (see below). After mentioning the death of Louis d'Outremer in 954 (Sept.10), and the succession of Lothaire (crowned 954 Nov.12), the passage reads:

"Cujus anno tertio, videlicet Dominicae Incarnationis DCCCCLVI, IV Nonas Septembris, Luna versa est in sanguinem; eodemque anno mense Junio signum mirabile apparuit in coelo, Draco scilicet magnus et sine capite. Sequuta statim est mors Hugonis Principis Francorum, Burgundionum atque Normannorum. Eodem anno eclipsis Solis facta est XV Kal. Januarii, et stellae a prima hora usque ad horam tertiam apparuerunt ..."

The date given for the lunar eclipse is 956 Sept.2. This was not a day of full moon, and the passage refers to the total lunar eclipse of 955 Sept.4-5 (i.e. II.Non. to Non. rather than IV.Non.). At Fleury, as at Greenwich, the eclipse began before midnight and ended after midnight. It is also recorded, in rather similar words, in Chronicon Floriacense (Bouquet 8, 1871, 254; SS 2, 1829, 255). It appears to have attracted little attention, but the identification here mooted was listed in Struyck 1740, 154 and thence in Struyck-Ferguson.

The abbey of Fleury or of Saint-Benoît-sur-Loire lay between Orléans and Gien. Hugh the Great died on 956 June 16 or 17. See F. Lot, Les derniers Carolingiens: Lothaire, Louis V, Charles de Lorraine (954-991), Paris 1891, pp.xxi, 16.

c.957 recte 961 — CONFUSED SOLAR RECORD

Misdating of solar eclipse (really 961).

The Fleury fragment quoted above under M.955 Sept.4-5 mentions, apparently for the third year of Lothaire, a solar eclipse on XV.Kal.Jan. (Dec.18), with stars visible from the first to the third hour (a manifest exaggeration). Such an eclipse is mentioned in:

Kepler, Astr. Opt. p.295 (Ges. Werke, ed. M. Caspar, 2, 1938, 256);
Riccioli 1653, 370; and, from Riccioli, Tycho-Curtius 1666, xxxiii.
Riccioli and Tycho-Curtius attribute the record to some "Vita Lotharii".

The date carries its own refutation. There was no solar eclipse in France during the third year of Lothaire (which cannot possibly fall outside 955-8), nor any solar eclipse anywhere on or close to Dec.18 in any of the years round about.

Thus identification is impossible unless conjectural emendation is allowed. One such emendation suggests itself. Much the largest solar eclipse in France around this time was the annular eclipse, S.961 May 17 (q.v.). For this, at Fleury, neither mid-eclipse about the second hour of the day nor some visibility of stars is unreasonable, though the third hour would be better still. Thus we may ponder some such emendations as the following:

(i) For "third year" of Lothaire read, say, "seventh year" (VII instead of III);

(ii) For "XV Kal.Jan." (Dec.18) read "XVI Kal.Jun." (May 17).

c.958 S.958 July 19 ECLIPSE NOT VISIBLE

This penumbral eclipse was not likely to have been noticed, but see Newt.1972, 72, 229, where this eclipse is considered, along with S.418 July 19 and S.939 July 19 (qq.v.), in discussing a record under "951" in the Annals of Blandigny.

961 S.961 May 17 ANNULAR IN GERMANY AND ITALY

"961. An eclipse of the Sun on 16.Kal.Jun. [= May 17]" is mentioned (e.g. SS 1, 1826, 624) by the anonymous continuator of Regino of Prüm, monk of Trier. The continuator has been conjectured to be Adalbert (monk of Trier; bishop at Kiev, 961-2; archbishop of Magdeburg, 960-81); see Camb. Med. Hist., 3, 202.

The eclipse information is also given in two later sources, Annalista Saxo (SS 6, 1844, 615) and Annals of Magdeburg (SS 16, 1859, 147). We have eleven other European sources giving year only (usually 961, though 962 twice), and specifying either "a sign in the Sun (six), an eclipse of the Sun (three), or a solar obscuration (two). Some of the sources are mentioned in Ginzel 1884 and Newton 1972.

It is appropriate that our primary source should be Regino's continuator, as Trier lies within, and Magdeburg almost within, the band of annularity shown in Schroeter, which runs from SW to NE by Madrid, Trier, Hamburg, Stockholm, and Finland. The curve of magnitude 0.75 to the SE runs by Palermo, the Balkans, and Moscow, so that the eclipse would be large throughout Central Europe.

We have not found the hour mentioned in any medieval record (unless a problem passage from Fleury refers to this eclipse, see above under circa 957). The "five hours before midday" quoted in T-C 1666 from Calv. is only roughly correct; annularity at Trier was rather less than four hours before midday.

The eclipse tends to be mentioned in the context of the crowning of Otto I's infant son, the future Otto II (961 May), the subsequent entry of Otto I into Italy (961 Aug.), and the crowning of Otto I in Rome as Emperor (962 Feb.2).

63 S.963 Sept.20 CONFUSED RECORD IN ITALY

This annular eclipse had noon point in the Atlantic, and proceeded via North-West Africa to an end in Central Africa. There is a reference in the Annals of the Bari region traditionally dubbed those of "Lupus protospatharius" (e.g. SS 5, 1844, 54-5). Having already (p.54) mentioned S.961 May 17 in a different context under its correct year, the Annals have "963. The emperor Romanus died, and Nicephorus succeeded, and reigned for 7 years; and king Otto entered Rome, and the Sun was obscured". A modest partial eclipse, of magnitude perhaps no more than 0.5, may have occurred in Southern Italy in the late afternoon. The words 'in his 6th year' may have been omitted and the eclipse of 968 may be meant.

964 S.964 Mar.16 SOLAR IN IRELAND NOT RECORDED

Totality ended at sunset approximately between South-West Ireland and North-West Spain. O'Connor 1952 found magnitude 0.76 at his standard mid-Ireland point, but we have encountered no Western record.

965 M.965 Aug.15 TOTAL IN GERMANY

"965. [xv]iii. Kal. Septembr. conver [sa est] luna quasi in sanguinem, [quae] et ipsa nocte erat 14." (i.e. the moon was changed as if to blood and that night was the 14th), Annales Prümienses (SS 15(2), 1888, 1292). The same annal includes the death of Bruno (925-965), archbishop of Cologne (and brother of Otto I).

This total eclipse on the night of 965 Aug.14-15 took place entirely after midnight at Greenwich, and a fortiori in the Rhineland. Since 18.Kal.Sept. = Aug.15, the Prüm date (as restored) is correct.

966 S.966 July 20 LARGE SOLAR IN SCOTLAND (AND ITALY?)

The track of totality is of the comparatively short and strongly curved kind only possible when part of it lies in high latitudes. Oppolz. shows a sunrise point in Northern Siberia, a midnight point in the Arctic Ocean, and then a track passing over Norway, Sweden and the Baltic for a sunset point in South Russia. Schroeter's band of totality includes Stockholm Helsinki and Riga, and ends a little short of the Sea of Azov. The curve of magnitude 0.75 to the west and south shows at least such magnitude in all Iceland, all Ireland, all France, Barcelona and Tunisia, and ends near Libya.

Anderson 1922, 473 accepted this eclipse as the one connected with the killing of King Dub (Dubh). He quotes the Chronicle of the Kings of Scotland, version D, from W.H. Skene, Chronicles of the Picts and Scots, 1867, p.151. "Dub, Malcolm's son, reigned for four years and six months; ahd he was killed in Forres, and hidden away under the bridge of Kinloss. But the sun did not appear so long as he was concealed there; and he was found and buried in the island of Iona". The maximum magnitude at Forres ($60^\circ$N, $3\frac{1}{2}^\circ$W) would occur in the late afternoon, and the table in Schroeter 1923, 31 makes it amount to about 0.86.

The latter eclipse is a more equal contender in connection with an apparently confused eclipse record mentioned in the Annals of Farfa in Italy; see below.

The next (967) eclipse would have been too small to be noticed in Scotland.

967 recte 966 & 968?

S.967 July 10 NO CERTAIN RECORD EVEN IN S.E. MEDITERRANEAN

A total eclipse. Schroeter's band of totality includes the eastern part of Crete and the southern part of the Caspian Sea.

There is probably some reference in the Annals of Farfa (e.g. SS 11, 1854, 589): "968. Otto II imperator. Sol defecit hora tertia, 14.Kal.Aug." It is difficult to know what to make of this. The year numbers in these annals are mostly about one unit too large, though Otto II was crowned in Rome on 967 Dec.25, which would count as the beginning of 968. There were solar eclipses on 966 July 20 (magnitude more than 0.8 about 6.20 p.m. in Central Italy) and 967 July 10 (magnitude 0.7 about 6.20 a.m. in Central Italy). The recorded date (14.Kal.Aug. = July 19) points to the first, the hour to the second. There was also the well-known solar eclipse on 968 Dec.22, discussed below, which was total at Farfa, and at about the third hour; but the month and day do not tally at all. Probably information from the two eclipses of 966 and 968 has been conflated.

968

M.968 Dec.7-8 (Mon.-Tu.) LARGE PARTIAL IN ISLAM

This large partial eclipse (Oppolz. magnitude 0.94) is clearly mentioned by Elias 1910, 136. He says that in AH 358 the Moon was eclipsed on the night of Tuesday the 14th (of Muharram), i.e. of AD 968 Dec.8 (Tu.). Most of the eclipse occurred before midnight at Babylon and points further west, but the whole night of Monday-Tuesday is taken as belonging to Tuesday. The Oppolzer date of mid-eclipse is naturally Dec.7. Elias gives his source for annal AH 358 as Tābit, son of Sinān; see under M.970 May 23-24.

There is a probable reference, too erroneous to be of much value, by the Spanish historian J.A. Conde, who in his History of the Arabs in Spain used primarily Moslem dating. To quote the editions we have seen, Conde 1840, 233 and Condé 1854, 467 say, apparently of AH 355 Rajab, "During the same month there were eclipses of the Sun and Moon. The eclipse of the Moon was on the night of the fourteenth, and that of the Sun on the twenty-eighth day commencing at the hour of sunrise". The dates mentioned are 966 July 6 and 20, and quite unsuitable. No lunar eclipse, umbral or penumbral, occurred at the full moon of 966 July 5, while S.966 July 20 (q.v.) was an afternoon, not a sunrise, eclipse in Spain. But there were a lunar eclipse on 968 Dec.7-8 (the Moslem night of AH 358 Muharram 14) and a solar eclipse at sunrise in Spain on 968 Dec.22 (AH 358 Muharram 28). Moreover no lunar and solar eclipses a fortnight apart occurred in Spain any nearer to the 966 summer dates mentioned by Conde, which are consequently over two years out.

968 S.968 Dec.22 TOTAL SOLAR IN EUROPE

This well-known total eclipse of the Sun had a short, very curved band of totality, including (in Schroe.) the far south of Ireland, Lyon, Rome, Constantinople and North-East Russia. We have encountered no record of it from the Irish and British end. We shall mention first some records from Europe (as far east as Italy), and then some more or less Byzantine records (as far west as Corfu). The eclipse is discussed in Ginzel 1883, 677 and in Newton 1972.

We have a dozen (mostly secondary) European references which mention the day, six giving correctly either Dec.22 or 11.Kal.Jan., and six by error giving 11.Kal.Dec. We also have nine references giving at most year and month (no day). See also Conde under M.968 Dec.7-8, and Farfa under S.967 July 10.

The four European sources which give the hour give also the correct 11.Kal.Jan., though the year, when given, tends to be wrong.

| | | | |
|---|---|---|---|
| 970 | hora diei 3. | Ann.Sangall.maior. (Hepidan.) | SS 1, 1826, 79 or Bouquet 9, 1874, 96 |
| | a 1.h. usque 3. | Chron.Floriacense (Fleury) | Bouquet 8, 1871, 254 |
| 969 | h.d.inter 3. et 4. | Ann. Cavenses (La Cava) | SS 3, 1839, 188 |
| | hora 3. | Leo Marsicanus Chr. Casin. | SS 7, 1846, 635 |

The eclipse occurred while Otto the Great was campaigning in Calabria (South Italy). No date or numbers, but some human interest, will be found in accounts of its effect on his army. The accounts are too long to quote, but the gist is as follows. The troops, though seasoned, were terrified, and took shelter as best they could; Everacrus (Evraclus), bishop of Liège, went round saying "Come out! This is quite natural! What will the enemy think of you?" After the eclipse, the soldiers laughed at their behaviour. The bishop is said in one version (Reiner) to have known about eclipses from reading Pliny, Macrobius, and Chalcidius.

The chief account appears to be in the *Gesta* of the bishops of Liège, by Anselm (c.1050), (SS *7*, 1846, 202, §24). This was abbreviated in the Chronicle of St. Laurence of Liège, by Rupert (d. 1129), (ibid. *8*, 1848, 263), and embellished in the Life of Evraclus, by Reiner (d. 1182 or later), (ibid. *20*, 1868, 563, ch.7). The account in Anselm is apparently copied in the *Gesta* of the bishops of Liège (to AD 1251) by Aegidius Aureaevallensis (Giles of Orval, in the diocese of Trier; ibid. *25*, 1880, 54).

In discussing the comparatively few Byzantine sources, we shall (like Newton 1972) find it convenient to include among them the contemporary account by Liudprand, bishop of Cremona, who saw the eclipse during his return to Italy from an embassy in 968 to Constantinople. Liudprand's Relatio de legatione Constantinopolitana (e.g. ed. G.H. Pertz, SS *3*, 1839, 362; Engl. tr., F.A. Wright, The Works of Liudprand, London 1930, 276), after mentioning his arrival at Corfu on Dec.18, says: "Four days later, on Dec.22, while I was breaking bread at table ... the Sun ... hid the rays of his light and suffered an eclipse". A note by Pertz (XV Kal.Jan. in place of XI) shows that the date is correct only by virtue of editorial emendation; but the identification is certain. Corfu lies slightly south of Schroeter's band of totality.

What appears to be the earliest surviving account in Byzantine Greek was not written until 992, by Leo the Deacon (Leo Diaconus, Hist., ed. C.B. Hase, Bonn 1828, 72; Ger. tr., F. Loretto, Graz etc. 1961, 71). His account reproduced in Stephenson and Clark 1978, 8 (cf. Newt.1979, 391 and 469) mentions the fourth hour of the day, and the visibility of stars. It may be taken, as in Newt.1972, 549-50, to refer to Constantinople, which lies just inside the southern edge of Schroeter's band of totality, and on that basis would have experienced brief totality around 11.15 a.m. (say about the end of the fifth hour of the day).

The 12th century Cedrenus 1839, 375 mentions Dec.22, "about the third hour", and visibility of stars.

The later 12th century Glycas 1836, 572 mentions "about the third hour of the day", and visibility of stars.

The hour in Cedrenus and Glycas is perhaps just tolerable if it refers to the very beginning of the eclipse (or if it is borrowed from Western Europe).

969 M.969 June 2-3 (Wed.-Th.) TOTAL IN IRAQ

This deep total eclipse (Oppolz. and Schroe. magnitude 1.82) is mentioned by Elias in the same annal for AH 358 (source Tābit, son of Sinān) as M.968 Dec.7-8 (q.v. for reference). He says that the Moon was totally eclipsed on the night of Thursday the fourteenth of Rajab, and set (eclipsed). The date given by Elias is correct, and the night is June 2-3 as we have indicated (Delaporte's editorial June 3-4 is to be disregarded).

At Babylon mid-eclipse was somewhat after 4 a.m., and moonset around 5 a.m., near the time at which the total phase ended and the second partial phase began. Further west (e.g. in Syria) the Moon may have set during the latter phase.

970 S.970 May 8 ANNULAR IN S.E. EUROPE NOT RECORDED

Schroeter has a band of annularity starting at sunrise near the border between Tunisia and Libya, and running via Greece and just south of Constantinople to the eastern part of the Black Sea and then Russia.

This eclipse is the identification given in Struyck 1740, 154 and Ginzel 1886, 980 of a slight reference in Glycas (1836, 576; 1866, col.575-6; Paris page 309). Glycas gives little help. It is surely inappropriate that his 'Chronicle' should have been latinized as 'Annals'; the form is in no way annalistic. The arrangement is by reigns, and there is a singular paucity of dates; even the reign changes, when one emperor is succeeded by another, are sometimes left vague. It is true that the laconic "such an eclipse of the Sun that the stars themselves were seen" occurs, as Ginzel says, after mention of S.968 Dec.22 and the murder of Nicephorus Phocas (969 Dec.11). But it also occurs after the comet of 975 Aug.-Oct. (also observed in China and Japan) and the death of the emperor John Tzimisces (976 January); it appears under the reign of Basil II (976-1025).

Near Constantinople Ginzel gives two computed tracks and Schroeter another. Struyck found for Constantinople a magnitude of about 0.95 at about 6.38 a.m. From Schroe. one may roughly estimate much the same. The solar eclipse of 970 May 8 was certainly very large at Constantinople, and one would not be surprised to find some report of it. But the passage in Glycas lacks enough good information, and not even the shrewdness of Struyck and Ginzel can provide an identification of good standing.

970 M.970 May 23-24 (Mon.-Tues.) PARTIAL IN IRAQ

This partial eclipse (Oppolz. magnitude 0.57) is mentioned by Elias 1910, 289: "Tābit, son of Sinān, son of Tābit, son of Qōri, another chronographer, indicates that the moon was eclipsed at Babylon, in AH 359, at the first hour of the night which began on Tuesday, 15 R; this was on the 24th [May] in the year of Alexander, ..." The night is here, as usually in Elias, counted as belonging to the following day. Oppolz. gives mid-eclipse on 970 May 23, UT $17^h4^m$ (Babylon MT $20^h2^m$), nearly enough at "one hour of the night", during what we should call Monday evening; the eclipse occurred entirely before Babylon midnight.

The source on whom Elias relies appears as a grandson of the famous astronomer Tābit ibn Qurra (826-901).

976 M.976 Jan.19 (p.m.) or M.976 July 14 (a.m.)
AMBIGUOUS TOTAL ECLIPSE

"976. Luna obscurata est." Annales Augustani (of Augsburg), SS 3, 1839, 124. There were two eclipses of the Moon, both total, in 976:

(i) entirely before Augsburg midnight on Jan.19-20, maximum magnitude about 1.18 around UT 9 p.m.,

(ii) entirely after Augsburg midnight on July 13-14, maximum magnitude about 1.27 around UT 3 a.m.

The Augsburg Annals may refer to either.

977 S.977 Dec.13 PARTIAL SOLAR IN EGYPT

The central line of totality runs from N.W. Africa to the Indian Ocean (noon point) and the South China Sea. One of the Ibn Yūnus eclipses, it was observed by astronomers at Cairo between 8.20 and 10.40 a.m., Cairo mean time, with a maximum magnitude about 0.67. The times follow from observed solar altitudes of the beginning and end of the eclipse. See Caussin 1799, 8; 1804, 164, Newcomb 1878 (§5, p.47f), and Newton 1970 (146, 242, 247). For earlier discussions, using manuscript material of W. Schickard (1592-1635), see T-C 1666, xxxiv, Str.1740, 122 and Dunthorne 1749, 164. See also Johnson 1874, 43 and Chambers 1902, 141.

The list of Chinese eclipses in Hoang 1925 gives this one as predicted but not observed.

978 S.978 June 8 ANNULAR IN EGYPT

The central line of annularity runs from Brazil to West Africa (noon point) and the Indian Ocean. This Ibn Yūnus eclipse was observed at Cairo between about 2.30 and 4.50 p.m., Cairo mean time. The times again follow from observed solar altitudes. See Caussin 1799, 8; 1804, 166, Newcomb 1878 (§5, p.48f), and Newton 1970 (146, 242, 247). Schickard's material is used in T-C 1666, xxxiv, Str.1740, 123, and Dunthorne 1749, 165. See also Johnson 1874, 43, quoting Vince. The observed maximum magnitude at Cairo is variously stated by modern authors as 7½ and 5½ digits (0.62 and 0.46).

~979 M.979 May 14-15 (Wed.-Th.) ANNULAR IN EGYPT

This partial lunar eclipse, of Oppolz. magnitude 0.73, occurred entirely before midnight throughout the Western world, during the evening of Wednesday (midnight to midnight reckoning) or Thursday (sunset to sunset reckoning). It is one of the Ibn Yūnus eclipses. At Cairo it was observed that the Moon rose eclipsed, and that the eclipse ended $1^h12^m$ after sunset. See Caussin 1799, 8; 1804, 168, Newcomb 1878 (§5, p.48f), and Newt.1970 (147, 215, 229). Schickard's material is mentioned in T-C 1666, xxxiv, Str.1740, 123, and Dunthorne 1749, 165, though the latter does not use the eclipse.

979 S.979 May 28 ANNULAR IN EGYPT

The central line of annularity runs from the Pacific via North America to Greenland (noon point), then by Iceland and Scandinavia to near the Caspian Sea. Schroe. shows the band of annularity running by Northern Ireland, south of Trondheim, north of Stockholm, south of Moscow, and ending just short of the Caspian.

This is another Ibn Yūnus eclipse. At Cairo the Sun was seen less than half eclipsed; Schroeter's table suggests a maximum magnitude about 0.40 there. From an observed solar altitude, it is computed that the eclipse began about 6.18 p.m., Cairo mean time. See Caussin 1799, 8; 1804, 168, Newcomb 1878 (§5, p.48f) and Newt.1970 (146, 150, 242, 247).

For Constantinople Str.1740, 145 found maximum magnitude about 0.72 (we find 0.65 from Schroe.). But Str. rightly preferred the identification S.990 Oct.21(q.v.), for an eclipse mentioned at around AD 979 by Cedrenus (e.g. Bonn ed., 2, 1839, 434). At Constantinople S.979 May 28 not only lacked sufficient magnitude for visibility of stars; it was also unquestionably an evening eclipse, whereas Cedrenus says "around midday" and AD 891 (sic) seems the only solution.

979 M.979 Nov.6-7 PARTIAL IN EGYPT

This partial eclipse, of Oppolz. magnitude about 0.86, occurred around Cairo midnight. From observed altitudes given by Ibn Yūnus, the beginning and end are computed to have occurred at roughly 10.10 p.m. and 1.20 a.m., Cairo mean time. See Causs.1799, 8; 1804, 168, Newcomb 1878 (§5, p.48f), and Newton 1970 (147, 215, 229).

980 M.980 May 2-3 DEEP TOTAL IN EGYPT

Oppolz. and Schroe. give this deep total eclipse as having maximum magnitude 1.60 at UT $0^h34^m$ and occurring entirely after Cairo midnight. Ibn Yūnus gives information about both beginning and end, but the lunar altitude at the beginning needs emendation (perhaps 47°40' is a corruption of 41°40', see E.B. Knobel, MN 39, 1879, 339). The end was observed about $36^m$ before sunrise, say around 4.36 a.m., Cairo mean time; Oppolz. and Schroe. yield about 4.29 a.m. See Causs.1799, 8; 1804, 170, Newcomb 1878 (§5, p.48f), and Newton 1970 (147, 150, 229).

981 M.981 Apr.21-22 SMALL PARTIAL IN EGYPT

This small partial eclipse, of Oppolz. magnitude 0.18 at UT $2^h20^m$, occurred entirely after midnight at Greenwich and therefore at Cairo, where Ibn Yūnus says it was observed. He gives maximum magnitude about 0.25, and his information about beginning and end leads to times around 3.30 and 5.08 a.m., Cairo mean time. For details see Causs.1799, 9; 1804, 170, Newcomb 1878 (§5, p.49f), and Newton 1970 (147, 215, 229).

981 M.981 Oct.15-16 SMALL PARTIAL IN EGYPT

This small partial eclipse, of Oppolz. magnitude 0.36 at UT $2^h54^m$, also occurred entirely after midnight at Greenwich and therefore at Cairo, where Ibn Yūnus says it was observed. He gives maximum magnitude about 0.42. His lunar altitude at "l'attouchement par dehors" is agreed to give about 4.18 a.m., Cairo mean time; Newt.1970, 150 ignores this time because of the phraseology, but Newc.1878, 52 found it less than a quarter of an hour later than his calculated beginning. See Causs.1799, 9; 1804, 170, Newc.1878 (§5, p.49f), and Newt.1970 (147, 150, 215, 229).

983 M.983 Mar.1-2 TOTAL IN EGYPT

According to Oppolz. and Schroe., this total eclipse had maximum magnitude about 1.09 some 20 minutes before Greenwich midnight. Ibn Yūnus says it was observed at Cairo. The stated lunar altitude at the beginning, namely 66°, is necessarily wrong, being more than the Moon's meridian altitude. E.B. Knobel (MN 39, 1879, 339) says it is almost certain that the original manuscript of Ibn Yūnus gave 62°. However, altitude is a poor guide to time when the object observed is nearly on the meridian. The reported altitude at the end yields about 3.40 a.m., Cairo mean time. See Causs.1799, 9; 1804, 172, Newc.1878 (§5, p.49f), and Newt.1970 (147, 150, 229).

983 recte 891 S.983 Mar.17 FALSE IDENTIFICATION

This identification of the solar eclipse mentioned by Cedrenus (cf. 979 May 28) at about AD 979 (see under S.990 Oct.21) was given by Muralt 1950, 566. We find S.983 Mar.17 invisible at Constantinople and was only penumbral in any case.

985 S.985 July 20 PARTIAL PHASE IN EGYPT

The central line of totality runs from New Mexico via the Atlantic and North Africa to an end in Equatorial East Africa. Ibn Yūnus says magnitude 0.25 was observed at Cairo, and from his solar altitudes at beginning and end Newcomb found times equivalent to about 5.02 and 6.23 p.m., Cairo mean time, respectively $24^m$ and $15^m$ later than Newcomb's tabular times. The eclipse is not used in Newton 1970. See Causs.1799, 9; 1804, 172, and Newcomb 1878 (§5, p.49f).

The eclipse was also computed in Str.1740, 144 for Messina, in an effort to identify an eclipse report (a mere "obscuratus est sol") under AD 987 in the Bari annals (to AD 1102) of 'Lupus protospatharius' (e.g. SS 5, 1844, 56). But Struyck inclined to the well-known S.990 Oct.21 (q.v.), which also provides a date under which Newton 1972, 449, 467 can put the 'Lupus' record, and dismiss it as uncertain, late and not usable. See also Struyck-Ferg. (e.g. 1773, 176). Another unlikely identification of the 'Lupus' record is S.988 May 18 (q.v.). 968 is a possibility.

986 M.986 Dec.19 (a.m.) LARGE PARTIAL IN EGYPT

A large partial eclipse, of Oppolz. magnitude 0.925 at UT $4^h11^m$. Ibn Yūnus gives magnitude about 0.83 at Cairo, and an initial lunar altitude from which Newcomb finds a beginning at 4.56 a.m., Cairo mean time, about $22^m$ later than Newcomb's tabular value. At Cairo the Moon set eclipsed.

There is also some problematical information, on which Causs.1799, 9 and Causs.1804, 172 differ, and E.B. Knobel (MN 39, 1879, 339) may be consulted. See Newcomb 1878 (§5, p.49f) and Newton 1970 (127, 147, 151); the latter author decides to ignore the record.

c.988 recte 990 & 993? S.988 May 18 CONFUSED RECORDS OF SOLAR ECLIPSE

Oppolzer's track of annularity goes from the U.S.A. via a noon point well north of Iceland to Scandinavia and Central Asia. Schroe. shows the band of annularity as passing near Hammerfest and well north of Archangel. We have encountered two possible, though unlikely, Western references.

The Annals of Magdeburg (e.g. SS 16, 1859, 158), besides giving the eclipse S.990 Oct.21 under the year 992, also mention a solar eclipse under the year 990. If this really meant 988, the reference would be to S.988 May 18 (Schroe. magnitude about 0.55 at Magdeburg). More probably, the passage may constitute only a confused double reference to S.990 Oct.21, and 1 point has been allocated accordingly.

The South Italian annals of 'Lupus protospatharius' (e.g. SS 5, 1844, 56) mention a solar obscuration under AD 987. As noted under 985, no solar eclipse was visible in South Italy in that year; and S.988 May 18 had Schroe. magnitude about 0.27 at Bari and 0.19 at Messina. These are small magnitudes indeed; but none of the other 10th century eclipses in 'Lupus' seems misplaced by more than one year. Possibly the correct eclipse is that of 993 and 2 points have been allowed. Only if the observation had been made in Scandinavia and transmitted to Italy by Norsemen could the 988 date be correct, but 1 point has been allowed for this, unlikely at such an early date.

990 M.990 Apr.12-13 (Sat.-Sun.) PARTIAL IN ISLAM

A partial eclipse, for which Oppolz. gives magnitude about 0.76 with mid-eclipse at UT $21^h43^m$ and a semi-duration of $1^h30^m$. Goldstine 1973 gives opposition (not quite the same thing as mid-eclipse) at UT $21^h39^m$.

The eclipse was observed at Cairo with magnitude 0.625, according to Ibn Yūnus, but the record has been mistrusted. From it, Newcomb deduced an observed beginning at about UT $19^h41^m$ and ignored the end entirely, while Newton deduced an observed end at about $21^h37^m$ and rejected the beginning. See Causs.1799, 9; 1804, 174, Newcomb 1878 (§5, p.49f), and Newton 1970 (127, 147, 151, 215, 229).

A calendrically detailed record appears in Elias (p.289). He correctly notes that the Moon was eclipsed in the middle of the night on which Palm Sunday began.

Making reference to a Fulda necrology, Str.1740, 154 computed maximum magnitude about 0.76 with (ignoring seconds) beginning at 8.44 p.m., mid-eclipse at 10.22 p.m., and end at 11.59 p.m. As Fulda is about $39^m$ east of Greenwich, the calculations of Str. and Oppolz. agree well. See also Struyck-Ferguson (e.g. 1773, 176).

990 M.990 Oct.6-7 (Mon.-Tues.) PARTIAL IN IRAQ

A large partial eclipse, for which Oppolz. gives magnitude about 0.96 with mid-eclipse at UT $2^h13^m$ and a semi-duration of $1^h37^m$.

Elias (loc.cit.) again gives full dating, and says that the Moon was eclipsed at the end of the night on which Tuesday began.

Repeating his reference to the Fulda necrology, Struyck 1740, 154 computed maximum magnitude about 0.93, with beginning, middle and end at about 1.32, 3.04 and 4.36 a.m. at Fulda, in reasonable agreement with Oppolzer. See also Struyck-Ferguson (e.g. 1773, 176).

990 S.990 Oct.21 ANNULAR IN EUROPE

Oppolzer's central line of annularity begins at sunrise in the Greenland Sea, proceeds via Southern Scandinavia to a noon point near Kaliningrad (formerly Königsberg), and ends at sunset near Lake Balkhash in Central Asia. Schroe.1923 shows a rather wide band of annularity which includes Bergen, Stockholm, Warsaw, and Astrakhan; his curve of magnitude 0.75 to the south-west goes through or near Dublin, Hampshire, Paris, Milan, Athens, and north of Jerusalem.

The eclipse is certainly reported from Central Europe, where some half-dozen annals or chronicles (mainly secondary) give the correct 12.Kal.Nov. (Oct.21), though sometimes under the wrong year. Four of these also give the hour, agreeing on the fifth hour of the day, which is about correct for Central Europe, where however the magnitude of the eclipse would be only about 0.8. Two early and important sources, both mentioning 12.Kal.Nov. and fifth hour of the day, are:

Thietmar (975-1018), bishop of Merseburg, Chronicle, wrongly under AD 989 (e.g., ed. J.M. Lappenberg, SS 3, 1839, 772; ed. F. Kurze 1889 and R. Holtzmann 1935 in SrG; ed. and Germ. tr., W. Trillmich, Berlin, no date, p.130; Bouquet 10, 1874, 123);

Annals of Quedlinburg (to 1025), under AD 990 (e.g. SS 3, 1839, 68). The eclipse has been treated in Calv.1605, 770; 1620, 602; 1650, 706, etc., Str.1740, 145 (see also Struyck-Ferguson), Ginz.1884, 553, and Newt.1972 (passim). To their references may be added the Annals of Magdeburg (e.g. SS 16, 1859, 158), though they put the correct 12.Kal.Nov. under 992 (ninth year of Otto III). See also S.985 July 20 and

S.988 May 18.

Newton 1979, 290 considered Hildesheim as a place of observation and used Ann. Saxonicum as a source from Saxony.

990 recte 891 S.990 Oct.21 (cont.) WRONG CENTURY IN CEDRENUS

Whether the eclipse was recorded in the Byzantine Empire is more questionable. A passage in Cedrenus (AD 1100) mentions (about AD 979, as far as one can judge) that "an eclipse of the Sun occurred around midday, and stars appeared" (1839, 434; PG 122, 1889, col.167-8; Paris page 694). The unsuitability of S.979 May 28 and S.983 March 17 as identifications has already been explained under those dates. Struyck (loc. cit.) computed for S.990 Oct.21 a maximum magnitude at Constantinople of about 0.84 at $45^m$ after noon, and adopted the identification, which was also used by Ginz. (loc.cit.). It may be noted that Shroe.1923 yields a maximum magnitude at Constantinople of about 0.83 at, roughly speaking, 12.40 p.m., local mean time. But Newt.1972, 551 justifiably doubts the identification, on the grounds of (i) insufficient magnitude for visibility of stars, (ii) wrong place in the narrative of Cedrenus, (iii) similarity of wording to the account by Cedrenus of S.891 Aug.8 (q.v.). Newt. implies that Cedrenus accidentally described S.891 Aug.8 twice, once in the right place and then again about a century later. Struyck (loc.cit.) refers also to Glycas, in whom however we do indeed find S.891 Aug.8 rather than S.990 Oct.21.

991 M.991 Apr.1-2 (Wed.-Th.) DEEP TOTAL IN IRAQ

A deep total eclipse, with Oppolz. and Schroe. magnitude 1.54 and mid-eclipse shortly before Greenwich mean midnight. The eclipse is mentioned by Elias (p.289). He fully dates it in several systems, and says it occurred at the ninth hour of the night on which the Thursday began.

991 M.991 Sept.26-27 (Sat.-Sun.) DEEP TOTAL IN IRAQ

A deep total eclipse, with Oppolz. and Schroe. magnitude 1.48, mid-eclipse around UT $14^h45^m$, and semi-duration of umbral eclipse $1^h50^m$. The eclipse is mentioned by Elias (loc.cit. p.289), who says the Moon was eclipsed at the beginning of Sunday night; in civil terms (midnight to midnight reckoning) this means the beginning of Saturday evening. In Mesopotamia the Moon would rise eclipsed.

993 S.993 Aug.20 TOTAL SOLAR S. MEDITERRANEAN (LARGE IN EGYPT)

Oppolzer's central line of totality begins in Morocco, and proceeds to a noon point near the coast of Pakistan and an end in Indonesia. Schroeter's band of totality passes just north of Alexandria and almost centrally over

Jerusalem.

Ibn Yūnus says the eclipse was seen at Cairo, with magnitude two-thirds of the surface (implying about 0.73 of the diameter). From solar altitudes at beginning and end, Newcomb and Newton find the equivalent of Cairo mean times about 7.43 and 10.25 a.m. See Causs.1799, 10; 1804, 174, Newcomb 1878 (§5, p.49f), and Newton 1970, 146, 247.

995 M.995 Jan.19-20 (Sat.-Sun.) TOTAL IN IRAQ

A total eclipse, with Oppolz. and Schroe. magnitude about 1.24, mid-eclipse about UT $15^h22^m$, and semi-duration of umbral eclipse $1^h46^m$. The eclipse is mentioned by Elias (p.289). He says it occurred at the first hour of the night on which Sunday commenced (in civil terms, this means the first hour of Saturday evening). See also the Augsburg reference below, under M.995 July 14-15. A Chinese record exists which may be a genuine documentation.

c.995 (Jan. or June) BLOOD MOON IN GERMANY

995 M.995 July 14-15 (Sun.-Mon.) TOTAL IN IRAQ

A total eclipse, with Oppolz. and Schroe. magnitude about 1.22, mid-eclipse about UT $22^h52^m$, and semi-duration of umbral eclipse $1^h45^m$. The eclipse is mentioned by Elias (p.290). He says the eclipse occurred in the middle of the night on which Monday commenced.

The primary Annales Augustani (i.e. of Augsburg; e.g. SS 3, 1839, 124) have "995. The Moon was turned to blood." If correctly placed in 995, this could refer to either M.995 Jan.19-20 (q.v.) or M.995 July 14-15. The July identification has been preferred in Str.1740, 154 and, with a question mark, in Newt.1972, 661. The Annals of Ratisbon use a very similar formula (Moon turned to blood) under 998, see M.998 Nov.6-7.

There is a Chinese record that clouds prevented visibility.

997 M.997 May 24-25 PARTIAL IN ASIA

A partial eclipse, for which Oppolz. gives maximum magnitude about 0.58 at UT $23^h31^m$, with semi-duration of umbral eclipse $1^h21^m$. It is given by R.R. Newton (Mem.Roy.Astr.Soc. 76(4), 1972, 114-5) as a likely identification of a lunar eclipse mentioned in a work of 1025 by al-Birūni. A lunar eclipse in the Muslim year 387 (beginning 997 Jan.14) was observed by al-Birūni in the Khwarizm district (near Khiva, between the Caspian and Tashkent) and by Abu al-Wafa in Bagdad; it was concluded that the longitude of Khwarizm is 15° east of Bagdad (Newton gives 16°23' as a modern value). Newton points out that since there is no way of confirming the year, other eclipses, particularly the total eclipse of M.998 Nov.6-7 (q.v.), must be considered possible. The latter eclipse fell in AH 388 (beginning 998 Jan.3).

998 M.998 Nov.6-7 (Sun.-Mon.) DEEP TOTAL IN IRAQ

A deep total eclipse, with Oppolz. and Schroe. magnitude about 1.62, mid-eclipse about UT $19^h 46^m$, and semi-duration of umbral eclipse $1^h 50^m$. The eclipse is mentioned by Elias (p.290). He says the Moon was eclipsed at the sixth hour of the night on which Monday began. For a possible Islamic record, see under M.997 May 24-25.

The Annals of Ratisbon or Regensburg (e.g. SS 17, 1861, 584) and their supplement (e.g. SS 30(2), 1934, 746) have "998. The Moon was turned to blood." If correctly placed in 998, this can only refer to M.998 Nov.6-7. As remarked under S.995 July 14-15 (q.v.), the verbal formulae of Augsburg for 995 and Regensburg for 998 are very similar.

Chinese eclipses in this century (listed in Hoang 1925) are again mainly those predicted and again include some dates that are not eclipse dates at all. However, the astronomers seem to have left some clues to eclipses that were observed, and an investigation on the lines of Cohen and Newton (1984) is desirable. We are told that eclipses on 965 Mar.6 and 967 Dec.19 "did not take place", suggesting that the intervening solar eclipse of 967 July 10 reported in Southern China was observed. Similarly, the Chinese state that the lunar eclipse of 974 Sept.3 and the solar eclipse of 977 Dec.13 did not take place, so that the important Eastern China eclipse of 975 Aug.10 (seen in Japan) was presumably observed. Nevertheless, for scientific purposes we need to have observations that come from a specific city or province and the Local Histories may provide some useful information.

Dr. Stephenson (Feb.1984) points out that a source not used by Hoang, the Astronomical Treatise of the Sung-shih (Chapter 52) states of the eclipse of AD 977 Dec.13: the Sun was eclipsed and it was complete. He adds "I suspect that this is another abortive prediction of a total eclipse". It was visible in Indo-China.

A complete list of the tenth century eclipses (AD 901-1000) in Hoang 1925, with their Oppolzer Greenwich dates, is:

S.904 Nov.10
S.906 Apr.26
S.909 Feb.23
M.911 Jan.17
S.911 Feb.2
M.913 Apr.24(a)
M.913 Oct.17(a)
S.921 July 8
S.923 Nov.11
M.925 Apr.11
S.925 Apr.25
M.925 Oct.4
S.926 Sept.10
S.927 Aug.30
S.928 Feb.24
M.929 Jan.27
M.929 July 24
M.930 Jan.17
S.930 June 29
S.931 Dec.12
S.937 Feb.13
M.937 Aug.24
S.938 Feb.3
S.939 July 19
M.940 Dec.17
S.942 May 17
S.943 May 7
M.944 Apr.11
S.944 Sept.30
M.944 Oct.4
S.945 Sept.9
S.946 Mch.6
M.948 Jan.28
S.948 July 9
S.949 June 28
S.950 Dec.12
S.952 Apr.26
S.955 Feb.25
M.957 Jan.18
S.958 May 21(a)
S.960 May 28
S.961 May 17
S.965 Mch.6(b)
S.967 July 10
M.967 Dec.19(a,b)
M.968 Nov.23(a)
S.968 Dec.22
M.969 Nov.26
S.970 May 8
M.970 May 23
S.971 Oct.22
M.972 Sept.25
S.972 Oct.10
S.974 Feb.25
M.974 Sept.3(a,b)
S.975 Aug.10
M.977 July 3
S.977 Dec.13(b)
M.977 Dec.28(a)
M.978 Nov.17(a,c)
M.980 Oct.26
S.981 Sept.30
S.982 Mch.28
S.982 Dec.18(a)
S.983 Mch.17
M.984 Feb.19
M.985 Aug.3(b)
S.986 Jan.13
S.986 July 9
M.987 June 14
M.989 Apr.24(a,b)
M.990 Apr.12
S.991 Mch.18
M.991 Sept.26
S.992 Mch.7
M.992 Mch.21
M.992 Sept.14(a,c)
S.993 Feb.24
S.993 Aug.20
M.994 July 25
S.994 Aug.9
S.995 Jan.4(c)
M.995 Jan.19
M.995 July 14(c)
M.996 Jan.8(a)
M.996 Nov.27(a)
S.997 June 7
S.998 May 28
S.998 Oct.23
M.998 Nov.6
S.999 Oct.12
M.999 Oct.27
M.1000 Mch.22(a)
S.1000 Apr.7
M.1000 Sept.16(a)

(a) As Hoang points out, no Oppolzer eclipse on this date (record spurious or corrupt?).

(b) Cette éclipse annoncée n'eut pas lieu.

(c) Cette éclipse fut dissimulée par les nuages.

S.975 Aug.10 INK-COLOURED SUN IN JAPAN

In the 7th century the eclipses in the Nihongi were evidently observed, but in the 8th and later centuries there are usually so many that it is impossible to tell which were observed and which calculated.

One total solar eclipse described at Kyoto, the capital (AD 794/1868) of Japan (35°02'N, 135°45'E) on 975 August 10 is of sufficient interest to be included by Stephenson and Clark 1978, 23, who give the following translation from the Nihon Kiryaku, a privately compiled history of Japan completed about AD 1028:

"The Sun was eclipsed. Some people say that it was entirely complete. During both the (double) hours mao (5-7 a.m.) and ch'en (7-9 a.m.) it was obscured. It was like the colour of ink and without brilliance. All the birds flew about in confusion and the various stars were all visible."

The tenth century Maya dates, unlike those of the ninth and eleventh centuries, show no clear associations with eclipses. The Eclipse Table would have continued to give the Maya fairly reliable predictions of such lunar and solar eclipses as they saw.

# APPENDIX A

## E C L I P S E   C H E C K - L I S T

The check-lists of Eclipses and Comets given in Appendices A and B provide our principal conclusions and assessments of probabilities for quick reference. In the text we do, however, try to indicate more precisely the exact place of observation and we consider there other possibilities not indicated here. The several headings are as follows:

| | |
|---|---|
| TYPE | S = Solar, M = Lunar, X = Non-eclipse, Y = Maya date. |
| DATE | Julian (Old Style) throughout. Midnight to midnight. |
| CHARACTER | Block capitals are reserved for eclipses believed to have been observed. |
| IDENTIFICATION | This score represents our assessment of the certainty that the eclipse-date has been correctly identified, using both historical and scientific judgements. 9 represents certainty, but scores of only 1 or 0 must sometimes be given for dates hitherto accepted. |
| PLACE | Historical and scientific considerations have again been taken into consideration in selecting our brief comments for this column. Numismatic or papyrus evidence might still be sought for some unrecorded solar eclipses for which places are specified in lower-case type. |
| CONTENT | For Content Analysis scores have been estimated. A score of 2 thus means inadequate or incorrect information; 5 implies complete information about date and time of day, and higher scores are given for precise or additional scientific information. |
| SOURCES | The apparent principal primary source. Other sources in the text may also be primary. |

## 1st CENTURY ECLIPSES

| TYPE | DATE | CHARACTER | IDENTIFICATION | PLACE | CONTENT | SOURCES |
|---|---|---|---|---|---|---|
| S | 5 Mar 28 | PARTIAL SOLAR | 8 | SPAIN or possibly ITALY | 3 | Dio |
| X | 14 (17) | Ghost Solar | 0 | See AD 17 | 2 | Dio 3rd Century |
| M | 14 Sep 27 | LUNAR (PANNONIA)<br>(Soon after death of Augustus) | 8 | YUGOSLAVIA | 4 | Tacitus, Dio |
| S | 17 Feb 15 | (<u>After</u> death of Augustus) | 6 | MEDITERRANEAN | 0 | See 14 (17?) cf. Eusebius |
| S | 29 Nov 24 | NEAR TOTAL IN BITHYNIA<br>(In same Olympic year as Crucifixion?) | 8 | ASIA MINOR | 3 | Phlegon 2nd Century |
| S | 45 Aug 1 | Birthday of Claudius | (5) | (Egypt) | 0 | Predicted and correct date<br>No record |
| M | 46 July 6 | (CLAUDIUS)<br>(ASSOCIATED WITH ERUPTION) | 8 | AEGEAN<br>(SANTORIN) | 3 | (Aurelius Victor, Seneca, Pliny) |
| S | 49 May 20 | (Nile Delta eclipse) | 0 | (S.E. Mediterranean) | 0 | No record |
| S | 59 Apr 30 | NERONIAN<br>SOLAR ECLIPSE<br>(TOTAL IN N. AFRICA, SYRIA) | 9<br>9 | ITALY: CAMPANIA<br>ARMENIA 40°N, 41°/45°E | 5<br>5 | Pliny, Tacitus, etc.<br>Pliny, etc. |
| M | 62 Mar 13-14 | LUNAR ECLIPSE OF HERO OF ALEXANDRIA | 3<br>3 | e.g. ALEXANDRIA or<br>ROME | 2<br>2 | Hero |
| S | 67 May 31 | 1st APOLLONIAN 'ECLIPSE'<br>ANNULAR | 4 | ROME | 2 | Philostratus |
| M | 68 Oct 29? | 1st ECLIPSE OF VITELLIUS | 6 | ITALY | 2 | Dio |

1st CENTURY ECLIPSES

| TYPE | DATE | CHARACTER | IDENTIFICATION | PLACE | CONTENT | SOURCES |
|---|---|---|---|---|---|---|
| M | 69 Apr 25? | 2nd ECLIPSE OF VITELLIUS | 6 | ITALY | 2 | Dio |
| M | 69 Oct 18? | ECLIPSE BEFORE CREMONA | 7 | N. ITALY | 2 | Dio |
| M | 71 Mar 4 | LUNAR ECLIPSE OF VESPASIAN | 8 | ROME-NAPLES | 2 | Pliny |
| S | 71 Mar 20 | SOLAR ECLIPSE OF VESPASIAN | 8 | ROME-NAPLES | 2 | Pliny (and Plutarch?) |
| S | 75 Jan 5 | PLUTARCH'S ECLIPSE ? (cf. 83) | 3 | (Rome?) | 1 | Plutarch |
| S | 80 Mar 10 | Annular Eclipse | 0 | (S.E. Mediterranean) | 0 | No record |
| S | 83 Dec 27 | PLUTARCH'S ECLIPSE ? (cf. 75) | 1 | (S.E. Mediterranean) | 1 | Plutarch |
| S | 88 Oct 3 | Annular Eclipse | 0 | (W. Mediterranean) | 0 | No record |
| X | 97 Oct 23 | Ghost Eclipse of Nerva (98) | 0 | (Calculated only?) | 0 | Victor |
| X | 98 Mar 21 | Ghost Eclipse of Nerva | 0 | (Calculated only?) | 0 | Victor |

2nd CENTURY ECLIPSES

| TYPE | DATE | CHARACTER | IDENTIFICATION | PLACE | CONTENT | SOURCES |
|---|---|---|---|---|---|---|
| S | 113 June 1 | (W. Europe) | 0 | | 0 | |
| S | 118 Sep 3 | TOTAL (Central Europe & Black Sea) | 8 | N. ITALY? | 2 | 'Ravenna' |
| M | 125 Apr 5-6 | PARTIAL LUNAR | 9 | N. EGYPT | 7 | Ptolemy No.1 |
| S | 125 Apr 21 | SOLAR (S.E. Mediterranean & Iraq) | 0 | (Alexandria) | 0 | |
| S | 131 June 12 | (Spain & N.W. Africa) | 0 | | 0 | |
| M | 133 May 6-7 | TOTAL LUNAR | 9 | N. EGYPT | 8 | Ptolemy No.2 |
| M | 134 Oct 20-21 | PARTIAL LUNAR | 9 | N. EGYPT | 8 | Ptolemy No.3 |
| M | 136 Mar 5-6 | PARTIAL LUNAR | 9 | N. EGYPT | 8 | Ptolemy No.4 |
| S | 138 Jan 28 | (Gaul) | 0 | | 0 | |
| S | 143 May 2 | (N. Europe) | 0 | (Rhineland) | 0 | |
| S | 145 Sep 4 | (W. Mediterranean) | 0 | (Rome) | 0 | |
| S | 164 Sep 4 | MEDITERRANEAN & PALESTINE | 8 | ATHENS | 2 | Sosigenes in Proclus |
| S | 172 Oct 5 | (Spain & N.W. Africa) | 0 | (Carthage) | 0 | |
| S | 174 Feb 19 | (Mediterranean, Syria, etc.) | 0 | (Athens) | 0 | |
| S | 176 July 23 | (Spain & N.W. Africa) | 0 | (Carthage) | 0 | |
| S | 185 July 14 | (W. Black Sea to Nineveh) | 0 | (Armenia) | 0 | |

## 2nd CENTURY ECLIPSES

| TYPE | DATE | CHARACTER | IDENTIFICATION | PLACE | CONTENT | SOURCES |
|---|---|---|---|---|---|---|
| S | 186 Dec 28 | PARTIAL SUNSET (N.W. Africa) | 4 | ROME? | 2 | Hist. Aug. |
| S | 197 June 3 | (Mediterranean) | 0 | (Rome) | 0 | (See AD 212) |
| S | 199 Oct 7 | (N.W. Europe) | 0 | (Gaul) | 0 | |

The only solar eclipses with extant Western records are those of 118 and 186. The possibility of papyrus or numismatic evidence has led us to include a number of others. The tracks in Ginzel 1899 (Plate XI) are accurate and tied to the evidence of Ptolemy's four lunar eclipses listed above.

3rd CENTURY ECLIPSES

| TYPE | DATE | CHARACTER | IDENTIFICATION | PLACE | CONTENT | SOURCES |
|---|---|---|---|---|---|---|
| S | 212 Aug 14 | W. Mediterranean, Italy | 8 | N. AFRICA (Near Carthage) | 4 | Tertullian |
| S | 218 Oct 7 | France, Italy, Asia Minor | 9 | ASIA MINOR (or ITALY) | 3 | Dio |
| S | 228 Mar 23 | N. Europe | 0 | None expected | | |
| S | 234 June 14 | Italy (Sunrise) | 0 | None apparently | | |
| S | 237 Apr 12 | Alps (Near Sunset) | 0 | None | | See 240 |
| S | 238 Apr 2 | N. Europe (Penumbral) | 0 | None | | See 240 |
| S | 239 Aug 16 | Siberia (See text) | 0 | None | | See 240 |
| S | 240 Aug 5 | Asia Minor (Total) | 8 | ASIA MINOR ? (and CHINA) | 3 | Hist. Aug. |
| S | 241 Jan 29 | Spain, Austria (Sunset) | 0 | None, probably | | See 240 |
| S | 265 Apr 3 | N.W. Africa | 0 | None expected | | |
| S | 266 Sep 16 | Syria | 0 | None | | |
| S | 272 Nov 8 | Egypt (Annular) | | None so far | | |
| S | 291 May 15 | N. Africa and Egypt | 6 | EGYPT (or N. AFRICA) ) ) | 3 | Hydatius and |
| | | or | | ) | | later |
| S | 292 May 4 | Portugal, France | 3 | N.W. MEDITERRANEAN ) | | Malalas |
| S | 295 Mar 3 | Mediterranean (Annular) | 0 | None apparently | | |

In this century the tracks of the solar eclipses in the Roman World are reliably mapped in Ginzel 1899, Plate XII. The elements of the marginal eclipses of 228, 238 and 262 June 4 are only noted in Ginzel p.83.

4th CENTURY ECLIPSES

| TYPE | DATE | CHARACTER | IDENTIFICATION | PLACE | CONTENT | SOURCES |
|---|---|---|---|---|---|---|
| S | 301 Apr 25 | Solar | 0 | Iran, Egypt | 0 | None. cf.Ginz.Plate XIII and pp.86-89 |
| M | 303 or 304 | Ghost (Felix) | | | | |
| S | 306 July 27 | ANNULAR (of Constantius?) | 9 | BYZANTIUM? | 3 | Hamartolus from lost source |
| S | 316 July 6 | ANNULAR | 0 | Iraq | 0 | None. See 320 |
| S | 316 Dec 31 | | 0 | East Africa | 0 | None. See 320 |
| S | 317 Dec 20 | | 0 | | 0 | None. See 320 |
| S | 319 May 6 | TOTAL in S.E. EUROPE | 6 | DANUBE? | 2 | Aurelius Victor |
| | | | 8 | CONSTANTINOPLE | 3 | Latin Annals |
| S | 320 Oct 18 | TOTAL. ALGERIA, SUDAN | 6 | ALEXANDRIA | 2 | Pappus |
| S | 324 Aug 6 | (Visible in S.W. Mediterranean) | 0 | N.W. Africa | 0 | No record |
| SX | 326 Dec 11 | Ghost (Hamartolus) | 0 | Africa | 0 | See 306 |
| X | c.331 | Non-eclipse | X | N. ARMENIA | X | Socrates |
| S | 334 July 17 | ANNULAR (Mediterranean) | 9 | SICILY? | 2 | Firmicus Maternus |
| S | 335 Jan 11 | (Visible in S.W. Mediterranean) | 0 | N.W. Africa | 0 | No record |
| X | 337 May 16? | Predicted but unobservable (Solar in Alaska, etc.) | 0 | - | 0 | No record |
| S | 346 June 6 | TOTAL | 9 | ALEXANDRIA or SYRIA | 5 | Theophanes from lost source |

4th CENTURY ECLIPSES

| TYPE | DATE | CHARACTER | IDENTIFICATION | PLACE | CONTENT | SOURCES |
|---|---|---|---|---|---|---|
| S | 346 or | AMBIGUOUS YEAR | ( 4 | SYRIA (approx.) | 2 ) | Jerome (cf. Ginzel, 89) |
| | 348 | | ( 2 | PALESTINE | 2 ) | |
| S | 348 Oct 9 | SOLAR, IMPORTANT NEAR CASPIAN SEA | 9 | CONSTANTINOPLE or ASIA MINOR | 3 | Theophanes from lost source |
| S | 349 Apr 4 | | 1 | PALESTINE? | 1 | Jerome (348 or 346?) |
| S | 355 May 28 | TOTAL | 0 | (Large at Constantinople) | 0 | No record found |
| S | 359 Mar 15 | Large eclipse | 0 | (Mediterranean) | 0 | No record |
| S | 360 Aug 28 | SIGNIFICANT ASIATIC SOLAR | 9 | PERSIA | 4 | Ammianus Marcellinus. Not in Ginzel's Map. |
| S | 364 June 16 | MINOR PARTIAL ECLIPSE OF THEON | 9 | ALEXANDRIA | 9 | Theon of Alexandria. Not in Ginzel's Map. |
| M | 364 Nov 25-26 | TOTAL LUNAR OF THEON | 9 | ALEXANDRIA | 9 | Theon |
| S | 374 Nov 20 or | Ghost Eclipse of Theon | ( 0 | | 0 ) | |
| | 378 Sep 8 | " " " " | ( 0 | | 0 ) | |
| S | 386 Apr 15 | | 0 | Mediterranean | 0 | |
| S | 393 Nov 20 | IMPORTANT ECLIPSE | 9 | N.E. ITALY | 3 | Fasti, from Ravenna? |
| | | | 9 | MEDITERRANEAN | 2 | Marcellinus Comes |

## 4th CENTURY ECLIPSES

| TYPE | DATE | CHARACTER | IDENTIFICATION | PLACE | CONTENT | SOURCES |
|---|---|---|---|---|---|---|
| S | 395 Apr 6? | Total Arabia | 1 | BETHLEHEM | 1 | Jerome, Ep. (c.398) |
| or S | 400 July 8? | or Total E. Arabia | 1 | BETHLEHEM (Sunrise) | 1 | contra Joannem |
| or X | c.398 Spring | or GHOST ECLIPSE (PROBABLE AURORA) | - | - | - | Jerome |
| M | 400 Dec 17 | CLAUDIAN'S ECLIPSE No.1 | 8 | ITALY | 2 | (See 5th C lists) |

5th CENTURY ECLIPSES

| TYPE | DATE | CHARACTER | IDENTIFICATION | PLACE | CONTENT | SOURCES |
|---|---|---|---|---|---|---|
| M | 400 Dec 17 | CLAUDIAN'S ECLIPSE No.1 | 8 | ITALY | 2 | Claudian |
| M | 401 June 12 | CLAUDIAN'S ECLIPSE No.2 | 8 | ITALY | 2 | Claudian |
| M | 401 Dec 6-7 | CLAUDIAN'S ECLIPSE No.3 | 8 | ITALY | 2 | Claudian |
| S | 402 Nov 11 | NEAR-TOTAL SOLAR | 9 | S. ITALY or N.E. PORTUGAL | 4 | Hydatius |
| X | c.410 | Later legendary solar | 0 | (Rome) | 0 | (14th C) |
| M | 412 Nov 4 | THE GOLDEN HORN LUNAR ECLIPSE | 6 | N. DENMARK | 2 | Hartner, 1969 |
| S | 413 Apr 16 | THE GOLDEN HORN SOLAR ECLIPSE | 6 | N. DENMARK | 2 | Hartner, 1969 |
| S | 418 July 19 | TOTAL | 9 | E. ASIA MINOR | 4<br>4 | Philostorgius<br>Marcellinus |
| | | | 9 | PORTUGAL or ITALY | 3 | Hydatius |
| | | | 9 | ITALY | 2 | Later Annals (Prosper?) |
| | | | 3 | SYRIA | | See 421 |
| S | 421 May 17 | SOLAR | 6 | SYRIA | 2 | Agapius (10th C) |
| X | 421 | False identification (Solar) | 0 | ITALY? | 0 | See 418 |
| X | 442 Oct 5 | False identification (Lunar) | 0 | | | |
| S | 443 Mar 17 | Sunset solar | 0 | (S.W. Europe) | 0 | No record |
| X | 445 July 20 | False identification (Irish) | 0 | | 0 | See 447 |

5th CENTURY ECLIPSES

| TYPE | DATE | CHARACTER | IDENTIFICATION | PLACE | CONTENT | SOURCES |
|---|---|---|---|---|---|---|
| S | 447 Dec 23 | TOTAL SOLAR | 9 | N.E. PORTUGAL | 4 | Hydatius |
| | | | 9 | S. FRANCE | 2 | Montpellier MS. |
| | | False location | 8 | Ireland, Wales, Sweden | 0 | Annals |
| XS | 449 May 8 | Sunrise eclipse | 1 | Syria | 1 | John of Asia |
| or | | | | | | |
| X | 450 Aug 15 | Meteorological darkness | | Constantinople | | John of Asia |
| XM | 451 Apr 2 | Ghost lunar eclipse | 0 | | | |
| M | 451 Sep 26 | LUNAR ECLIPSE | 9 | N.E. PORTUGAL | 4 | Hydatius |
| | 453/4 | Sunspot? | | Portugal | | Hydatius |
| S | 458 May 28 | PARTIAL SOLAR | 9 | PORTUGAL | 6 | Hydatius |
| M | 462 Mar 1-2 | LUNAR 'BLOOD' | 9 | PORTUGAL | 3 | Hydatius |
| S | 464 July 20 | PARTIAL SOLAR | 9 | PORTUGAL | 6 | Hydatius |
| | " | (Near) TOTAL SOLAR | 9 | SYRIA (near) | 2 | Agapius |
| X | 467 | Comet? | 1 | PORTUGAL | 1 | (Hydatius) |
| S | 472 Aug 20 | Total | 0 | (S. Egypt) | 0 | No record |
| X | 473 | Darkness (Cold front?) | 0 | CONSTANTINOPLE | 0 | Cedrenus (Late source) |
| S | 484 Jan 14 | (SUNRISE SOLAR ECLIPSE<br>( | 9 | ATHENS | 6 | Marinus |
| | | (SOLAR OF KING PIRUZ | 9 | PERSIA | 4 | Elias 11th C |

5th CENTURY ECLIPSES

| TYPE | DATE | CHARACTER | IDENTIFICATION | PLACE | CONTENT | SOURCES |
|---|---|---|---|---|---|---|
| X | 485 May 29 | Improbable year (see 497) | 2 | | | See 497 |
| S | 486 May 19? | Prediction | | (Greece) | | (Marinus) |
| S | 486 May 19 | SOLAR | 9 | SYRIA or ARABIA | 5 | Elias 11th C |
| X | 492 Jan 15 | False identification | 0 | (Ireland) | 0 | |
| S | 493 Jan 4 | Sunrise eclipse | 1 | Syria | | See 497 (Marc.) |
| S | 496 Oct 22 | Possible alternative | 1 | (S.E. Europe) | 0 | See 497 (Marc.) |
| S | 497 Apr 18 | SOLAR | 8 | E. MEDITERRANEAN | 2 | Marcellinus |
| S | 497 Apr 18? | 1st SOLAR OF GREGORY (PARTIAL) | 7 | VANDAL N. AFRICA | 2 | Gregory of Tours |

6th CENTURY ECLIPSES

| TYPE | DATE | CHARACTER | IDENTIFICATION | PLACE | CONTENT | SOURCES |
|---|---|---|---|---|---|---|
| S | 511 Jan 15 | Sunset | 1 | (Constantinople) | | See 512 |
| S | 512 June 29 | MEDITERRANEAN | 9 | S.W. ITALY | 5 | P. Campanum |
| | | | 8 | S.E. EUROPE | 3 | Marcellinus, Lydus |
| | | | 9 | SYRIA | 5 | Syriac Chronicle |
| | | | 8 | SYRIA, ARMENIA | 3 | Elias and Samuel |
| S | 526 Sep 22 | African | 1 | NEAR RED SEA | 3 | Elias 11th C |
| S | 527 Sep 11 | Atlantic | 0 | Portugal | | No record found |
| S | 534 Apr 29 | ANNULAR | 9 | SYRIA | 5 | Agapius 10th C |
| | | | 4 | ITALY? | 2 | Consularia Italica |
| | 536 May/<br>537 June | Volcanic darkness | | Mediterranean | | Various |
| S | 538 Feb 15 | E. MEDITERRANEAN | 9 | ASIA MINOR?<br>or ITALY? | 5 | 'Bede' |
| S | 540 June 20 | TOTAL | 9 | ROME? | 5 | 'Bede' |
| S | 541 Dec 3 | E. MEDITERRANEAN | 0 | (N.W. Syria) | | No record found |
| S | 547 Feb 6 | PARTIAL | 9 | N. EGYPT, etc. | 3 | Cosmas |
| M | 547 Aug 17 | Correct prediction | 2 | N. Egypt | 0 | Cosmas |
| M | 560 Nov 18? | DEEP TOTAL | 8 | SWITZERLAND | 5 | Marius of Avenches |

6th CENTURY ECLIPSES

| TYPE | DATE | CHARACTER | IDENTIFICATION | PLACE | CONTENT | SOURCES |
|---|---|---|---|---|---|---|
| S | 563 Oct 3 | PARTIAL<br>& COMET (565) etc. | 9 | FRANCE | 4 | Gregory of Tours |
| S,M | c.565 [563?] | Confused portents | 1 | (Constantinople?] | 1 | Sigebert (Late source) |
| S | 566 Aug 1 | TOTAL | 9 | ARABIA?<br>EGYPT or<br>SYRIA | 4 | Syriac 10th C |
| M | 567 Dec 31 | DEEP TOTAL | 9 | S.E. EUROPE? | 5 | St. Gall 9th C |
| M | 577 Dec 11?<br>(&/or 574/5?) | LUNAR ECLIPSE(S) | 8 | FRANCE | 2 | Gregory of Tours |
| M | 581 Apr 5 | LUNAR ECLIPSE<br>& COMET (581 Jan 20?) | 5 | FRANCE | | Gregory of Tours |
| M | 582 Mar 25<br>or Sep 17 | DEEP TOTAL | 2<br>or 7 | FRANCE | | Gregory of Tours |
| S | 590 Oct 4 | PARTIAL | 9 | FRANCE | 4 | Gregory of Tours |
| S | 590 Oct 4 | SOLAR | 7 | CONSTANTINOPLE | 3 | Simocatta, etc. |
| M | 590 Oct 18 | PARTIAL | 8 | N.E. FRANCE? | 3 | Burgundian Chron. |
| S | 591 Mar 30 | False identification | 0 | (CHINA) | | No Western record |
| S | 591 Sep 23 | False identification | 0 | | | No genuine record |
| S | 592 Mar 19 | PARTIAL PHASE | 9 | SYRIA | 4 | Agapius 10th C |
| | | | 9 | BURGUNDY | 4 | Fredegar |

6th CENTURY ECLIPSES

| TYPE | DATE | CHARACTER | IDENTIFICATION | PLACE | CONTENT | SOURCES |
|---|---|---|---|---|---|---|
| | | NEAR TOTAL PHASE | 7 | CONSTANTINOPLE | 3 | Michael and Theophanes |
| S | 594 July 27 | TOTAL SOLAR | 9 | N.W. BRITISH ISLES | 3 | Irish Annals |
| S | 596 Jan 5 | Annular | 0 | (S.W. Europe) | | No record |

7th CENTURY ECLIPSES

| TYPE | DATE | CHARACTER | IDENTIFICATION | PLACE | CONTENT | SOURCES |
|---|---|---|---|---|---|---|
| S | 601 Mar 10 | NEAR-TOTAL | 9 | EGYPT (THEBES)<br>and SYRIA | 5<br>5 | Coptic ostracon<br>Syriac chronicle |
| S | 603 Aug 12 | | 9 | E. FRANCE<br>(Track of totality:<br>S.W. Ireland to N.E. Spain) | 4 | Burgundian Annals |
| M | 604 July 16-17 | LUNAR | 9 | MESOPOTAMIA<br>(EDESSA = URFA) | 5 | Syriac chronicle |
| S | 606 June 11 | ANNULAR | 0 | (Asia Minor) | 0 | No clear record |
| M | 610 Mar 15 | Calculated Lunar | 0 | | 0 | No contemporary source |
| S | 612 Aug 2 | | 0 | (N.W. Africa &<br>e.g. Seville) | | Irish 'Ghost'<br>(from S. Spain?) |
| S | 616 May 21 | | | (Arabia) | 1 | See 617 |
| S | 617 Nov 4 | SOLAR | 7 | ALEXANDRIA or<br>CONSTANTINOPLE<br>(Vienna to Babylon) | 2 | Stephanus of Alexandria<br>(at Constantinople?) |
| M | 618 Oct 9 | LUNAR | 7 | ALEXANDRIA or<br>CONSTANTINOPLE | 3 | Stephanus of Alexandria |
| M | 622 Feb 1-2 | LUNAR eclipse of<br>Persian Expedition<br>Later calculation | 2 | ARMENIA<br>Islamic | 2<br>0 | George of Pisidia<br>Later astrological sources |
| S | 624 June 21 | | 0 | (e.g. Seville) | 0 | No Spanish record traced |
| | | 'Ghost' eclipse of Edwin | | | 1 | Irish 'Ghost' |

7th CENTURY ECLIPSES

| TYPE | DATE | CHARACTER | IDENTIFICATION | PLACE | CONTENT | SOURCES |
|---|---|---|---|---|---|---|
| M | 625 Nov 19-20 | LUNAR | 1 | (Islamic?) | 0 | No certain Islamic record |
| XS | 626-9 | 'Ghost' eclipses of Agapius | 0 | | 0 | See 632 |
| M | 630 Aug 28 | Lunar (calculated) | 0 | Invisible | 0 | Later calculations only |
| S | 632 Jan 27 | ANNULAR | 8 | MECCA | 2 | Early Arabian sources |
| | | Agapius | 1 | SYRIA? | 2 | True date? See 626/9 |
| M | 634 June 16-17 | Lunar (calculated) | 0 | | 0 | Later Islamic calculation |
| S | 639 Sep 3 | Total | 0 | (N. Europe) | 0 | No Western record |
| M | 644 May 27 | Lunar (calculated) | 0 | (Invisible in Near East) | 0 | Later Islamic calculation |
| S | 644 Nov 5 | ANNULAR | 7 | SYRIA (Constantinople, Smyrna, Bagdad) | 5 | Syriac source in Theophanes (Later Syriac sources corrupt) |
| S | 646 Apr 21 | | 4? | EDESSA? (and CHINA) (Constantinople & Asia Minor) | 1 | Sebokht? |
| XS | 650 Feb 6 | Scottish 'Ghost' | | | | No genuine record |
| S | 655 Apr 12 | TOTAL in SPAIN | 7 | SPAIN | 2 | Continuation of Isidore (Perhaps AD 666) |
| M | 655 Oct 20-21 | LUNAR (calculated?) | 3? | ISLAMIC? | 1? | Horoscope |
| S | 659 Jan 28 | LARGE ANNULAR - TOTAL | 0 | (Italy and Balkans) | 0 | No record |

## 7th CENTURY ECLIPSES

| TYPE | DATE | CHARACTER | IDENTIFICATION | PLACE | CONTENT | SOURCES |
|---|---|---|---|---|---|---|
| M | 660 Dec 22 | Lunar (calculated) | 0 | Invisible in Islam | 0 | Later calculation |
| S | 661 July 2 | Important in N & W of British Isles | 0 | (Recorded only in CHINA) | 0 | No Western record |
| S | 664 May 1 | FIRST GENUINE ENGLISH ECLIPSE | 9 | N.E. IRELAND/IONA and | 5 | Annals of Ulster |
| | | | | N.E. ENGLAND | 3 | Bede |
| S | 666 Sep 4 | TOTAL in SPAIN | 2 | (SPAIN?) | | Probable 'ghost', but see 655 |
| M | 670 Jan 11-12 | LUNAR (calculated?) | 3 | (Islam) | 1? | Later astrological sources |
| S | 671 Dec 7 | ANNULAR in ARABIA | 9 | SYRIA | 4 | Michael's Syrian source |
| | | | 9 | ARABIA | 3 | Arabian sources |
| X | 672 Nov 11-12 | Ghost Lunar (Irish aurora?) | | | | |
| M | 679 Dec 22-23 | IMPORTANT LUNAR (calc.?) | 5? | ISLAM | 1? | Later astrological sources |
| M | 680 June 17-18 | TOTAL LUNAR | 9 | ITALY | 5 | Papal annals |
| M | 683 Apr 16-17 | TOTAL LUNAR | 9 | ITALY | 5 | Papal annals |
| S | 688 July 3 | OBSERVED PARTIAL IMPORTANT IN N. SCOTLAND (Less than 75% at Iona) | 9 | SCOTLAND or IRELAND | 3 | Annals of Ulster |
| M | 691 Nov 11 | LUNAR (PARTIAL) | 9 | IONA or N.E. IRELAND | 4 | Annals of Ulster |
| S | 693 Oct 5 | IMPORTANT from Constantinople | 9 | S.E. EUROPE | 5 | Theophanes |
| | | to Basra | 9 | MESOPOTAMIA | 5 | James of Edessa |
| S | 698 Dec 8 | Important Annular | 0 | (N.W. Europe to Syria) | 0 | No record found |

## 8th CENTURY ECLIPSES

| TYPE | DATE | CHARACTER | IDENTIFICATION | PLACE | CONTENT | SOURCES |
|---|---|---|---|---|---|---|
| M | 716 Jan 13 | TOTAL | 9 | ITALY | 3 | Papal |
| M | | TOTAL | 1 | IRELAND | | (See 718) |
| S | 718 June 3 | TOTAL SOLAR | 9 | SPAIN | 5 | Cont. Hisp. |
| M | 718 Nov 12-13 | SMALL LUNAR | 6 | IRELAND | 4 | Annals (AU) |
| S | 720 Oct 6 | SOLAR | 4 | N. AFRICA? | 1 | Lost Islamic? |
| M | 725 Dec 24 | | 1 | IRELAND | | (See 726) |
| M | 726 Dec 13-14 | | 8 | IRELAND | 2 | Annals (AU) |
| S | 733 Aug 14 | TOTAL SOLAR | 9 | N.E. ENGLAND | 5 | Cont. Bede |
| | | | 9 | CAUCASUS | 3 | 'Albanian' Chronicle |
| M | 734 Jan 24 | | 9 | N.E. ENGLAND | 5 | Cont. Bede |
| M | 734 Jan 24 | | 9 | IRELAND | 4 | AU |
| M | 735 Jan 13 | | 0 | IRELAND | | See 734 |
| M | 744 | Probable Comet (Jan 745) | 1 | SYRIA | | Agapius |
| M | 752 July 30-31 | DEEP TOTAL | 9 | N.E. ENGLAND | 4 | N. Annals (SD) |
| S | 753 Jan 9 | ANNULAR SOLAR | | (Especially Central Europe) | | |
| | | | 9 | N.E. ENGLAND | 4 | N. Annals (Cont. Bede) |
| | | | 9 | IRELAND | 3 | Annals (AU, AT) |

8th CENTURY ECLIPSES

| TYPE | DATE | CHARACTER | IDENTIFICATION | PLACE | CONTENT | SOURCES |
|---|---|---|---|---|---|---|
| M | 753 Jan 23 | LARGE | 9 | N.E. ENGLAND | 4 | N. Annals (Cont. Bede) |
| | | | 9 | IRELAND | 2 | I. Annals (AT) |
| M | 755 Nov 23 | LUNAR with OCCULTATION | 9 | N.E. ENGLAND | 6 | N. Annals (SD) |
| S | 758 Apr 12 | Solar | 0 | (N. Europe) | 0 | No record |
| S | 760 Aug 15 | ANNULAR SOLAR | | (E. Europe) | | |
| | | | 9 | CONSTANTINOPLE | 5 | Theophanes |
| | | | 3 | CENTRAL EUROPE | 1 | Annals? |
| M | 762 Jan 14 | PARTIAL | 1 ) | | | |
| M | 763 Jan 4 | TOTAL | 2 ) | | | |
| | | | ) | IRELAND | 2 | AU, AT |
| M | 763 June 29-30 | TOTAL | 1 ) | | | |
| M | 763 Dec 26 | PARTIAL | 5 ) | | | |
| S | 764 June 4 | ANNULAR SOLAR | | (N.W. to E. Europe) | | |
| | | | 9 | GERMANY? | 1 | Carolingian |
| | | | 9 | IRELAND | 3 | AU |
| | | | 0 | (Constantinople) | 0 | No Greek source |
| M | 773 Dec 3 | TOTAL | 9 | IRELAND | 4 | AU |
| M | 774 Nov 22-23 | NEAR-TOTAL | 8 | ITALY | 2 | La Cava (?) |

8th CENTURY ECLIPSES

| TYPE | DATE | CHARACTER | IDENTIFICATION | PLACE | CONTENT | SOURCES |
| --- | --- | --- | --- | --- | --- | --- |
| | 777/778 | Calculated solar of Roncevaux | 0 | Spain | 0 | Fictitious (Calculations) |
| S | 779 Aug 16 | TOTAL S. SPAIN | 6? | SPAIN | 3 | Escorial MS. |
| S | 786 Apr 3 | Total | 0 | (S.E. Mediterranean) | 0 | No record |
| S | 787 Sep 16 | ANNULAR | | (S.W. Europe to Syria and China) | | |
| | | | 9 | RHINELAND | 3 | Ann. Lorsch. |
| | | | 9 | CONSTANTINOPLE & NICAEA | 3 | Theophanes only |
| | | | 4 | ITALY? | 3 | Lucca. MS. |
| M | 788 Feb 26 | NEAR TOTAL | 9 | N. FRANCE (Autun) | 3 | Ann. Flav. |
| | | | 8 | IRELAND | 3 | AU |
| S | 791 July 6 | Sunrise | 0 | (Constantinople) | 0 | No record |
| M | 795 Apr 9 | TWO LUNAR | 9 | FRANCE | 2 | Ann. Flav. |
| M | 795 Oct 3 | TWO LUNAR | 9 | FRANCE | 2 | Ann. Flav. |
| M | 796 Mar 28 | TOTAL LUNAR | 8 | N. ENGLAND | 4 | N. Annals (SD) = ASC MS. D(E) |
| S | 796/8 | VOLCANIC DARKNESS | | | | (Theophanes, etc.) |
| M | 800 Jan 15-16 | MAJOR LUNAR | 9 | ENGLAND | 5 | ASC MSS. D,E |

9th CENTURY ECLIPSES

| TYPE | DATE | CHARACTER | IDENTIFICATION | PLACE | CONTENT | SOURCES |
|---|---|---|---|---|---|---|
| M | 800 Jan 15-16 | (See 8th C chapter) | 9 | ENGLAND | 5 | ASC (D, E) |
| M | 802 May 21 | TOTAL | 9 | ENGLAND | 5 | ASC (D, E) |
| M | 803 Nov 2 | | 9 | E. FRANCE | 2 | Ann. Flav. |
| M | 806 Sep 1 | TOTAL | 9 | FRANCIA | 6 | RFA |
| | | | 9 | ENGLAND | 5 | ASC (D, E) |
| S | 807 Feb 11 | ANNULAR | 9 | FRANCIA | 6 | RFA |
| | | | 9 | ITALY | 5 | Ann. Farfa |
| | | | 9 | WALES | 3 | Brut |
| | | | 5 | (AUSTRIA)? | 5 | Ann. Juvavenses maior. |
| M | 807 Feb 26 | TOTAL | 9 | FRANCIA | 6 | RFA |
| | (Probable identification) | | 6 | IRELAND | 3 | Irish Annals |
| M | 807 Aug 21 | | 9 | FRANCIA | 6 | RFA |
| S | 809 July 16 | PARTIAL | 9 | ENGLAND | 5 | ASC MS.F |
| | | | or 9 | or N.E. FRANCIA | | from Ann. S. Col. Senon? |
| M | 809 Dec 25 | TOTAL | 9 | FRANCIA | 5 | RFA |
| | | | 9 | WALES | 5 | Brut |
| M | 810 June 20 | TOTAL | 9 | FRANCIA | 5 | RFA |

## 9th CENTURY ECLIPSES

| TYPE | DATE | CHARACTER | IDENTIFICATION | PLACE | CONTENT | SOURCES |
|---|---|---|---|---|---|---|
| X | 810 July 5 | Calculated (S) | | | | |
| S | 810 Nov 30 | TOTAL | 9 | AUSTRIA | 5 | Ann. Juv. maior. |
| | | | 9 | GERMANY | 5 | RFA |
| M | 810 Dec 14 | TOTAL | 9 | FRANCIA | 5 | RFA etc. |
| S | 812 May 14 | TOTAL | 9 | SYRIA | 5 | |
| | | ASIA MINOR & N.E. SYRIA | 9 | W. ASIA or CONSTANTINOPLE | 5 | Theophanes |
| S | 812 May (14) | PARTIAL PHASE | 7 | FRANCIA | 5 | RFA etc. (15th May) |
| X | 812 June | Spurious date | | | | |
| M | 812 Oct 23 | Small, possibly observed | 3 | (FRANCIA)? | 1 | cf. Einhard |
| S | 813 May 4 | TOTAL SUNRISE | 9 | S.E. EUROPE | 6 | Theophanes |
| | | (PARTIAL PHASE) | 2 | GERMANY | 1 | cf. Einhard |
| S | 814 (Sep 17) | Calculation or invention | | | | |
| Y | 815 Mar 14 | Important Maya solar eclipse | 0 | MESOAMERICA | | None found |
| M | 817 Feb 5 | TOTAL | 9 | FRANCIA | 5 | RFA etc. |
| S | 818 July '8' (7) | SMALL SOLAR (Calculated?) | 2 | FRANCIA? | 5? | RFA etc. |

9th CENTURY ECLIPSES

| TYPE | DATE | CHARACTER | IDENTIFICATION | PLACE | CONTENT | SOURCES |
|---|---|---|---|---|---|---|
| M | 820 Nov '24' (23) | LUNAR | 5 | FRANCIA? | 5 | RFA etc. |
| M | 824 Mar '5' (18) | LUNAR | 9 | FRANCIA | 5 | RFA etc. |
| M | 828 July 1 | DAWN LUNAR | 9 | FRANCIA | 5 | RFA etc. |
| X | 828 July 15 | False identification | 0 | | | |
| M | 828 Dec 25 | TOTAL | 9 | FRANCIA | 5 | RFA etc. |
| | | | 9 | ENGLAND | 5 | ASC etc. |
| S | 829 Nov 30 | PARTIAL | 9 | BAGDAD? | 5? | (Ibn Yūnus, 10th C) |
| Y | 830 May 25 | Minor Maya Solar | 2 | (AMERICA) | | (Hieroglyphic stairway?) |
| M | 831 Apr 30 | Near total | 0 | FRANCIA? | | (No source traced) |
| X | 831 May '16' (15) | Calculation only (Solar) | 0 | FRANCIA? | | (No source traced) |
| M | 831 Oct 24 | TOTAL | 9 | FRANCIA | 4 | Ann. Xanten |
| | | | 9 | WALES | 5 | Brut |
| M | 832 Apr 18 | TOTAL | 9 | N.E. FRANCE | 5 | Ann. St. Bertin |
| | | | 9 | W. GERMANY | 5 | Ann. Xanten |
| X | 832 May 4 | Prediction of Solar Eclipse | 0 | (Visible Southern Hemisphere) | | (Ann. Fulda) |

## 9th CENTURY ECLIPSES

| TYPE | DATE | CHARACTER | IDENTIFICATION | PLACE | CONTENT | SOURCES |
|---|---|---|---|---|---|---|
| X | c.833 (832) | Confused predictions cf. Solar & Lunar | 0 | | | Ann. Xanten |
| M | 835 Feb (17) | TOTAL | 9 | W. GERMANY | 4 | Ann. Xanten |
| S | 836 July 17 | E. Europe Solar | 0 | (E. Europe) | | No record |
| Y | 837 July 6 | Sunrise Solar | | MAYA | | Yaxchilan date? |
| M | 838 Dec 5 | TOTAL | 9 | N.E. FRANCE | 5 | Ann. St. Bertin |
| S | 840 May 5 | IMPORTANT SOLAR (TOTAL) | 9 | FRANCE, GERMANY | 6 | Ann. Fulda etc. |
| | | | 9 | SWITZERLAND & ITALY | 5 | Chr. Ven. etc. |
| | | | 9? | CONSTANTINOPLE | 5? | Continuator of Theophanes |
| S | 841 Oct 18 | PARTIAL (0.5) SOLAR | 9 | N. FRANCE | 4? | Nithard (Observed?) |
| | | | 5 | MEDITERRANEAN | 3 | See 840 (Theoph. Cont.) |
| M | 842 Mar 30 | LUNAR | 9 | GERMANY | 5 | Ann. Fulda |
| | | | 9 | N.E. FRANCE | 6 | Fontanelle |
| Y | 842 Mar 30 | MAYA: LUNAR | 8 | MAYA: COPAN | 5 | Dresden Codex |
| M | 843 Mar 19 | TOTAL LUNAR | 9 | FRANCIA | 5 | Nithard |
| Y | 847 July 2 | MAYA: LUNAR | 6 | MAYA | 4 | Quirigua dates |
| X | 848 Oct 1 | (Calculated (later) as Solar) | 0 | Invented for N.E. England | | (Roger of Wendover 13th C) |

## 9th CENTURY ECLIPSES

| TYPE | DATE | CHARACTER | IDENTIFICATION | PLACE | CONTENT | SOURCES |
|---|---|---|---|---|---|---|
| S | 849 May 25 | Scottish Solar | 0 | (Scotland) | | No record |
| Y | 849 Nov 4 | MAYA: LUNAR | 4 | MAYA | 4 | Copan date (Nov 2) |
| Y | 850 Oct 24 | MAYA: LUNAR | | | | Quirigua date |
| S | 852 Mar 24 | Annular N.W. Europe | 0 | (N.W. Europe) | | No record |
| M | 854 Feb 16 | LARGE PARTIAL | 9 | BAGDAD | 5 | Ibn Yūnus |
| M | 854 Aug 11-12 | TOTAL LUNAR | 9 | BAGDAD | 5 | Ibn Yūnus |
| S | 856 Jan 11 | Total Solar | 0 | S.W. Europe & N.W. Africa | | No record |
| M | 856 June 21-22 | Partial | 3 | BAGDAD | 5 | Ibn Yūnus |
| Y | 857 June 11 | MAYA: LUNAR (NODAL) | | MAYA | | Piedras Negras date |
| M | 861 Mar 30 | TOTAL N.W. EUROPE | 9 | FRANCE | 5 | Ann. St. Bertin |
| S | 863 Aug 18 | Especially S. Europe (Solar) | 0 | S. Europe | 0 | No record |
| S | 865 Jan 1 | N. British Isles | 9 | IRELAND | 5 | AU etc. |
| M | 865 Jan 15 | TOTAL | 9 | IRELAND | 4 | AU |
| | | | 9 | GERMANY | 4 | Ann. Xanten |
| S | 866 June 16 | PARTIAL SOLAR | 9 | BAGDAD | 5 | Ibn Yūnus |
| M | 866 Nov 26 | Good calculation? (Small Lunar) | 2 | Bagdad | 6 | Ibn Yūnus |

9th CENTURY ECLIPSES

| TYPE | DATE | CHARACTER | IDENTIFICATION | PLACE | CONTENT | SOURCES |
|---|---|---|---|---|---|---|
| M | 867 Nov 15 | TOTAL | 9 | BAGDAD | 7 | at-Tabarī |
| M | 871 Sep 2 | Calculated (& observed) | 4 | Basra | 2 | at-Tabarī |
| S | 873 July 28 | ANNULAR | 9 | IRAN | 8 | al-Birūni |
| M | 878 Oct 15 | TOTAL | 9 | IRELAND | 4 | AU |
| | | | 9 | GERMANY | 3 | Ann. Fulda |
| S | 878 Oct 29 | TOTAL SOLAR | 9 | IRELAND | 5 | AU |
| | | | 9 | ENGLAND | 3 | ASC & Charter |
| | | | 9 | GERMANY | 4 | Ann. Fulda, Regino, Asser's source |
| X | 879 Mar 26 | False identification (Solar) | 0 | | 0 | No record |
| X | 879 Oct 4 | False identification (Lunar) | 0 | e.g. Visible England | 0 | No good record |
| X | 880 Mar 14 | False identification (Solar) | 0 | | 0 | No record |
| X | 881 (Aug 28) [Recte 891] | False identification | 0 | | 0 | No record |
| M | 882 Feb 7 and/or Aug 3 | TOTAL ECLIPSES | 9 | SWITZERLAND | 3 | Ann. Sangall. |
| M | 882 Aug 3 | TOTAL | 9 | BAGDAD | 6 | at-Tabarī |

## 9th CENTURY ECLIPSES

| TYPE | DATE | CHARACTER | IDENTIFICATION | PLACE | CONTENT | SOURCES |
|---|---|---|---|---|---|---|
| S | 882 Aug 17 | Partial (0.17) Sunset Solar (Calculated?) | 1 | Bagdad | 0 | at-Tabarī |
| X | 883 Jan 27 | False identification | 0 | | | |
| M | 883 July 23-24 | LARGE LUNAR | 9 | SYRIA | 7 | al-Battanī |
| S | 885 June 16 | SOLAR | 9 | SCOTLAND | 5 | Pictish Chron. |
| | | (NEAR TOTAL) | 9 | N. IRELAND | 3 | AU |
| X | 887 Oct.20 | False identification (Solar) | 0 | | | |
| S | 891 Aug 8 | ANNULAR | 9 | N.E. SYRIA | 6 | al-Battanī |
| | | | 7 | CONSTANTINOPLE | 3 | Symeon et al. |
| | | | 9 | N.E. FRANCE | 4 | Folcwin |
| | | | 9 | SWITZERLAND | 3 | St. Gall (Ann. Alam.) |
| | | | 9 | GERMANY | 3 | Various: Corvey? |
| | | | 9 | BELGIUM | | Ann. Laub. |
| M | 893 July 2 or Jan 6 | ST. GALL '893' | 7 | SWITZERLAND | 3 | St. Gall (Ann. Alam.) |
| M | 893 Dec 26-27 | CORRECTED DATE | 8 | N.W. IRAN | 3 | Elias |
| | | ISLAMIC | 9 | ARDABIL | 5 | at-Tabarī (Contemp.) |

9th CENTURY ECLIPSES

| TYPE | DATE | CHARACTER | IDENTIFICATION | PLACE | CONTENT | SOURCES |
|---|---|---|---|---|---|---|
| M | 893 Dec 26 or 894 June 21-22 | ST. GALL '894' | 7 | SWITZERLAND | 3 | St. Gall (Ann. Alam.) |
| S | 894 June 7 | Extreme N. Europe | 0 | | | No record |
| M | 894 (893) | False dates | 0 | | | No true records |

## 10th CENTURY ECLIPSES

| TYPE | DATE | CHARACTER | IDENTIFICATION | PLACE | CONTENT | SOURCES |
|---|---|---|---|---|---|---|
| S | 901 Jan 23 | PARTIAL SOLAR | 9 | SYRIA, ARMENIA | 8 ) | Albategnius (al-Battanī) |
| M | 901 Aug 2-3 | TOTAL | 9 | SYRIA | 7 ) | |
| S | 903 June 27 | S.W. EUROPE SOLAR | 8 | S. SPAIN | | Ibn Hayyān |
| M | 904 May 31 or Nov 25 | 'MERCIAN' LUNAR | 9 | ENGLAND | 3 | ASC |
| S | 906 Apr 26 | Annular Islamic Solar | 0 | | 0 | No Western record |
| M | 908 Mar 20 | LUNAR | 8 | S.E. EUROPE | 2 | Symeon |
| M | 912 Jan 7 | TOTAL | 1 | IRELAND? | 1 | AU |
| S | 912 June 17 | TOTAL SOLAR (SUNSET) | 9 | S. SPAIN | 6 | Ibn Adarī and Ibn Hayyān |
| | | | 5 | IRELAND? | 1 | AU |
| S | 913 June 7 | Total in Egypt | 0 | (Egypt) | 0 | No record |
| X | 916 Apr 5 | False identification | 0 | (Near East) | 0 | No record |
| (S) | 917 Sep 19 | No record | 0 | (Bagdad) | 0 | No record |
| M | 921 Dec 17 | TOTAL | 9 | IRELAND | 5 | AU |
| M | 923 June 1 | PARTIAL | 9 | BAGDAD | 8 | Ibn Yūnus |
| S | 923 Nov 11 | PARTIAL SOLAR | 9 | BAGDAD | 8 | Ibn Yūnus |
| M | 925 Apr 11 | TOTAL | 9 | BAGDAD | 6 | Ibn Yūnus |
| M | 926 Apr 1 | TOTAL | 9 | FRANCE | 5 | Ann. Flodoard |

10th CENTURY ECLIPSES

| TYPE | DATE | CHARACTER | IDENTIFICATION | PLACE | CONTENT | SOURCES |
|---|---|---|---|---|---|---|
| M | 927 Sep 13-14 | SMALL LUNAR (0.2) | 9 | BAGDAD | 8 | Ibn Yūnus |
| S | 928 Aug 18 | SMALL SOLAR (0.2) SUNRISE (0.2; observed on water) | 9 | BAGDAD | 8 | Ibn Yūnus |
| M | 929 Jan 27-28 | TOTAL | 9 | BAGDAD | 6 | Ibn Yūnus |
| M | 932 Nov 16 | NEAR TOTAL | 9 | SYRIA & BAGDAD | 6 | al-Hashimi in al-Birūni (fl. AD 1025) |
| S | 932 Nov 30 | N.E. Atlantic Solar | 0 | (N.W. Ireland) | | No record |
| M | 933 Nov 4-5 | TOTAL | 9 | BAGDAD | 7 | Ibn Yūnus |
| S | 934 Apr 16 | ANNULAR SOLAR (Late afternoon) | 1 | GERMANY (Aurora?) | 2 | Widukind |
| | | (Conspicuous in NW Europe) | 2 | BRITISH ISLES? | | No clear record |
| M | 936 Sep 3-4 | 'BLOOD-MOON' | 9 | FRANCE | 2 | Flodoard |
| S | 939 July 19 | TOTAL SOLAR | ( 9 | S. SPAIN | 4 | Masʿūdi's source |
| | | | ( 9 | N. SPAIN | 2 | Sampiro? |
| | | | ( 9 | ITALY | 5 | Various |
| | | PARTIAL SOLAR | ( 9 | SWITZERLAND & | 5 | Ann. Sang. Mai. |
| | | | ( 9 | GERMANY | 5 | Ann. Corb. |
| M | 940 Dec 17 | TOTAL | 9 | SYRIA | 5 | Tābit (Lost) in Elias |

## 10th CENTURY ECLIPSES

| TYPE | DATE | CHARACTER | IDENTIFICATION | PLACE | CONTENT | SOURCES |
|---|---|---|---|---|---|---|
| S | 946 Mar 6 | (Visible S. Egypt & Arabia) | 0 | | 0 | No record found |
| M | 955 Sep 4-5 | TOTAL | 9 | W. FRANCE (NEAR ORLEANS) | 4 | Fragment from Fleury |
| S | 961 May 17 | ANNULAR SOLAR | 9 | W. GERMANY | 5 | Adalbert at Trier |
| | | Misdated c.957 | 5 | W. FRANCE | 3 | Fleury fragment |
| | | | | (CHINA) | | |
| | | | 3? | ITALY | 1 | Lupus |
| S | 963 Sep 20 | Small (0.5) Solar | 4 | S. ITALY | 2 | Lupus |
| S | 964 Mar 16 | Sunset | 0 | (S.W. Ireland) | | No record |
| M | 965 Aug 15 | TOTAL (LIKE BLOOD) | 9 | W. GERMANY | 5 | Ann. Prüm. |
| S | 966 July 20 | LARGE SOLAR (0.86) | 9 | SCOTLAND | 2 | Pictish Chron. |
| | | (0.8) | 4 | CENTRAL ITALY | 2 | Ann. Farfa |
| S | 967 July 10 | (Total) | 0 | (Asia Minor) | | No record |
| M | 968 Dec 7-8 | LARGE PARTIAL LUNAR | 9 | SYRIA or IRAQ | 5 | Tābit (Lost) in Elias |
| S | 968 Dec 22 | TOTAL SOLAR | 9 | ITALY | 5 | Anselm, Ital.Ann.(Var.) Ann. Beneventum |
| | | | 9 | CORFU, ADRIATIC | 5 | Liudprand |
| | | | 9 | CONSTANTINOPLE | 5 | Leo the Deacon |

10th CENTURY ECLIPSES

| TYPE | DATE | CHARACTER | IDENTIFICATION | PLACE | CONTENT | SOURCES |
|---|---|---|---|---|---|---|
| | | | | S. SPAIN | 5 | (Lost) Islamic |
| | | | 9 | W. FRANCE | 5 | Chron. Fleury |
| | | | 9 | SWITZERLAND | 5 | Sangall. |
| | | | 6 | S. GERMANY | 1 | Herimannus |
| | | | | (CHINA) | | |
| M | 969 June 2-3 | TOTAL | 9 | IRAQ or SYRIA | 5 | Tābit |
| S | 970 May 8 | Annular Solar (0.95 at Constantinople) | 1 | S.E. Europe | 1 | No clear record |
| M | 970 May 23 | PARTIAL | 9 | BABYLON | 5 | Tābit. See Elias |
| | | | | (CHINA) | | |
| M | 976 Jan 19 or July 14 | ALTERNATIVE LUNAR | 9 | S.W. GERMANY (AUGSBURG) | 3 | Ann. Augustani |
| S | 977 Dec 13 | PARTIAL PHASE (0.67 at Cairo) | 0 | CAIRO | 7 | Ibn Yūnus |
| S | 978 June 8 | ANNULAR (0.5 or 0.6) | 9 | CAIRO | 7 | Ibn Yūnus |
| M | 979 May 14 | PARTIAL | 9 | CAIRO | 8 | Ibn Yūnus |
| S | 979 May 28 | ANNULAR | 0 | ( (N.E. Europe) | | No record there |
| | | (0.4 in Egypt) | 9 | ( CAIRO | 8 | Ibn Yūnus |
| M | 979 Nov 6-7 | PARTIAL | 9 | CAIRO | 7 | Ibn Yūnus |

10th CENTURY ECLIPSES

| TYPE | DATE | CHARACTER | IDENTIFICATION | PLACE | CONTENT | SOURCES |
|---|---|---|---|---|---|---|
| M | 980 May 2-3 | DEEP TOTAL | 9 | CAIRO | 6 | Ibn Yūnus |
| M | 981 Apr 21-22 | SMALL PARTIAL | 9 | CAIRO | 6 | Ibn Yūnus |
| M | 981 Oct 15-16 | SMALL PARTIAL | 9 | CAIRO | 6 | Ibn Yūnus |
| M | 983 Mar 1-2 | TOTAL | 9 | CAIRO | 6 | Ibn Yūnus |
| X | 983 Mar 17 | False identification (Solar) | 0 | | | |
| S | 985 July 20 | PARTIAL PHASE | 9 | CAIRO | 7 | Ibn Yūnus |
| M | 986 Dec 19 | LARGE PARTIAL | 8 | CAIRO | 6 | Ibn Yūnus |
| X | 987 | False year | 0 | S. ITALY | | Bari Annals (i.e. Lupus) |
| S | 988 May 18 | Confused records (Solar) | 1<br>1 | GERMANY<br>N. EUROPE | 1 | Ann. Magdeburg<br>Lupus |
| M | 990 Apr 12-13 | PARTIAL | 9 | IRAQ | 5 | See Elias |
| | | | 9 | CAIRO | 6 | Ibn Yūnus |
| M | 990 Oct 6-7 | PARTIAL | 9 | IRAQ | 5 | See Elias |
| S | 990 Oct 21 | ANNULAR | 9 | CENTRAL EUROPE | 5 | Thietmar, Quedlinburg or Hildesheim |
| M | 991 Apr 1-2 | DEEP TOTAL | 9 | IRAQ | 5 | See Elias |
| M | 991 Sep 26-27 | DEEP TOTAL | 9 | IRAQ | 5 | See Elias |

10th CENTURY ECLIPSES

| TYPE | DATE | CHARACTER | IDENTIFICATION | PLACE | CONTENT | SOURCES |
|---|---|---|---|---|---|---|
| S | 993 Aug 20 | TOTAL SOLAR S. MEDIT. | 9 | CAIRO (Partial) | 6 | Ibn Yūnus |
| | | | 2 | ITALY | | Lupus |
| M | 995 Jan 19-20 | TOTAL | 9 | IRAQ | 5 | See Elias |
| M | c.995 (Jan or July?) | (BLOOD-MOON) | 4 | S.W. GERMANY | 3 | Ann. Augustani |
| M | 995 July 14-15 | TOTAL | 9 | IRAQ | 5 | See Elias |
| M | 997 May 24-25 | PARTIAL | ( 5 | E. CENTRAL ASIA | 4 | al-Birūni |
| | | | ( | | | |
| M or | 998 Nov 6-7 | Ambiguous date | ( 4 | BAGDAD | 4 | Abu al-Wafa |
| M | 998 Nov 6-7 | DEEP TOTAL | 9 | IRAQ | 5 | See Elias |

# APPENDIX B

## C O M E T  C H E C K - L I S T

| | |
|---|---|
| TYPE | True comets with scores of 3 or more are indicated by C, Novae by N, Supernovae by S and Halley's Comet by HC. O indicates a minor object (mistakenly identified with a European comet) with score less than 3, and X is an erroneous date. |
| THE YEAR AND MONTHS | up to 630 usually based on Chinese sources (Ho 1962). All Western comets at <u>other</u> dates in the standard catalogues (e.g. Hasegawa 1971) are (except as noted) believed to be wrongly dated or to relate to either meteors or the aurora. Korean comets before AD 1000 were rejected as fictitious. |
| THE NAMES | of the comets are, except for Halley's, a personal choice. Comets described as post-Halley may have followed Halley's comet in a similar orbit. |
| IMPORTANCE | The score is based on the amount of space devoted to it in the Chinese and other sources. Scores below 3 relate to comets (some were called Novae because their tails were not seen) seen only by experienced Chinese or Japanese observers and, if included at all, are bracketed. |

The scale of marks adopted can be interpreted as follows:

9 Created terror. Remembered for generations.
8 Created consternation. Long remembered.
7 Noted as remarkable even in short annals.
6 Noted as remarkable in most chronicles.
5 Noted by most chroniclers.
4 Noted by some chroniclers.
3 Noted by at least one contemporary chronicler.
2 Not noted by the general public.
1 Noted only by experienced sky-watchers.

PLACE — This is indicated by initials as follows:

| | |
|---|---|
| C = CHINA | J = JAPAN |
| M = MEDITERRANEAN | I = ISLAM |
| E = EUROPE | B = BRITISH ISLES |

HO'S NUMBER — The comet-number in Ho's Far Eastern Catalogue (Ho 1962).

SELECTED SOURCES — Primary Western sources are given wherever possible.

## 1st CENTURY COMETS

| TYPE | YEAR AND MONTHS | | NAME | IMPORTANCE | PLACE | HO'S NUMBER | SELECTED SOURCES |
|---|---|---|---|---|---|---|---|
| C | 13 | Nov/Dec | COMET | 4 | C(M?) | 65, 69 | |
| C | 22 | Nov/Dec | MINOR COMET | 3 | C | 66, 69 | |
| O | 29 | (?) | Minor Comet/Nova | (2) | (C?) | 67 | |
| C | 39 | Mar | MODERATE COMET | 4 | C | 68 | |
| C | 54 | June/July | BRIGHT COMET (Claudius) | 5 | CM | 70 | Suetonius, Seneca |
| C | 55/56 | Dec/Mar | BRIGHT COMET (1st Neronian) | 5 | CM | 71 | Pliny, Dio |
| C | 60 | Aug | 2nd NERONIAN COMET | 4 | CM | 73, 72 | C. Siculus, Tacitus, Seneca |
| O? | 61 | Sep/Nov | 3rd Neronian Comet? | (2?) | C(M?) | 74 | 'Octavia' (60?) |
| O? | 64 | May/July | 4th Neronian Comet? | (2?) | C(M?) | 75 | See 65 |
| C | 65 | July/Sep | PRE-HALLEY COMET | 5 | CM | 76 | (Tacitus?) |
| HC | 66 | Jan/Apr | HALLEY'S COMET | 7 | CM | 78, 77 | Josephus, Dio |
| O? | 70/71 | Dec/Jan | ?NOVA IN LEO | (2) | C ) )(One comet?) | 79 | |
| O | 71 | Mar/May | Minor Comet | (2) | C ) | 80 | |
| O | 75 | July/Aug | Minor Comet | (2) | C | 81 | |
| O | 76 | Oct/Nov | Minor Comet | (2) | C(M?) | 82 | See 77 |
| C | 77 | Jan/May | VESPASIAN'S COMET | 4 | CM | 83, 84 | Pliny, Dio, Titus |
| C | 84 | May/July | COMET | 3 | C | 85, 86 | No Western record |

## 2nd CENTURY COMETS

| TYPE | YEAR AND MONTHS | NAME | IMPORTANCE | PLACE | HO'S NUMBER | SELECTED SOURCES |
|---|---|---|---|---|---|---|
| C | 101 Jan 12/21 | COMET | 3 | | 87 | |
| O | 101 Dec 30 | Minor Comet/Nova | (2) | | 88 | |
| C | 104 May 30/June | COMET | 3 | | 89 | |
| O | 107 Sep 13 | Minor Comet/Nova | (1) | | 90 | |
| C | 116 Jan 15/Feb | COMET | 3 | M | 93 | Juvenal (c.115) |
| C | 132 Jan | HADRIAN'S COMET & STAR OF ANTONINUS | 5 | M | 97 & 98 | Various |
| O | 133 Feb 8 | Comet (or aurora?) | (1) | | 99 | |
| HC | 141 Mar/Apr | HALLEY'S COMET | 9 | ARABIA, M & PALESTINE | 100 | (Hist. Aug.) Jewish sources? |
| C | 149 Oct 19/22 | COMET | 3? | | 101 | |
| C | 161 June 14/Aug? | COMET | 4 | M | 105 | Eusebius |
| C | 178 Sep/Dec | GREAT COMET | 5 | | 106 | No Western record |
| O | 182 Aug/Sep | Minor Comet | (2) | | 108 | |
| S | (185 187 Dec/ July | SUPERNOVA | 9 | M | 109 | Herodian, etc. |
| O | 188 Mar/at least July | ?Two Comets | (2) | | 111, 112 | |
| C | 191 Oct | ?COMET OF COMMODUS | 4 | M | 113 | Herodian c.192 |

3rd CENTURY COMETS

| TYPE | YEAR AND MONTHS | NAME | IMPORTANCE | PLACE | HO'S NUMBER | SELECTED SOURCES |
|---|---|---|---|---|---|---|
| X | 204 | Invented | (0) | Korea | 116 | Based on China (No 117) |
| C | 204 Dec/205 Jan | WINTER | 4 | CM | 117 | Dio's first comet |
| HC | 218 May/June | HALLEY | 6 | CM | 122 | Dio's second comet |
| C | 236 Nov/Dec | WINTER | 3 | C | 126 | |
| C | 238 Aug/Sep | IMPORTANT | 5 | C | 127, 128 | No other record |
| C | 240 Nov/Dec | WINTER | 4 | C | 129 | No other record |
| N | 247 Jan/June | PROBABLE NOVA | 3 | C | 131 | |
| C | 248 Apr/Sep | SUMMER | 3 | C | 132, 133 | No other record |
| C | 251/2 Dec 21/Apr | WINTER | 3 | C | 134, 135 | No other record |
| C | 253/4 Dec/June | TWO? | 3 | C | 136 | No 137 and a 255 event may have been aurorae |
| C | 262 Dec/263 Jan | WINTER | 3 | C | 142 | |
| C | 276 June/Sep | SUMMER | 4 | C | 147 | |
| C | 277 Mar/Aug | IMPORTANT | 6 | C | 148 | Not a supernova, but seen for five months |
| O | 290 May | Unimportant | (1) | C | 155 | Conceivably a nova, but ... |
| HC | 295 May/June | HALLEY | 5 | C | 156 | |

4th CENTURY COMETS

| TYPE | YEAR AND MONTHS | NAME | IMPORTANCE | PLACE | HO'S NUMBER | SELECTED SOURCES |
|---|---|---|---|---|---|---|
| C or CN | 305 Sep/Nov | COMET or NOVA plus COMET (conceivably) or Comet plus Aurora (Nov 21) | 4 | CM | 164, 165 | Seen in Rome? |
| C | 329 Aug/Sep | SUMMER COMET (Nova?) | 3 | C | 167 | |
| C | 336 Feb | COMET OF CONSTANTINE'S DEATH | 3 | CM | 168 | Eutropius |
| C | 349 Dec/350 Mar | WINTER COMET(S) (One or two?) | 4 | C | 171 | |
| C | 363 Aug/Sep | SUMMER COMET | 4 | CM | 173 | Ammianus in Italy |
| S or N | 369 Apr/Aug | SUPERNOVA or NOVA? (Not the radio source suspected) | 4 | C | 174 | 'North' |
| HC | 374 Mar/Apr | HALLEY'S COMET | 7 | CM | 175 | Ammianus (Not 373) |
| C | 390 Aug/Sep | COMET OF PHILOSTORGIUS | 6 | CM | 178 | Not 389. Also Marcellinus. |
| S or N | 393 Mar/Oct/Nov | SUPERNOVA or NOVA Radio source | 4 | C | 179 | Scorpio (Constellation) |
| C, N or S | 396 Aug/Dec approx. | COMET, NOVA or SUPERNOVA | 4 | C | 182 | Taurus (Constellation) |

## 5th CENTURY COMETS

| TYPE | YEAR AND MONTHS | | NAME | IMPORTANCE | PLACE | HO'S NUMBER | SELECTED SOURCES |
|---|---|---|---|---|---|---|---|
| C | 400 | Mar/May | GREAT SPRING COMET | 5 | C, Byz. | 183 | Claudian |
| O | 401 | Jan | Minor Comet | (1) | C | 185 | |
| C | 402/3 | Nov/Jan | WINTER COMET | 3 | C | 186 | Synesius? |
| C | 415 | June | SUMMER COMET | 3 | C | 188, 187 | |
| C | 418 | June/Sep | GREAT SUMMER COMET | 5 | Byz., C, Italy | 191, 189, 190 | Marcellinus Philostorgius |
| X | 419 | Feb 17 | Minor (Ghost) Comet | (0) | C | 192 | |
| O | 420 | May | Minor Comet (?) | (1) | C | 193 | |
| O | 421 | Jan/Feb | Minor Comet | (1) | C | 194 | |
| C | 422 | Mar/Apr | SPRING COMET | 3 | C, Byz. | 195 | Marcellinus? |
| O | 422 | Dec 18 | Minor Comet | (2) | C | 196 | |
| C | 423 | Feb | WINTER COMET | 3 | C, Byz. | 197 | Marcellinus? |
| C | 423 | Mid-Dec | WINTER COMET | 4 | C, Byz. | 197 | Marcellinus? |
| O | 428 | Mar 3? | Comet/Meteor | | Ravenna? | - | |
| O | 436 | June 21 | Minor Comet | (1) | C | 199 | |
| O | 437 | Feb 26 | Meteor (or Nova?) | (1) | C | 200 | |
| C | 442/3 | Nov 4 | GREAT WINTER COMET | 5 | C, Portugal, Byz. | 201 | Hydatius |

5th CENTURY COMETS

| TYPE | YEAR AND MONTHS | | NAME | IMPORTANCE | PLACE | HO'S NUMBER | SELECTED SOURCES |
|---|---|---|---|---|---|---|---|
| O | 449 | June & Nov | Minor Comets | (2) | C | 202, 203 | |
| C | 451 | June/Aug | HALLEY'S COMET | 6 | Portugal, C, Italy | 204 | Hydatius |
| O? | 460 | Nov | Aurora or Minor Comet | (1) | C | 207 | |
| X | 461 | Apr 20 | Aurora (?) | (0) | C | 208 | |
| O? | 464/5 | Dec/Jan | Minor Comet or Aurora | (2) | C | 209 | |
| C | 467 | Feb | GREAT COMET | 5 | C, Portugal, Italy, Byz. | 210 | Hydatius Victor |
| C | 483 | Nov/Dec | COMET | 3 | C (Athens?) | 211 | |
| C | 500 or 499 | Jan | COMET | 4 | Syria, C | 212/4 | Edessa |

## 6th CENTURY COMETS

| TYPE | YEAR AND MONTHS | NAME | IMPORTANCE | PLACE | HO'S NUMBER | SELECTED SOURCES |
|---|---|---|---|---|---|---|
| X | (501) | (Aurorae?) | | C? | 213, 214 | (Chinese gaps 450-520) |
| X | (504) | (Legend) | | (Scotland) | | Veremundus (c.1020±) |
| C | 520 Oct | GREAT COMET | 5 | CM | 216 | Malalas |
| C | 530 Aug/Sep | HALLEY'S COMET | 7 | CM | 217 | Malalas |
| O | (537 Also 541 & 561) | Conceivable Novae | (1) | (C) | 224 | (Probably small comets) |
| C | 539 Nov/Dec | GREAT COMET | 6 | CM | 221 | Procopius, Lydus |
| C | 560 Oct | LOTHAIR'S COMET | 3 | CEM | 223 | Gregory of Tours (V 21) (China: Good sources 560+) |
| C | 565 July/Oct | GREAT COMET | 6 | CEM | 226 | Gregory (IV 31), Marius, Agnellus, Olympiodorus, Agapius |
| C | 568 July/Nov | GREAT COMET(S) | 5 | C | 227, 228 | John of Biclaro? |
| C | 574 Apr/June | SIGEBERT'S COMET | 4 | CE | 229, 230 | Gregory of Tours, John of Asia |
| C | 582 Jan | GREGORY'S COMET (& AURORAE) | 3 | CE | 233 | Gregory (VI 14) (Not 580 or 584) |
| C | 595 Jan | COMET OF SECUNDUS | 5 | CE | 236 | Paul the Deacon, Fredegar (Not 594) |

7th CENTURY COMETS

| TYPE | YEAR AND MONTHS | | NAME | IMPORTANCE | PLACE | HO'S NUMBER | SELECTED SOURCES |
|---|---|---|---|---|---|---|---|
| X | 602 | Apr | False date | (0) | (Byzantine) | - | See 607 |
| HC | 607 | Apr/Dec<br>Nov/Dec | HALLEY and<br>POST-HALLEY | 7 | ITALY<br>BYZANTINE | 237-239 | Paul the Deacon, Theophanes<br>Sui shu and Pei shih |
| C | 615± | July | MAJOR COMET? | 4 | CE<br>MIDDLE EAST | 240 | Sui shu, Irish Annals<br>Syriac sources |
| O | 617 | July/Nov | Minor Comet | (2) | C (only) | | Sui shu |
| C | 626 | Late Mar | SPRING COMET | 3 | C<br>SYRIA (& CONST ?) | 243 | Chiu T'ang shu (CTS)<br>Agapius |
| C | 634/5 | Sep/Feb | IMPORTANT COMET(S) | 5 | J(Two?) & C<br>BYZANTINE | 244 | Nihongi & CTS<br>Theophanes, etc. |
| C | 636 | Mar/Apr | MAJOR COMET | 4 | JC | 245 | Nihongi & CTS |
| O | 642± | Aug | Minor Comet | (2) | JC | 246, 247 | Nihongi & CTS |
| O | 663 | | Minor Comet | (2) | C | 250 | CTS |
| C or N | 668 | May/June | MAJOR COMET or NOVA | 4 | C, SYRIA | 251, 252 | CTS, Theophanes |
| C | 676 | Sep/Oct | MAGNIFICENT COMET | 9 | JC<br>MIDDLE EAST<br><br>E | 255 | Nihongi & CTS<br>Elias (from James of<br>Edessa), al-Khwarizmi<br>Liber Pontificalis |
| C | 681 | Oct/Nov | AUTUMN COMET | 3 | CJ | 256 | CTS, Nihongi |
| O | 683 | Apr/May | Minor Comet | (2) | C (only) | 257 | CTS |
| HC | 684 | Sep/Jan | HALLEY and<br>POST-HALLEY | 7 | ITALY<br>CJ | 258 | Liber Pontificalis etc.<br>CTS & Nihongi |

## 8th CENTURY COMETS

| TYPE | YEAR | MONTHS | NAME | IMPORTANCE | PLACE | HO'S NUMBER | SELECTED SOURCES |
|---|---|---|---|---|---|---|---|
| C | 707 | Nov/Dec | CHINESE COMET | 3 | C | 261 | |
| C | 712 | July/Aug | SUMMER COMET | 4 | CM | 264 | Later Islamic writers (misdated 710 & 716) from Syriac source? |
| | (729 | Jan) | | | | | Bede's year is 729 |
| C | 730 | June/July | BEDE'S COMET (N. ENGLAND) | 5 | CMB | 268 | Chinese date (Bede) |
| C | 738 | Apr | SPRING COMET | 5 | CM | 269 | |
| C | 745 | Jan | 'JAPANESE' WINTER COMET | 5 | JM | 271 | Often misdated |
| HC | 760 | May | HALLEY'S | 7 | CM | 273-275 | Theophanes & al-Khwarizmi Often misdated 761 or 762 Denys, Theophanes |
| ?C | 767 | Jan/Feb | UNCERTAIN WINTER | 3 or 0 | C? | 277 | China only |
| C | 770 | May/July | GREAT COMET | 6 | CJM | 279 | Perihelion June 5 Denys, Theophanes |
| C | 773 | Jan | FAR EASTERN WINTER | 7 | CJ | 280 | No Western record |

The best Western records in this century were kept by a (Nestorian-Syrian?) Bishop of Hirta (760/770) and are preserved in Denys (now often called Dionysius). No comets of importance 774/816.

9th CENTURY COMETS

| TYPE | YEAR AND MONTHS | NAME | IMPORTANCE | PLACE | HO'S NUMBER | SELECTED SOURCES |
|---|---|---|---|---|---|---|
| C | 817 Feb | FRANKISH/CHINESE COMET | 4 | CEM | 283 | RFA |
| C | 821 Feb/Mar | CHINESE SPRING COMET | 3 | C | 284 | |
| C | 834 Oct/Nov | CHINESE AUTUMN COMET | 3 | C | 289 | |
| HC | 837 Mar/Apr | HALLEY'S COMET | 8 | CJEI | 291 | Life of Louis (Ennin's Diary) |
| C | 838/9 Nov/Apr | POST-HALLEY GREAT WINTER COMET | 8 | CJEMI (incl. Egypt) | 292 a b | (Ennin's Diary) Life of Louis |
| O | 840 Mar & Dec | Small Comet/Nova in China | (2) | (C) | 292 d | |
| X | 840 | International Aurorae | (0) | (CJME) | | |
| O | 841 Summer | Small Nova/Comet in China | (2) | (C) | 295? | |
| C | 841/2 Dec/Feb | COMET OF NITHARD | 4 | CEJ | 295 | Nithard, Fontanelle |
| O | 852 Mar | Small (in Japan only) | (2) | | 297 | |
| O | 853 Feb | Small (in China only) | (2) | | | |
| C | 857 Sep/Oct | AUTUMN COMET | 3 | CI | 299 | |
| C | 864 Apr/June | FRANKISH SPRING COMET | 3 | JCEM | 300 | Ann. Flor. |
| C | 867/8 Nov/Sept | GREAT WINTER COMET | 5 | JECBM Afghanistan | 302 | Sens |

9th CENTURY COMETS

| TYPE | YEAR AND MONTHS | NAME | IMPORTANCE | PLACE | HO'S NUMBER | SELECTED SOURCES |
|---|---|---|---|---|---|---|
| C | 875 June | SUMMER COMET OF FULDA | 3 | JE | 306 | Fulda (Pre Death of Lothair) |
| CN | 877 Feb/Mar | COMET/NOVA SEEN e.g. IN ITALY? | 3 | JME | 307 | |
| C | 891 May/July | ALFRED'S COMET | 5 | CJBEI | 313 | ASC |
| C | 894 Mar | FAR EASTERN COMET | 3 | CJ | 316 | ASC, Saxo, Ibn al-Jawzi |

## 10th CENTURY COMETS

| TYPE | YEAR AND MONTHS | | NAME | IMPORTANCE | PLACE | HO'S NUMBER | SELECTED SOURCES |
|---|---|---|---|---|---|---|---|
| C | 905 | May/June | GREAT SUMMER COMET | 6 | CJEBMI | 321 | |
| C | 907 | Apr | JAPAN & EUROPE | 4 | JE | 322 | Corvey Annals |
| C | 912 | Apr/May | PRE-HALLEY | 4 | CMI | | Symeon, Mas'udi, etc. |
| HC | 912 | July | HALLEY'S COMET | 7 | JEBI | 325 | Perihelion in July Symeon. Often misdated |
| C | ?923 | Oct | PROVISIONAL DATE | 4 | CI ) | 327-329 | Chronology confused 913/940 |
| C | ?938 | Jan | PROVISIONAL DATE | 3 | CE ) | | 928 in Ho = 938 |
| C | 941 | Sep/Oct | BRIGHT COMET | 5 | JCEMB India Afghanistan | 335 | Widukind; Cordova |
| C | 947 | Sep | AUTUMN COMET | 3 | CI | 338 | Iraq and Egypt |
| C | 959/62 | Jan/Apr | YEAR UNCERTAIN | 4 | CMJI | 341, 342 | |
| C | 975 | Aug/Oct | GREAT COMET | 6 | CJMIEB N. Africa Armenia | 346 | (Not Iceland!) |
| HC | 989 | Aug/Sep | HALLEY AGAIN | 5 | CJEMBI ARMENIA | 349 & 350 | 995 in ASC |
| C | 998 | Feb/Mar | LATE WINTER | 4 | CJE | 352 | St. Gall (NOT 1003) |

## APPENDIX C

## MAYA ECLIPSES

The medieval Maya book known as the Dresden Codex contains a table for the prediction of eclipses. In the accompanying text glyphs for solar and lunar eclipses can be recognized. The dates in this Codex are expressed in the Long Count form, and such dates can readily be converted (Aveni 1980, 211) to Mayan Day Numbers (M). However, the problem has been to find the difference between this floating chronology and the Climatic chronology as expressed in Julian Day Numbers (J). The Correlation problem was to find the constant A (the 'Ahau' number) in the formula

$$J - M = A$$

In the sixteenth century we have some 'double' dates - dates expressed in both the Short Count Maya form and the AD Christian manner. Such dates are not all consistent, but the conventional correlation for this period, i.e. A = 584,283 (Modified Thompson Correlation) works well in the Spanish Colonial period.

In the Classic Maya period it is often assumed that this formula can still be applied, but the 'Original Thompson' version with its addition of two days, 584,285, improves the fit with eclipse dates. However, astronomers who have attempted to determine the constant A from the astronomical evidence of the Classic inscriptions and codices have found many correlations: a list of such alternatives has been given in Jnl. Hist. Astron. (Schove 1984b Table 1, cf. also Kelley 1983). A new correlation A = 615,824 is there shown to give a better fit with both planetary and seasonal phenomena. In another article (Schove 1984a) this is shown to give a better fit also with lunar and solar eclipses.

### THE DRESDEN CODEX AND ECLIPSES

The Dresden Codex (Thompson 1972) contains over 30 Maya dates and the assumption that there are some connections with eclipses has made it possible (Schove 1984a) to fix positions of the node or eclipse half-year in the floating Maya chronology. Division of these Maya dates by 173.3098 revealed that the remainders were found to cluster within a 38-day section of the eclipse half-year. Eclipses cluster within 18 days of node passage and the deduction was clear. This test is satisfied by both the 'Original Thompson' 584,285 and the new 615,824, both of which leave decimal remainders 0.32±0.01 after division (see Schove 1984b Table 1). Nevertheless, it eliminates many hypotheses which place the node cycle a fortnight earlier, that is, correlations which leave remainders about 0.44. This means, as we explain below, that the Eclipse Table in the Codex must be an Eclipse Warning Table. The pattern of the Eclipse Table is explained by Aveni (1980, 174-179): the first three bases and their implied expansions can now be interpreted as follows:

The first (12 Lamat) base represents the New Moon at the beginning of an eclipse month and of the Maya eclipse half-year. At such dates a solar eclipse was possible but difficult to predict.

The second (1 Akbal) base represents the Full Moon in the middle of an eclipse month. At such dates (as in 842) lunar eclipses were frequently observed and easily predicted.

The third (3 Etznab) base represents the end of the Eclipse Moon, another New Moon where a solar eclipse could occur.

This table implies that the Maya had a long series of dated observations of lunar eclipses when its prototype was completed. These led to the second column in the table below, and the observation that solar eclipses were possible a fortnight earlier or later to the first and third columns.

With the conventional correlation the second base becomes AD 755 Nov 21-23 when a lunar eclipse occurred with the centre of the shadow over India; with the new correlation the date 842 March 30 is that of an eclipse visible on both sides of the North Atlantic and one which was recorded in what is now France and Germany.

# ECLIPSE MOONS TABLE

| | | Commencement New Moon (Occasional Solar Eclipses) | Full Moon (Frequent Lunar Eclipses) | End New Moon (Occasional Solar Eclipses) |
|---|---|---|---|---|
| 0 | 0 = | 842 Mar 15 | 842 Mar 30 | 842 Apr 4 |
| 1 | + 177 = | - | - | (+1) |
| 2 | + 177 = | (-1) | 843 Mar 19 | - |
| 3 | + 148 = | - | (-29) | - |
| 4 | + 177 = | - | - | |
| 5 | + 177 = | - | - | - |
| 6 | + 177 = | - | (-1) | - |
| 7 | + 178 = | - | (+1) | - |
| 8 | + 177 = | - | - | - |
| 9 | + 177 = | - | - | - |
| 10 | + 177 = | (-1) | 847 July 2 | - |
| 11 | + 177 = | - | (-½) | - |
| 12 | + 177 = | (-1) SUNRISE | - | - |
| 13 | + 148 = | - | - | 848 Oct 1 |
| 14 | + 178 = | - | (-1) | (+1) |
| 15 | + 177 = | - | - | 849 May 25 (+1) |
| 16 | + 177 = | Month No.1 | 849 Nov 4 | (+1) |
| 17 | + 177 = | (-1) | 850 Oct 24 | - |
| 18 | + 177 = | - | - | - |
| 19 | + 148 = | - | (-29) | - |

| | | | | |
|---|---|---|---|---|
| 20 | + 177 = | - | - | (+1) |
| 21 | + 177 = | - | - | 852 Mar 2 |
| 22 | + 177 = | - | - | - |
| 23 | + 178 = | - | - | (+1) |
| 24 | + 177 = | - | (+1) | - |
| | | | | |
| 25 | + 177 = | - | 854 Feb 16 | - |
| 26 | + 148 = | - | 854 Aug 11 (-29) | (+1) |
| 27 | + 177 = | - | - | (+1) |
| 28 | + 177 = | - | - | (+1) |
| 29 | + 178 = | - | - | 856 Jan 11 (+1) |
| | | | | |
| 30 | + 177 = | - | 856 June 21-22 | - |
| 31 | + 177 = | - | (+1) | - |
| 32 | + 177 = | - | 857 June 11 | (+1) |
| 33 | + 177 = | - | - | - |
| 34 | + 177 = | - | - | - |
| | | | | |
| 35 | + 177 = | - | - | - |
| 36 | + 148 = | - | - | - |
| 37 | + 178 = | - | - | (+2) |
| 38 | + 177 = | - | (+1) | (+1) |
| 39 | + 177 = | - | (+1) | (+1) |
| | | | | |
| 40 | + 177 = | - | 861 Mar 30 | - |
| 41 | + 177 = | - | (+1) | - |
| 42 | + 148 = | - | - | (+1) |
| 43 | + 177 = | - | - | - |
| 44 | + 177 = | - | - | (+1) |

# ECLIPSE MOONS TABLE
(continued)

| | | | | |
|---|---|---|---|---|
| 45 | + 177 = | - | - | - |
| 46 | + 177 = | - | - | - |
| 47 | + 177 = | - | - | - |
| 48 | + 177 = | 865 Jan 1 (-1) | 865 Jan 15 | - |
| 49 | + 148 = | - | (-30) i.e. 865 July 12 | (+1) |
| 50 | + 177 = | 866 June 16 | - | or 866 June 16 |
| 51 | + 177 = | - | - | - |
| 52 | + 178 = | - | 866 Nov 26 | - |
| 53 | + 177 = | - | - | - |
| 54 | + 177 = | - | 867 Nov 15 | - |
| 55 | + 177 = | (-1) | 868 May 10 | - |
| 56 | + 177 = | - | 868 Nov 9 (-1) | - |
| 57 | + 177 = | (-1) | - | 869 Apr 15 |
| 58 | + 148 = | - | - | - |
| 59 | + 177 = | - | 870 Mar 21 (-1) | 870 Apr 4 |
| 60 | + 178 = | - | - | (+1) |
| 61 | + 177 = | - | - | (+1) |
| 62 | + 177 = | - | (+1) | - |
| 63 | + 177 = | - | 872 Feb 28 (-1) | - |
| 64 | + 177 = | 872 Aug 7 | 872 Aug 22 | - |
| 65 | + 148 = | - | - | - |
| 66 | + 177 = | - | - | - |
| 67 | + 177 = | - | - | - |
| 68 | + 177 = | - | (-1) | - |
| 69 | + 177 = | - | 874 Dec 26 | (-1) |

The structure of the above table is that of the Eclipse Moons Table in the Dresden Codex. The dates are those given by the 615,824 correlation. The dates completed are (mainly European) eclipses included in our check-list. Other eclipses visible to the Maya could be added. The errors in the Maya table as far as the period 842-874 is concerned are indicated in brackets - ± 29 or 30 days where there is an error of 1 moon, ± 1 or 2 where the error is only 1 or 2 days. Eclipse No.40 is 19 Tropical Years after the 842 eclipse. Eclipse No.69 is 32¾ or 11,959 days after the original eclipse and completes the '11,960-day cycle'.

## THE VENUS TABLE

The 'Eclipse Overtones' of the Venus Table in the Dresden Codex were first noted by H.J. Spinden (in 1928). The main base is dated 9.9.9.16.0. and two implied bases seem to be respectively 11,960 days later and 9,360 days before it and both intervals are eclipse intervals. C.H. Smiley (*Nature*, 188, 1960, 215-216, and 1970, see L. Satterthwaite in *Estudios de Cultura Maya*, Mexico, 2, 1962, 271) developed the idea further. With the 615,824 correlation all three Venus bases (AD 709, 742 and 683) occur at the New Moons 15-17 days after visible lunar eclipses. These eclipses were central in longitudes 84° to 169°W so that they were striking in Mesoamerica. Moreover, if we expand the table adding the first increment of 33,280 days (91 years) we get three more visible eclipses (i.e. 774, 800 and 833; the last, although central at 89°W, had a magnitude of only 0.1). The Venus Table was in short useful for both eclipse and Venus predictions, and the three base-dates marked the ends of Eclipse Months in which lunar eclipses had been seen, corresponding in this respect to dates in the third set of the Eclipse Table.

In the Dresden Codex there are eclipse associations with many of the Long Count dates. The 11,960-day eclipse cycle - the length of the Eclipse Table - thus connects the dates in AD 447 (roughly), 709, 742, 774 and others, suggesting that some dates in this Codex may have been adjusted slightly from the original dates of the observed eclipses. The centre of the cluster in the eclipse cycle is presumably near the node: in any correlation it is near the *second* set of the Eclipse Table, confirming its connection with *lunar* eclipses. Possibly the 447 eclipse was used in the prototype of the Venus Table. The last date in the Codex is an unsuccessful prediction adding 12 x 11,960 days to a date 393 years earlier.

## THE 819-DAY CYCLE

Certain Maya dates between AD 750 and 900 specify connections with an 819-day cycle, suggesting an interest in lunar longitude: after 30 sidereal months the moon is back to the same position among the stars. Each return was associated by the Maya with one of the cardinal points so that the complete cycle was 3,276 days or 120 sidereal months, that is 111 lunations or one fortnight less than 19 eclipse half-years. One set of these Maya dates leaves a remainder of 816 when the Mayan Day Number is divided by 819, but it would seem that they began to record the cycle when there had been a lunar eclipse two weeks earlier, as four cycles would then lead to further eclipses. Thus the first 819-dates of a set sometimes correspond, like the Venus bases, to the third set of the Eclipse Table, and after the 3,276 days cycle the date would correspond to the Lunar Eclipse date of the second set.

Some of the 819-dates at Palenque are very close to node passage, but there was also a Venus association and the surviving samples are too few to be certain yet that the eclipse associations are not accidental. Further investigations would seem worthwhile.

## THE MOON NUMBERING

The secret of Maya moon-numbering (Glyph C) was solved by Teeple 1930, 93; he found that certain dates were considered as the New Moon at the start of eclipse half-years of 5 or 6. Dr. Kudlek kindly supplied me with a list of eclipses at Tikal, and this confirmed in each case that the lunar eclipses 14-16 days later were indeed visible to the Maya in 842 Mar.30, 847 July 2, 849 Nov.4 and 850 Oct.24. Thus these dates correspond to the first (12 Lamat) set of the Eclipse Table, indicating that the Maya tried to begin their half-years with an eclipse moon.

## OTHER ECLIPSE ASSOCIATIONS

The dates on the monuments are associated mainly with historical events, but Thompson 1950, 240, discussing irregular forms of Glyph D (cf. Kelley 1976, fig.7 Numbers 2 to 8), noted "two or three unusual glyphs which seem to refer to disappearance(s) of moon or perhaps to conjunction". Five out of six cases do indeed precede eclipses (AD 762 Solar, 847, 857, 872 Lunar). These cases are discussed more fully in Schove 1984a.

The paper 'Maya Eclipses and the Correlation Problem', although dated 1982, arrived in June 1984 (Estudios de Cultura Maya, 14, 1982, 241-260). Two tables from that paper (pp. 248 and 260) are reproduced below with brief comments.

TEEPLES MOON NUMBERING AND LUNAR ECLIPSES

| | | | |
|---|---|---|---|
| 9.16.8.6.3 | + 15 days | = | 842 Mar 30 |
| 9.16.9.16.9 | + 14 days | = | 847 Jul 2 |
| 9.16.11.14.7 | + 15 days | = | 849 May 11 |
| 9.16.12.5.4 | + 15 days | = | 849 Nov 4 |
| 9.16.13.4.7 | + 16 days | = | 850 Oct 24 |

Later astronomical dates in the same class include e.g.

| | | | |
|---|---|---|---|
| 9.17.12.17.5 | + 16 days | = | 870 Mar 21 |

The accuracy of moon numbering at Copan in this period suggests that the Maya waited for the eclipses to be observed before Glyph C was inscribed.

ECLIPSES AND GLYPH D

The Maya dates and precise eclipse dates referred to above are as follows:

| | | Maya date | + 615 824 = | Lunar Eclipses |
|---|---|---|---|---|
| 1 | Copan I | 9.12.3.14.0 | 762 July 26 | (Solar August 5) |
| 4 | Quirigua F | 9.16.10.0.0 | 847 July 19 | 847 July 2 |
| 3 | Quirigua D | 9.16.15.0.0 | 852 June 23 | None |
| 2,5 | Quirigua E and Piedras Negras | 9.17.0.0.0 | 857 May 28 | 857 June 11 |
| 6 | Quirigua Alt O | 9.17.14.16.18 | 872 Feb 16 | 872 Feb 28 |

Illustrations of 2, 4 and 6 are given by Thompson (1950, fig. 36, Nos. 29-33; fig. 37, Nos. 63-66 and Nos. 26-31 and see his p. 269).

The eclipse associations of Maya dates explained above suggest that at least the Maya dates selected for inclusion in the sections for the fifth to ninth centuries were mainly dates of observed eclipses.

## MAYA CORRELATIONS

The conventional chronology is rejected on astronomical grounds for several reasons explained in Schove 1984b. The conventional correlation fares nearly as well in the standard Moon, Eclipse and Venus tests, but the position of Midsummer is then clearly wrong, as is demonstrated also by Kelley 1983. Moreover, various planetary tests reveal that the correct correlation is 615,824, that is 86.4 years further on than the conventional view and 130 years 5 days less than Kelley's 1983 alternative. Kelley's correlation was made to fit the solstices and equinoxes (not the Venus observations). The new correlation indicates nevertheless that the dates on the monuments relating to the four solar stations are five days ahead astronomically. The Chinese, from whom the Maya may indirectly have gleaned their knowledge of the 11,960-day eclipse cycle, began their celebrations three days ahead. Perhaps in Mesoamerica the rituals of the earlier date, like the Aztec sacrifices, were assumed by the people to produce the changes that were to follow (cf. Needham 1959, 188 note).

In the later Middle Ages the Eclipse Table would have ceased to function and the Maya seem to have changed their Short Count dates (used by the people) slightly so that the Moon, Eclipse and Venus tables would function almost as before. This slight change in the Short Count necessitated an 86.4 year change in the Long Count.

## AZTEC ECLIPSES

The Aztec picture-chronicles or comic-strips of the fifteenth and sixteenth centuries include records of the solar eclipses of 1426, 1477, 1492, 1496, 1504, 1507, 1510 and 1531 (cf. Schove 1961, and Aveni 1980, 26 and figs. 9a, b and c). Earlier Mixtec chronology is being studied, but so far no accepted eclipse dates have been found.

These Aztec dates include some eclipses in the period 1504-10 when the magnitude was only 0.4, and this suggests that they had been predicted and were studied in reflection in water.

Table Some Maya dates of Eclipse interest (615 824 correlation)

| PAGE & COLOUR<br>a | DATE<br>b | | EQUIVALENT<br>c | DIFFERENCE<br>d | e |
|---|---|---|---|---|---|
| | 8.6.16.7.14 | 7 Mac | 262 July 4 | (-30) | Calculated Solar Eclipse? (See fig.1) |
| 63b &<br>31b | 8.16.3.12.3 | 13 Akbal | 447 Jan 23 | -21 (S) | Important<br>Eclipse cycle and glyphs |
| | 8.16.14.11.5 | | 457 Nov 8 | +25 (S) | (See our fig.1) S Pacific only |
| 70b | 8.16.19.0.12 | 4 Eb | 462 Mar 14<br>(380 years before AD 842) | + 3 (S)<br>-12 (L) | E Hemisphere only<br>Total (28°W 7°N) Eclipse glyphs pp.69 & 71 |
| 45b | 8.17.11.1.0 | 13 Oc | 474 Jan 18 | + 1 (L)<br>-14 (S)<br>+15 (S) | Total (118°W 21°N)<br>Solar eclipses but partial only |
| 24b | 9.8.3.16.0 | 3 Xul<br>Venus Table base | 683 Oct 28 | -17 (L) | Lunar (169°W)<br>(Kelley 1977 pp.59-62) |
| 24b | 9.9.9.16.0 | 1 Ahau<br>Venus Table base | 709 June 13 | -16 (L)<br>-30 (S) | Total (84°W 22°N) |
| 24b | 9.11.3.2.0 | 13 Ix<br>Venus Table base | 742 Mar 12 | -16 (L) | Lunar (116°W)<br>(Kelley 1977 pp.59-62) |
| 58b | 9.12.10.16.9 | 13 Muluk | 769 Aug 6 | +16 (L)<br>+30 (S) | Small (32°W 11°N)<br>Eclipse glyphs |
| 24b | (9.12.16.6.0)<br>Implied Venus date by addition of 33,280 days or just over 91 years | | 774 Dec 9 | -16 (L)<br>-31 (S) | Near total |
| 24b | (9.14.2.6.0) | | 800 July 25 | -15 (L) | Visible |

| a | b | | c | d | e |
|---|---|---|---|---|---|
| 24b | (9.15.15.10.0) | | 833 Apr 23 | -15 (L) | Almost unobservable |
| 51b | 9.16.4.10.8 | 12 Lamat | 842 Mar 15 | +15 (L) | |
| | | 1st New Moon base in Eclipse Table | | | |
| 52r | 9.16.4.11.3 | 1 Akbal | 842 Mar 30 | 0 (L) | Total (54°W 5°S)<br>Eclipse glyphs |
| | | Full Moon base in Eclipse Table | | | |
| 52b | 9.16.4.11.8 | 3 Etznab | 842 Apr 14 | 0 (S) | Partial, invisible |
| | 9.17.0.0.0<br>(Not in Dresden) | 13 Ahau | 857 May 27 | 0 (S)<br>+15 (L)<br>+29 (S) | Eclipse glyphs. Partial<br>(Kelley 1977 p.63) |
| 69r | 9.17.15.6.14 | 9 Ix | 872 July 21 | +18 (S) | |
| | 9.17.19.13.16<br>(Not in Dresden) | | 876 Nov 19 | 0 (S) | Eclipse of Sun glyphs |
| | 9.19.7.2.14 | 9 Ix | 903 Nov 16 | +21 (L) | Visible lunar. Solar invisible |
| 62b | 10.4.6.15.14 | 3 Ix | 1002 Mar 1 | 0 (L)<br>-14 (S) | Total (14°E 5°N) Visible<br>Partial only |
| 62b | 10.6.1.1.5 | 3 Chicchan | 1035 Nov 14 | + 4 (L)<br>-12 (S) | Small (78°W 21°N)<br>Eclipse glyphs |
| 61b | 10.7.4.3.5 | 3 Chicchan | 1058 Aug 25 | - 3 (S) | Visible in S Pacific<br>Eclipse glyphs |
| 70b | 10.11.4.0.14 | 9 Ix | 1137 May 11 | +10 (S) | Visible<br>Eclipse glyphs |
| 51r | 10.19.6.1.8 | 12 Lamat | 1297 Jan 25 | +87 (S) | New Moon but no eclipse<br>Calculated from 904 date. (See Introduction) |

Dates are taken from

1) the explicit Dresden Codex dates. There are only 35 in all
2) the implicit dates of the Venus Table bases and its first expansion (In brackets)
3) two dates with eclipse glyphs found on inscriptions (AD 857, 876)

a = Page and colour of date in the Codex
b = Maya Long Count date
c = Maya Short Count (1st part only)
d = Eclipse date - Maya date
e = Notes e.g. if Eclipse Glyphs are found on the same page. Central position of lunar eclipse is given.

Note longitudes all 169°W/14°E

For fuller details see Schove 1984a.

# APPENDIX D

## EYEWITNESS SOURCES

The observations of eclipses, comets and of other ephemeral phenomena collected for the Spectrum of Time project provide clues to sources that incorporate contemporary eyewitness descriptions. Aurorae, thunderstorms, earthquakes and heat-waves for instance seem important at the time, but they will not be recorded in a chronicle at all unless the writing is 'continuous with the flow of Time'. In the first millennium AD it was usual for a chronicler to incorporate and abridge the works of his predecessors - the Anglo-Saxon Chronicle of king Alfred's time in the mid-ninth century is surely just such an abridgment - the author summarized earlier annals of the Danish invasions of England and the records of comets and eclipses may have been omitted. Indeed, the primary "selected Western sources" in our Appendixes A and B were often written in later centuries, but they are named because contemporary eyewitness accounts are embedded in them. In this Appendix we include the selected sources but attempt to specify the earliest authors for each decade: those no longer extant are given in brackets, the later writers have 'pre-' affixed to their names; this prefix is used to indicate that there must be an earlier primary source which is probably no longer extant.

We needed a definitive list of this kind at the beginning of our enquiry, and it is only because historians have never provided one that I have an excuse for the publication of the notes that follow. This list is based especially on the sources used in this book and is neither complete nor definitive; additions and corrections or further information about eyewitness writers and <u>where</u> they wrote would be appreciated. In the meantime, where we have mentioned several sources in the text this Appendix should help to indicate not only who copied from whom but also where an eclipse was originally seen.

1st CENTURY



AD 1-9 pre-Dio.

10s pre-Dio and pre-Tacitus (e.g. Aufidius?).

20s pre-Phlegon AD 29, ASIA MINOR.
Velleius wrote a History of ROME to AD 29.

30s pre-Gospels of New Testament, PALESTINE. ROMAN annals lost.

40s pre-Jerome and pre-Orosius (re-Famine),
Seneca, Pliny I.

50s Pliny the Elder,
pre-Tacitus (Prodigies from AD 51).

60s pre-Tacitus 'Annals' 51-64 and History 69-70 (written c.100),
Seneca, Pliny the Elder, Josephus, C. Siculus, Hero,
Philostratus, pre-Dio.

70s Pliny the Younger (Volcano AD 79),
pre-Dio.

80s Eyewitness sources few in this decade.

90s Aurelius Victor, Tacitus (Germania).

2nd CENTURY

100-109 Eyewitness sources few again.
(Dio extant only in epitome; Tacitus wrote about the past.)

110s Phlegon of Tralles = pre-Africanus (Fragments only survive),
(Annals evidently kept in ASIA MINOR c.105-129),
pre-'Ravenna' Annals, Juvenal.

120s Ptolemy at ALEXANDRIA (127-151),
Theon I probably at SMYRNA (127-132).

130s Arrian in CAPPADOCIA, ASIA MINOR: he wrote on the past.

140s Aelius Aristides (Autobiography 140s to 160s),
pre-Historia Augusta (some genuine phenomena 117-161).

150s (Latin literature mainly AFRICAN 150-250.) Sources few.

160s Sosigenes (= pre-Proclus), pre-Eusebius (c.168-189).
A Latin flood inscription in GAUL.

170s No extant Western records of comets or eclipses.

180s Herodian, pre-Historia Augusta, pre-Eusebius.
A Latin inscription near SALZBURG (Pestilence).

190s pre-Herodian.
Dio writes his History 193-219.

3rd CENTURY

200s Dio. Evidence of Annals in SYRIA (EDESSA flood 201).

210s Dio Cassius came from NICAEA in BITHYNIA and wrote in Greek. Herodian, a SYRIAN, wrote in Greek.

220s (Julius Africanus contemporary, but lost from 221.) Herodian. Decline of sources.

230s Absence of eyewitness sources.

240s pre-Historia Augusta preserves some items from 240/242.

250s pre-Eusebius 245/252 (from Dexippus etc.), pre-Aurelius Victor.

260s Absence of eyewitness sources.

270s pre-Zosimus (Pestilence, Famine etc.).

280s Consular Annals, pre-George the Syncellus (to 285, but he died 800) and pre-Theophanes, Mamertinus (RHINELAND).

290s pre-Hydatius and pre-Malalas, Mamertinus.

4th CENTURY

300s pre-Hamartolos, Eusebius.

310s Aurelius Victor, Eusebius, Latin Annals.

320s Eusebius, Pappus.

330s Firmicus Maternus (Astrologer), pre-Socrates (Eutropius).

340s Jerome, pre-Theophanes.

350s Absence of eyewitness sources.

360s Ammianus Marcellinus of ANTIOCH (in IRAN, 350/365), Jerome, Theon.

370s Ammianus again (settled in ROME 378) writes in Latin.

380s RAVENNA Annals (379-572).

390s Jerome, pre-Zosimus (Eunapius), Marcellinus Comes.

5th CENTURY



400s Claudian, Synesius, Hydatius (PORTUGAL).

410s Philostorgius, Olympiodorus, Hydatius, pre-Prosper, pre-Marcellinus Comes, Orosius (SPAIN). The Golden Horn eclipse record (?) in DENMARK.

420s pre-Count Marcellinus, Sozomen, pre-Agapius (SYRIA).

430s Absence of eyewitness sources.

440s Hydatius, MONTPELLIER MS. John of ASIA.

450s Hydatius, Prosper. (Chinese sources decline.)

460s Hydatius -468, Agapius, pre-Consular Annals (c.-468), pre-Gregory of TOURS (FRANCE).

470s pre-Cedrenus, pre-Gregory of TOURS (FRANCE).

480s pre-Elias (SYRIA), Marinus.

490s EDESSA sources, pre-Gregory of TOURS (FRANCE), pre-Marcellinus, pre-Denys.

6th CENTURY

500s Absence of eyewitness sources.

510s Lydus, pre-Elias (SYRIA), Samuel of ANI (ARMENIA).

520s pre-Elias, Malalas.

530s Cassiodorus, pre-Bede, Procopius, Annals in ITALY, SYRIAC sources (Edessa Chronicle), Lydus, Marcellinus (up to 534 and continuation to 548).

540s Cosmas, pre-Bede, Procopius, pre-Sigebert 540-565 (incl. Johannid of CORIPPUS 546/8).

550s IRISH Annals, Procopius -554, Malalas.

560s CHINESE sources recover, Gregory of TOURS (FRANCE), Marius (ALPS), Agnellus, Olympiodorus?, pre-Agapius, John of BICLARO, pre-Sigebert, SYRIAC sources, Victoris Tonnennensis -567.

570s Gregory of TOURS (FRANCE), John of ASIA (EPHESUS), Marius (SWITZERLAND), Fasti Vindobonenses (RAVENNA at this date, not Vienna).

580s Gregory of TOURS (FRANCE), Evagrius. Easter Chronicle (c.585).

590s Gregory of TOURS (FRANCE), pre-Fredegar (BURGUNDIAN Annals 584/603), Paul the Deacon, IRISH Annals (AI), Evagrius -593, St. Gallen, Simocatta, pre-Agapius, pre-Michael, pre-Theophanes.

7th CENTURY

600s BURGUNDIAN Annals (pre-Fredegar to 603), SYRIAC Chronicles, COPTIC eclipse ostracon, Paul the Deacon, pre-Theophanes.

610s Stephanus of ALEXANDRIA, George of PISIDIA.

620s George of PISIDIA, pre-Agapius, CTS in CHINA begins, Isidore -624.

630s Early ISLAMIC sources, pre-Theophanes (from SYRIA), JAPANESE sources (Nihongi).

640s Sebokht?

650s IRISH Annals.

660s IRISH Annals (AU), Anglo-Saxon Chronicle (ASC), pre-Theophanes.

670s pre-Michael (SYRIA), James of EDESSA (= pre-Elias), pre-al-Khwarizmi, Liber Pontificalis (Book of the Popes), ASC MS C.

680s 'Book of the Popes', IRISH Annals.

690s pre-Theophanes, James of EDESSA.

8th CENTURY

700s Continuator of James of EDESSA.

710s IRISH Annals (incl. Brut), SPANISH continuator, pre-at-Tabari, pre-Hamza.

720s IRISH Annals.

730s Continuator of Bede (etc. NE ENGLAND), CAUCASUS, IRISH Annals, pre-Agapius (Theophilus).

740s pre-Agapius, Annals (NE ENGLAND AND IRELAND), pre-Theophanes, pre-al-Khwarizmi, Continuator of Fredegar.

750s Annals (IRELAND AND NE ENGLAND).

760s pre-Theophanes, IRISH Annals, pre-Denys.

770s IRISH Annals, ITALIAN Annals, pre-Denys (Henanjésu), ASC (MS C = SE ENGLAND).

780s pre-Theophanes, Royal FRANKISH Annals (both versions), Paul the Deacon -787, Annals of FLAVIGNY, IRISH Annals, Fragm. Annals CHESNII.

790s Ann. FLAVIGNY, pre-Simeon of DURHAM (NE ENGLAND), Ann. Laureshamensis (-803) at LORSCH.

800s ENGLISH Chronicle (ASC MSS, ASC MS F (= pre-Ann. S. Maximini Trevirensis, TRIER), Royal FRANKISH Annals (RFA), Ann. Juvavenses maiores, Brut (WALES), Ann. Farfa (ITALY), Syncellus -810.

810s RFA (& Einhard) etc. e.g. Ann. Juv. maiores again, Theophanes -813, Ann. Loiselianos -814.

820s RFA, ASC, pre-Ibn Yūnus. RFA = Ann. Laurissenses -829.

830s Ann. Xanten (XANTEN), Ann. St Bertin (N FRANCE), Ennin (JAPANESE traveller in CHINA), Nithard, Flavigny (NW of DIJON) -840, Vita Hludowici Imperatoris (Life of Emperor Louis) -840.

840s Ann. Fulda (GERMANY), Fragm. Fontanelle (NE FRANCE), Nithard, Agnellus at RAVENNA -841, Denys (ends 842), Ann. St Bertin (at TROYES) 843-861, Georgios Hamartolos -842, Ann. Lugdunenses -841, Ann. Weissemburgenses -846. Maya records MESOAMERICA.

850s Ibn Yūnus (BAGDAD), Ann. Xanten (LOWER RHINE), Ann. Fulda (GERMANY), John the Deacon (VENICE), Ann. St Bertin (Prudentius, W of RHINE).

860s Ibn Yūnus, Ann. St Bertin, Hincmar at REIMS 862-882, AU, at-Tabari.

870s at-Tabarī, al-Birūni, AU, ASC, Ann. Fulda (FULDA), pre-Regino, pre-Asser, ASC Wessex, Ann. St Bertin again, Andreas Bergomatis (ITALY), Ann. Xanten -c.873.

880s at-Tabarī (BAGDAD), Ann. Sangall. (ST GALL), al-Battanī (RAAY-on-EUPHRATES), Pictish Chron. (SCOTLAND), pre-Elias (SYRIA), Ann. Fulda (Continuator at RATISBON), pre-Dandalus (ITALY).

890s al-Battanī, ST GALL (SWITZERLAND = Ann. Laubacenses -926), ISLAMIC (as 880s) and pre-Ibn-al-Jawzi, pre-Elias, Folcwin I, Saxo, FULDA -901, Ann. Vedastini -900.

10th CENTURY

900s Ann. Corb. (CORVEY, PADERBORN), al-Battanī, Ibn Hayyān, Leonis Imperatoris Imperium -912, MERCIAN Register and ASC, Symeon, Elias (lost sources), at-Tabarī, Ann. Mettenses -905 (also 678-831), Regino -c.906.

910s Symeon, pre-Mas'udi, at-Tabarī (BAGDAD -915), Ann. S. Neots -914, AU, AI, AC all very good, pre-Ann. Aug. (AUGSBURG), Ibn Adarī and Ibn Hayyān, Cont. Regino (near TRIER).

920s AU, Symeon, Ibn Yunus, pre-Abu-l-Faraj, Flodoard (REIMS) 919-961, ST GALL -926, pre-Ibn-al-Jawzi.

930s al-Hashimi (= al-Biruni), Ibn Yūnus, pre-Widukind, Flodoard, Mas'ūdi, Ann. Sang. mai., Ann. Corb., Ann. Weingartenses -936.

940s pre-Widukind, CORDOVA Chronicle, INDIA, AFGHANISTAN sources, Tābit, (pre-Elias), Continuator of Eutychius (Yahya), Mas'ūdi, Cont. of Symeon by George the Monk II (-948), ST GALL (Ann. Sang. mai.), Hamza.

950s IRISH Annals, Fleury, Brut (SW WALES), Folcwin II -962, Simeon of DURHAM (NE ENGLAND), Hamza (IRAN), Mas'ūdi (CAIRO etc.), Symeon (BYZ.), Continuator of Theophanes -961, Ibn Miskawaih (Tābit), Ann. Augienses -954.

960s Lupus (ITALY), Widukind (WESER), Ann. Prum (MAGDEBURG), Pictish Chron. ( SCOTLAND), Richer (REIMS), Ann. Farfa (ITALY), Tābit (= pre-Elias), ITALIAN Annals, Liudprand (MEDIT.) -961, Chron. Fleury, ST GALL, Leo the Deacon 957-976, Yaḥya of ANTIOCH, Cont. Regino - c.967 (TRIER).

970s ASC, MS D (SW ENGLAND), ASC MS C (WINCHESTER then ABINGDON), Tābit, Widukind, Leo the Deacon 957-976, ARMENIA, Tābit (in Elias), Ann. Augustani (AUGSBURG), Ibn Yūnus (CAIRO), Ann. Juvavenses -974.

980s IRISH Annals (several), ASC (S ENGLAND: ABINGDON), Richer of REIMS, ARMENIA, Ibn Yūnus (CAIRO), Observations 977/1007, ITALIAN Annals (BARI?), Ann. Flavinienses -985, Leo the Deacon -990, Pictish Chronicle -990.

990s ST GALL (SWITZERLAND), Richer of REIMS, IRISH Annals, Brut (WALES), Ibn Yūnus (CAIRO), pre-Elias, Thietmar, QUEDLINBURG or HILDESHEIM, Lupus, Ann. Augustani (AUGSBURG), al-Birūni, Abu-al-Wafa.

## APPENDIX E

## ECLIPSES AND COMETS AFTER AD 1000

AD 1000-1300 The eclipses recorded in chronicles after AD 1000 are usually based on observations near the monasteries or cities from which the chronicles were named. For this period the lists in Newton 1971 (Table XVIII: 7 to 9 for Solar Eclipses and Table A I for Lunar Eclipses) are most useful, but we need to take note of his later revisions in Newton 1979. Clues as to the locality of the observations are given by the meteorological information e.g. in P. Alexandre 1976, Le Climat au Moyen Age en Belgique et dans les régions voisins (i.c. 1000-1400), Liège. See pp.66 ff. (Centre Belge d'Histoire Rurale. Pub. No.50.) The solar eclipse tracks of Schroeter 1923 are accurate for Europe and the Mediterranean but omit some eclipses recorded in European Russia.

### COMETS AFTER AD 1000

In this millennium comet dates can usually be determined reliably from independent records in Japan, Korea, China and Europe. We have, however, again selected comets scoring 3 or more points, mentioning in brackets lesser comets only when these have been wrongly identified with novae or European comets in the catalogues. The fifteenth-century table was previously published by Schove 1975 in JBAA. These lists will be published elsewhere.

## APPENDIX F

## CHRONOLOGICAL SYNTHESIS

The date-list includes selected notes on

(a) SOLAR(S) and LUNAR(M) ECLIPSES, together with the marks allocated in Appendix A for identification and information

(b) COMETS(C), together with the marks allocated in Appendix B for magnitude

(c) items from the SPECTRUM OF TIME collections (Schove 1960, 1961, 1983a, p.106). Some years can be confirmed from the revised dates given in (a) or (b). Others might be checked from ice-core, tree-ring and varve analyses; South Germany drought-sensitive oak thus has narrow rings in years of spring or summer dryness; hot summers lead to high density wood as tested by X-rays; frosts in spring may cause damage or frost rings and so on (Schove and Lowther 1957, cf. Schove 1983c and d)

An attempt has been made to combine documentary and tree-ring evidence to identify global weather upsets (often of the type experienced in 1450, 1782/3, 1877/8 and 1982/3)

(d) approximate or <u>inconsistent</u> dates that might be confirmed when e.g. Tabarī and other Islamic writers become available

(e) Sunspot and Auroral information based on <u>documentary</u> evidence (Far Eastern and European combined) or on tree-ring <u>radiocarbon</u>.

Page-references to Schove 1983a are indicated by the prefix SC

Some sources (bracketed) are indicated; many eclipses and comets were recorded by only one contemporary chronicler, and borrowed by others who indulged in the international trade in marvels. Many quotations derived in this way can be found e.g. in Newton 1972, who gives the dates of his sources. The meteorological facts often supply the key to the real eye-witness eclipse-observer, who is then selected for inclusion also in our source-index.

# CHRONOLOGICAL SYNTHESIS

## 1st CENTURY

| | | |
|---|---|---|
| 2 | | Solar eclipse in China (fig.7 and 1st Century) |
| 5 Mar 28 | S | 8, 3 (Dio from a lost source) |
| c. 7 or 8 | | Very cold winter in SE Europe (Velleius) |
| c. 9 to 14 | | Auroral displays and an inferred sunspot maximum (Many displays in period 30 BC/AD 20) SC 133, 166, 373 |
| 13 Nov/Dec | C | (No certain Western record found) |
| 14 Sep 27 | M | 8, 4 (Dio, Tacitus) |
| c.15 | | A cold winter in SE Europe (Dio) and a series (AD 12-19) of poor Nile floods and presumed scarcity in Egypt |
| 17 Feb 15 | S | 6, 0 (Dio, etc) |
| 20 | | Sunspot seen in China; possible year of the 'Tiberius' Aurora at Rome and Ostia (SC 53, 155, 166) |
| 22 Nov/Dec | C | (No Western record found) |
| 29 Nov 24 | S | (Lost Asia Minor source) |
| 27/36, especially 30 | | Possible years for the Crucifixion (cf. Addenda) |
| 39 Mar | C | 4 (No Western record found) |
| c.45 | | Weather upset. Abnormally high Nile flood and 'Universal Famine' |
| 46 July 6 | M | 8, 3 (Seneca, etc) |
| 54 June/July | C | 5 (Suetonius, Seneca) |
| 55/56 Dec/Mar | C | 5 (Pliny, Dio) |
| c.58 | | Heath fires - drought? Germany (Tacitus) |
| 59 Apr 30 | S | 9, 5 |
| c.53/63 | | Poor Nile floods (Tacitus, etc) |
| 60 Aug | C | 5 (Tacitus, etc) |
| 62 Mar 13/14 | M | 8, 2 |
| 65 July/Sep | C | (For Chinese eclipse of AD 65 see fig.5) |
| 66 Jan/Apr | C | (Dio, etc) |
| 67 May 31 | S | 4, 2 |
| 68/69 | M | Lunar eclipses of Dio (See text) |
| 69 | | Drought in Rhineland (Tacitus). Tree-rings suggest dryness lasted from 67 to 69 |
| 71 & 75 (83) | | Eclipses of Pliny and Plutarch (See text) |
| 77 Jan/May | C | 4 (Pliny, Dio, etc) |
| 84 May/July | C | 3 (No Western record found) |
| 90 (about) | | Epidemic with spots ascribed to malicious needle-pricks (Dio) |

| | | |
|---|---|---|
| 111-124 | | Poor Nile floods and scarcity in Egypt (e.g. 113, 120, 129/130, 134/5) SC 285 (after Bonneau) |
| | | Auroral activity was weak from AD 80 to 130 (SC 138, 349) |
| c.124-128 | | Long drought in N Africa |
| 144 | | No documentary evidence of drought, but narrow tree-ring in S Germany |
| 147 | | Cold winter recorded in the 'diary' of Aristides |
| 150-170 | | Poor Niles e.g. 156/7, 166 (SC 285) |
| 165 | | Pandemic spreads from Asia (Smallpox?) and in: |
| 166 | | reaches Italy and by 171 'the whole world' (Orosius, etc) |
| 171 | | Epidemic in China |
| 160/195 | | Strong sunspot and auroral activity (SC 53, 133, 196)<br>Radiocarbon minimum follows c.AD 210 (SC 373) |
| 180/9 | | Pestilence in Europe and China |
| 185 Dec | C | (9) Supernova (Herodian, etc) |
| 186 Dec 28 | S | (4, 2) (Hist. Aug.) |
| 187/191 | | Other comets e.g. 191 (4) (Herodian) |

## 3rd CENTURY

| | |
|---|---|
| 204, 218 | Comets (and solar eclipse, also in 218) of Dio |
| 212 | Solar eclipse and Aurora of Tertullian |
| 233-6 | Probable global weather upset. In the period 230-250 prolonged drought in S Mediterranean |
| 240 | Solar eclipse in Asia Minor is a genuine record in the Historia Augusta |
| 220-280 | Only a few aurorae (SC 54, 109, 133, 138, 166, 349, 373) |
| 251 | After long pestilence-free period since c.190, the 1st Pestilence temp. Gallus (Numismatic evidence, etc) |
| 254-265 | Weak Niles and prolonged scarcities in Egypt |
| 261 | Second wave of pestilence temp. Gallienus (Zosimus) |
| 271 | Third wave temp. Claudius II (Zosimus, etc) |
| 272-286 | No more pestilence but famines instead (Zosimus, etc) |
| 277 | Great comet (6). (Halley's comet came in 295) |
| 280-297 | Nile floods improve |
| 289, 291 | Mild and cold winters respectively (Mamertinus) NW Europe (cf. Solar eclipses of Hydatius 291-2) |
| c.299 | Ice on the Rhine (Uncertain year) (cf. Weiss) |

| | |
|---|---|
| c.302 | Many sunspots and aurorae (299 to 305) SC 138, 161, 208 |
| c.302 | Drought-famine N Africa (Also c.366/7 and c.388) |
| 358 | Cold but short winter in NW Europe (Ammianus Marcellinus) |
| 4th Cent. | Many sunspots and aurorae (SC 54, 111, 166, 379), especially c.350-380 (SC 161, 196, 349, etc) |
| c.372 | Sunspot cycle no. -124 active 370-375 (SC 208, etc) |
| 374 | Halley's Comet (Ammianus Marcellinus) |
| 377 | Cold February NW Europe (Ammianus Marcellinus) |
| 390 | Comet of Philostorgius |
| c.394/6 | Earthquake and aurorae in Mediterranean (Marcellinus Comes) |

## 5th CENTURY

| | |
|---|---|
| 400 | Spring Comet (5). The 'armed angels' that frightened Gainas? |
| 401 | Cold winter: ice on the sea near Constantinople (Marc., Paschale) |
| 402 | Eclipse and Comet (3) |
| 407/411 | Famine in the Mediterranean |
| 409 Feb 1 | Astrologers recalled from exile |
| 422 | Spring Comet (3). Prosper's 'miraculous sign in heaven'? (Marc., etc) |
| 431 | Rain-famine Antioch and locusts in E Mediterranean |
| 432 | Cold winter in Europe |
| 442/3 | Comet (5) and cold winter: snow lasted nearly six months SE Europe (Marcellinus, Hydatius) |
| 430/445 | Includes an undated severe drought in Jerusalem |
| 440/449 | Few sunspots and aurorae and thus radiocarbon maximum (SC 161, 196, 349, 373) |
| 450/3 | Aurorae and possible sunspot (SC 116, 121, 371 and text 453) |
| 447/474 | Evidence of Maya interest in eclipses |
| 452/5 | World weather upset and famine Middle East (SC 268) |
| c.463 | Another global climate upset. Nile failure and famines in Eurasia (SC 268) |
| 467 | Comet (5). The 'cloud shaped like a trumpet'? (Pasc., Victor Tonn., etc) |
| 467/8 | Heavy rains follow in Constantinople and Asia Minor |
| c.469 | Cold winter in SE Europe (Jordanes) |
| 472 Nov 6 | Eruption of Vesuvius and rain of cinders in Constantinople. (Volcanic ash from Iceland also in this century, but no confirmed ice-dates from Greenland ice-cores yet) (SC 344) |
| 479 | Auroral peak and new sunspot dates 478/482 from China (SC 116, 121 and 371 revision) |
| 489/490 | Auroral maximum and new sunspot dates (SC 121, 155 and 371 revision) |

| | |
|---|---|
| 502 | Aurora as far south as Iraq (Joshua the Stylite). Many more in period 500-530 (SC 116, 138, 373) and many sunspots seen in China (SC 54, 63) |
| 507-511 | Droughts in Italy and SE Europe. Narrow tree-rings in S Germany 509-511 and fires in Constantinople 509, 510 and 512 |
| 529 | Cold winter SE Europe |
| 530 | Halley's comet blamed for the drought, famine and mortality that followed |
| 530 | Meteor showers Europe and China (cf. fig.12) |
| c.535-537 | Unknown volcanic eruption: sun obscured: global weather upset and harvest failures pave way for Plague |
| c.538-9 | Mediterranean Famine follows. Comet year is 539, not 540 |
| c.542/3 | Plague spreads from Ethiopia and Middle East to the British Isles |
| 546 | Narrow tree-ring suggests drought in Germany |
| c.548 | Cold winter in France (recalled by Gregory of Tours) |
| 556 Nov | Unconfirmed 'comet' of Malalas (Meteor, perhaps) |
| c.558 | 2nd wave of Plague, affects children (Agathias, John of Nikiu, etc) |
| 562 | Drought leads to public prayers: north winds had lasted from August to November (Malalas) |
| 563 | Drought leads to closure of public baths and murders near the fountains. Abnormal drought also in American Southwest (tree-rings) |
| 561, 564 and 565 | Narrow tree-rings (drought?) in S Germany (Becker) |
| c.567 | Strange dust and darkness (Volcanic?) (John of Biclaro) |
| 565 and 568 | Correct dates of great comets (Gregory of Tours, etc) 6 and 5 |
| 569-575 | Smallpox pandemic in Europe |
| c.574 | 3rd wave of Plague e.g. Gaul (Gregory) and the Turks (Simocatta) |
| 580s | Wet, warm decade (Gregory). 587, 589 Wide tree-rings (Becker) |
| 581 | 4th wave of Plague. Spring frost in France (Gregory) |
| 582-587 | Aurorae (e.g. Blood rains) in Gregory (SC 118-120) |
| c.590 | 5th wave of Plague (Evagrius, Secundus) |
| 591 and 592 | Narrowest tree-rings since 565 in S Germany confirm the long drought (Jan-Sep) reported for Italy and France |

| | |
|---|---|
| c.600 | 6th wave of Plague (e.g. Aleha-zeka in Elias) |
| 607 | Halley's Comet (7). This is wrongly dated as 602 in Muralt (cf. Proudfoot) |
| c.608-610 | Mortality and cold winter in SE Europe |
| 617 Nov 4 | Eclipse wrongly dated as 618 |
| 620s | Dark sun affected by eruption(s). Years uncertain |
| 626 Mar | Comet (3), presumably the 'shining star in the west' (cf. Paschale) |
| c.627/8 | Famine and drought in Arabia |
| 634/5 | Comet (5) is probably the comet that later historians (Theophanes, etc) backdated to precede the Arab victories |
| c.638 | Prolonged drought in Arabia and Syria, and |
| c.636-645 | world weather upset (Arabia, Nile, Far East and tree-rings in U.S.) |
| c.648 | Famine-cannibalism in Iraq and Armenia (Michael) |
| c.652-4 | 'Dust fell from sky' (Theophanes, etc). Either aurora, meteor shower or volcanic eruption |
| 676 | Great Comet (evidently the 'sign in the sky' of Theophanes misdated 675) |
| 630-680 | Few aurorae and high radiocarbon production (SC 8, 24, 138, 374) |
| 680s | Severe droughts in Syria, Libya, etc<br>Narrower tree-rings in Germany suggest dryness (680/9) there too<br>Famine and emigration from Syria (Theophanes) |
| 684 | Cold winter precedes Halley's Comet |

## 8th CENTURY

| | |
|---|---|
| c.706 | Frost in April in Middle East (Michael after lost source) |
| 706/7 | Nile fails in Egypt. Too much rain in China |
| c.707 | Auroral maximum |
| c.709 | Cold winter in NW Europe (Mosell.) |
| 709/710 | Planetary conjunctions (Schove 1977) |
| 706, 709, 711 | Narrow tree-rings and presumed droughts in S Germany (Becker) |
| 685/785 | Meteor showers every twenty years (E-group Jan 1-4) |
| c.710 | Epidemic like polio affects Ireland and Egypt (Makrisi) (Schove 1974) |
| 717 | Long, cold winter affects all Europe between eclipses of 716 and 718 (e.g. Nicephorus) |
| c.721 | Hot summer, probably in Wales or Ireland (Annales Cambriae) |

8th CENTURY (continued)

| | |
|---|---|
| 722 | Very good harvest (Ann. Mosell., etc)<br>Wide tree-rings 721 on (S Germany) |
| 724 | Auroral maximum |
| 730 or 729 | Bede's Comet (See Addenda) |
| 730s | Nile failures |
| 737 and 741 | Droughts reported in N England in Continuation of Bede<br>(The fire of York on 23 Apr 741 may be linked) |
| 738-743 | World weather upset |
| c.743 | Long, cold winter in SE Europe (Misc. Hist.)<br>Price of vegetables very high afterwards |
| c.743 June | Aurora seen in Syria (Agapius, Michael)<br>Many in this decade (SC 8, 374) |
| 745 Jan 1 | Comet and (E-group?) meteor shower<br>(Theophanes, Elias, 'Simeon', AT) |
| 746 | Plague and great darkness Aug 10/14 (Volcanic?)<br>(Theophanes, etc) |
| 763/4 | Long, cold winter with a meteor shower (seen also in China) possibly on Dec 31. Thaw-floods followed, then droughts in both NW and SE Europe |
| 764 | Droughts in England with fires suggests a narrow tree-ring likely (Schove and Lowther 1957, but not yet confirmed) |
| 760 and 770 | Comets fix the chronology of Henanjésu (in Denys), who notes the aurorae of c.766 in Middle East |
| c.766/7 | More droughts in E Mediterranean (Nicephorus, etc)<br>Prolonged period of poor Nile floods 758-787<br>(See SC 234, 285) |
| 769 | Acorn crop in Ireland suggests preceding summer was hot<br>(Schove and Lowther 1957) |
| c.770 | Famine in Fertile Crescent. Mortality throughout Middle East (Henanjésu) in 770/1 |
| c.772 | Drought in W Europe (Ann. regn. Francor.) |
| c.773 | World weather still upset |
| c.779 | Auroral maximum: in England from 776 |
| 784 | Wet in NW Europe |
| 786/788 | Auroral maximum (Ann. Lauresham, Chesnii, etc) |
| c.793 | Famine general in Europe. In 793 aurorae are seen in N Europe ('fiery flying dragons' and 'rain of blood')<br>Great drought, but a good harvest according to Ann. Mosell. |
| c.797 | Auroral maximum further south |

| | |
|---|---|
| 800 | Summer frost (Ann. regn. Francor.) |
| 801 | Warm winter ( " " " ) |
| c.803 | Locusts in NW Europe |
| 807 | Sunspot seen in Europe (SC 47, 317 and cf. 32), but few aurorae 770-830 (SC 8, 24, 374) |
| 808 | Famine - Iraq, Armenia and Byzantine Empire |
| 811 | Cold winter in which two sons of the Emperor died |
| 815, 819, 822 | Narrow tree-rings S Germany (Drought?) |
| 820 | Cold summer. Wet (RFA) |
| 822 and 824 | Long, cold winters. 824 followed by drought. (RFA) Tree-rings Germany narrow 822-840 |
| 827-840 | especially 832-842 Weak Niles. Nile frozen one winter c.828/832 |
| 840 | Important sunspot and auroral year. Many spots in period 830/850 (SC 8, 43, 55, 134, 138, 317, 374) |
| 841 | Dry, late summer (Nithard) |
| 843 | Long, cold winter (Nithard); 844 mild; 845 cold (S. Bertin) |
| 846 & 849 | More cold winters (Bertin, Fontanelle) |
| 850 & 852 | Great heat (Xanten) 2-year cycle? (Narrow rings 850, 851, 855, 856 and 858) |
| 856 | Cold, dry winter (Prudentius in S. Bertin) |
| 860 | Long, very cold winter Germany and Italy (Fulda, Andreas, John the Deacon, etc) |
| 863 | Very wet, mild winter (Xanten) |
| 870 | Great summer heat; great drought June to August (Fulda, Folcwin I) |
| 873 | Cold winter. Locusts fly west into NW Europe in hot summer and destroy crops in Germany (Abnormally narrow tree-ring in 874) |
| 874 | Another very cold winter and dry, hot summer |
| 880 | Cold winter; rivers frozen (Xanten) |
| 886 | Very wet May/July (Fulda, John the Deacon) |
| 887 | Long, cold winter (Fulda) |
| 891 | Correct date of Alfred's comet (May/July 5 points); in Egypt comet followed by a poor Nile flood. In NW Europe drought and spring frosts spoilt the wine in 891 or 892 (Chron. Andregav., Ann. Nivern.) |
| 893 | Another cold winter. Characteristic of odd years 875/893 |
| 894 | Cold again so that 2-year cycle ends in 893 |
| 890s | Droughts in Islam generally (Elias, etc) |
| 897 | Sky extremely red at night (Volcano, Sandstorm, Aurora ?) in Egypt<br>Few aurorae in period 880-910 (SC 56, 63, 374) |

| | |
|---|---|
| 900/902 Oct | Spectacular meteor showers in Europe, Islam and the Far East (cf. fig.12 and Schove 1972) The Leonids |
| c.902 | A drought in Constantinople |
| c.904 | Auroral maximum, with a great comet in May 905 |
| 900/930 | Wet in Islam, but cold in N Europe and narrow tree-rings commence in Germany |
| 912 | Correct date of a very great (7-point) comet (Symeon, etc) Called a wet and a dark year in Ireland (See text) |
| 913 Feb 6 or 13 | Meteor shower seen in Europe and Islam |
| 913/921 | Some severe winters in Europe and Islam, but years not certain |
| 915/6, 926 and 929 | Approximate years of droughts in Spain |
| 917-919 | Aurorae in Europe |
| 921 | Long drought of Flodoard confirmed dendrochronologically, but comet years (918, 922/3, 928) claimed are uncertain |
| c.925-7 | Sunspots and aurorae (SC 374) |
| 927 | Probable date of very cold winter and of Iraq floods |
| c.928 | Dry summer; fits narrow tree-ring |
| 931-4 | On Oct 20-23 spectacular Leonids (cf. fig.12) |
| 931-950 | Weak Niles; global weather upsets with cold winters and famines in Europe and Islam up to 947. Tree-rings in Germany narrow |
| 941/2 | The comet was supposed to have led to high prices and cannibalism in Tukaristan (E.S. Kennedy in Centaurus, 1980) |
| 948/9 | Droughts in Europe and large acorn crop follows in Ireland in 950 |
| 951 | Mortality of bees in Ireland (Cold summer?) |
| 957 | Approximate year of hot summer in Wales (Brut) |
| 953 and 963-9 | Nile failures |
| c.965/8 | Aurorae in Mediterranean and China (SC 138) Leonids in 967 |
| 974 | Great drought in Europe (Ann. Corb.) and Middle East (Leo diaconus) |
| c.978/9 | Aurorae in England and N Africa |
| c.977-980s and 990s | Hot summers continue. Good acorn crops in Ireland. Droughts and some narrow tree-rings; returning again in c.988-998 |
| 994 | Very cold winter lasts from previous Oct to May (Europe) |
| 997 | Scarcity in Iraq and many rulers in SW Asia and Fertile Crescent die (as-Suyuti from a lost source), presumably through pestilence |
| c.999 | A cold winter causing the death of palm trees in Baghdad |

# ADDENDA AND CORRIGENDA

Introduction: p.xi

1131 BC and 1406/7 ECLIPSE OF GIBEON: SUN STANDING STILL

Some scholars believe the Gibeon incident (Joshua 10: 12-13) might belong to a later century: there was, however, no important eclipse after 1131 BC until 831 Aug 15, which was 0.97 in magnitude at an altitude of 72%. (See F.R. Stephenson, Pal. Expl. Qly., 107, 1975, 107-120. See p. 112.)

The repetition of this miracle about 1407 is referred to by Dowling (17 C) whose claim that the sun stood still so that the Anglo-Norman Earls could defeat O'Carrel is mentioned in a footnote to the Annals of Ulster (Vol. 3, 56, n.1. Reference kindly supplied by Dr A.P. Smyth). This defeat, however, took place in 1407; the eclipse on 1406 June 16 probably coincided with some other battle and the two events have been assimilated. Aurorae occurred in the Far East as late as 1406 but the year of the battle, 1407, as our table at the end of the Introduction implies, is too near the sunspot minimum for a display bright enough to continue a 'daylight' battle.

Introduction: p. xi VOLCANIC DARKNESS

Darkness not due to total solar eclipses can be due either to

(a) Cold fronts (e.g. 331?), as happens today in modern NW Europe (and the USA),

(b) Dust-storms, notably in the Middle East, e.g. 418?

or (c) Volcanic eruptions as in 44/43 BC, c. 472, 536, 797, etc. Earlier such dates would include the dates of the eruptions of Thera (Santorin). Dates for this have been suggested (Schove 1983d), from summer temperature indications based on tree-ring densities at 1479 BC and (a later eruption, perhaps from somewhere else?) 1349 BC. No documentary evidence is available: the stories of the plague of Egypt in the time of Moses seem to reflect folk-memories following a series of Nile Flood failures, perhaps in the thirteenth century. A minor eruption in Thera is dated in our text as AD 46.

536 Stothers' Nature reference (304, 1984, 344-345) is now available.

1st Century VELLEIUS

The History of Rome by Velleius Paterculus is not lost as Dr Peter Bicknell has kindly pointed out. It is indeed cited (from Weiss 1914) in our Appendix F as the source for the cold winter of AD 7/8.

S 29 Nov 24 'ECLIPSES' ASSOCIATED WITH THE CRUCIFIXION

L 33 April

A claim that the lunar eclipse date of AD 33 was the date of Christ's death has been made by C.J. Humphreys and W.G. Waddington in 'Dating the Crucifixion' (Nature, 306, 1983, 743-746). They start with the admission 'The only certainty about the date of the Crucifixion is that it occurred during the 10 years that Pontius Pilate was procurator of Judea (AD 27-36)'. They proceed to show, as we did, that the Gospel evidence, if accepted literally, leads to either AD 30 Friday 7 April or AD 33 Friday 3 April. They reject the tradition going back to Tertullian (AD 200) - again as we did - that the date was AD 29 March 25. They then argue that the lunar eclipse of AD 33 explains why, in Acts 2, 14-21, at Pentecost, Peter is said to have used Joel's phrase 'The Sun will be turned to darkness and the Moon to blood ...'. They suggest that this quotation was a topical reference to both a duststorm (not a solar eclipse) and a lunar eclipse, at the time of the Crucifixion seven weeks earlier. Additional evidence for a lunar eclipse is based on their hope that the so-called Report of Pilate was not entirely a late Christian forgery. However, I feel that the lunar eclipse at Jerusalem on April 3 was so small that any redness would have been forgotten by Pentecost and certainly by the time that the Acts were written. The AD Solar eclipse in Asia Minor was presumably assimilated to the eclipse date, but this would surely have been some time after AD 45 when Christians there began to reconcile their memories with the local records - the statement that the Sun was eclipsed now occurs in some translations of Luke. Of the two dates AD 30 is thus still slightly preferred to AD 33 for the Crucifixion year. Astronomical attempts to fix the exact date of both the birth and death of Christ have thus all been unsuccessful.

622, 644 and 695 THEOPHANES

In the new translation (Turtledove) the eclipses of 644 and 655 are given on pp. 42 and 65 and correctly dated. In 622 a 'lightless night' is mentioned on pp. 14-15.

## 718 and 720 TOTAL ECLIPSES IN SPAIN

Important solar eclipses occurred in Spain in this century (706, 710, 718, 720, 779 and 787). The 'Cronica mozarabe de 754' with Latin and Spanish translations has been edited by J.E. Lopez Pereira and published as No. 58 in the series Textos Medievales by Anubar at Zaragoza. This includes under the Spanish year 758 (38 years ahead of the AD year):

At that time, at the beginning of the year 758 (i.e. AD 720), in the 100th year of the Arabs (i.e. beginning August 718), in Spain the sun was eclipsed (deliquium solis in the Latin) from the seventh to the ninth hour (13.30 to 14.30 in the Spanish), making stars visible.

The timing and the general indications that the information came from Toledo in Central Spain indicate that this is from the same source as ascribed by us to Isidore's Continuator i.e. it relates to the total eclipse of 718 June 3, but the Spanish year 758 (= 720) again suggests that the annalist conflated the reference to the two eclipses, 718 in Central Spain and 720 in South Spain.

## 729 or 730 THE YEAR OF BEDE'S COMET

Mr K. Harrison does not think Bede's comet (Eccl. Hist. 5.23) is a genuine record of the year 729. I prefer the Chinese date of 730. No independent West European reference to a comet (e.g. preceding the 732 battle of Poitiers) has been noticed.

## 752, c.760, 848, 879 ROGER OF WENDOVER

Roger of Wendover does preserve some details from a lost version of the Anglo-Saxon chronicle; these deserve further consideration but the 848 quotation probably refers to the 878 solar eclipse.

## 934/6 DARKENED BY THE BLOOD OF MEN

Dr D.N. Dumville points out that the A-text is a derivative of an ancestor of B but that the later MS. B (written probably 977/8) gives a better text for 924-46, and he refers to the edition of The Battle of Brunanburh by Alistair Campbell (London 1938). The reading in B, C and D 'blood of men' is thus to be preferred to 'bold warriors'. There is no convincing evidence that the scribe of A 924-46 was writing at Winchester

The poem was written specifically as a written poem (Campbell op.cit.) for the chronicle. Dr Dumville notes that dunnade is a preterite of dunnian, a verb meaning 'darken', 'obscure', a verb not commonly met with, but concludes 'I doubt the basis for the inclusion of this item at all'.

I looked for possible eclipse inspiration in the Anglo-Saxon poem 'The Phoenix' with its many references to the Sun (ed. N.F. Blake, 'The Phoenix', Manchester University Press, 1964: See Sunne in Index). Line 288 has certainly been amended by the editor who comments (p.76) 'It is unlikely that the poet would refer to the renewal of the sun ...'. The terminus ad quem is indeed c.940 (p.23) but no evidence of any topical references to the 934 eclipse has been traced in the poem, which could indeed be a half-century earlier.

934 thus remains thc 'Missing Eclipse' in British sources. There is likewise no reference to any after-effects of the Icelandic eruption of this decade.

# BIBLIOGRAPHY

In the accounts of eclipses, references have usually been given in sufficient detail as they occur. But some works are mentioned so often that a brief name and year form has been used (e.g. Oppolzer 1887, Ginzel 1899, Schroeter 1923), or even abbreviated (e.g. Ginz.). The chronicle sources for Europe from the 1st to the 12th century are listed in place-name order in Newton 1972.

Further details of the sources are available in Potthast (1896, 1962-70 etc.) or in The Guide to the Sources of Medieval History by R.C. van Caenegem, 1978, Amsterdam (North Holland Publishing Co.) and in the references there cited.

---


Agapius of Manbij (Hieropolis). (10th C but useful for even 6th C.) Ed. A.A. Vasiliev, Universal History PO 8(3). 1912.

Ahnert, P. 1960, 1961. Astronomisch-chronologische Taflen für Sonne, Mond und Planeten (Astronomical-chronological tables for the sun, moon and planets). 1st ed. Leipzig, Barth. 42pp., 43 tables.

Aimoin of Fleury. Floruit 960s. 1869, 1880. History of the Franks. Useful editions of Books 1-4 in Bouquet, 3, 1869 and PL 139, 1880.

Albategnius (Albatenius, al-Battānī). (c.920.) Astronomy (in Arabic), formerly often latinized as De Scientia Stellarum, now usually called Opus Astronicum, see Nallino 1903.

Anderson, A.O. 1922. Early Sources of Scottish History, AD 500 to 1286. Vol.1. Edinburgh and London, Oliver and Boyd.

Annals of Clonmacnoise (AC). 1896. Ed. D. Murphy. Dublin University Press.

Annals of Inisfallen (AI). 1951. Ed. Sean MacAirt. Institute of Advanced Studies, Dublin.

Annals of Lorsch (Annals Laurissenses). (See RFA.)

Annals of Saint-Bertin (abbey of that name at Saint Omer, Pas-de-Calais, France), 830-882 (continuing RFA). Latin texts in:

1826, ed. G.H. Pertz, SS 1, 419-515.
1871, ed. C. Dehaisnes, Paris, Soc. de l'Hist. de France.
1883, ed. G. Waitz, in S & G.
1964, ed. F. Frat, Jeanne Vielliard & Suzanne Clémencet, with Introduction by L. Levillain, Paris, Klincksieck.

Annals of Tigernach (AT). 1895-6. Ed. W. Stokes.

1895. Revue Celtique, XVI, 374-419.
1896. Revue Celtique, XVII, 6-33, 116-263, 337-420.
1897. Revue Celtique, XVIII, 9-59, 150-303, 374-391.

Annals of Ulster (AU). 1887-1901. Ed. W.M. Hennessy and B. MacCarthy, Dublin (HMSO).

Aveni, A.F. 1980. Skywatchers of Ancient Mexico. University of Texas Press. Austin and London.

Bar Hebraeus (13th C). The Chronology of Bar Hebraeus. Ed. E.A.W. Budge, I. Oxford, 1932.

Barrett, A.A. 1978. Observations of comets in Greek and Roman sources before AD 410. Journ.Roy.Astr.Soc. of Canada, lxxii, 81-106.

Bede. See Colgrave and Mynors 1969 and C.W. Jones 1943. There are various other editions and translations.

Bekker, I. Editor of Byzantine Bonn editions. (See e.g. Cedrenus.)

van den Bergh, G. 1955. Periodicity and Variations of Solar (and Lunar) Eclipses. Haarlem.

Bickerman, E.J. 1968. Chronology of the Ancient World. London. Thames Hudson. 2nd ed. c.1980.

Boll, F. 1909. Article 'Finsternisse' in RE, 6, cols.2329-2364. Stuttgart.

Bouquet, M. Recueil des historiens des Gaulles et de la France. Paris. 2nd ed. 1869-1880.

Brady, J.L. 1982. Halley's Comet AD 1986/2647 BC. JBAA, 92(5), 209-215.

Bricker, H.M. and Bricker, V.R. 1983. Classic Maya prediction of solar eclipses. Current Anthropology, 24, 1-23.

Bunting (Bünting), H. 1590. Chronologia ............. Magdeburg
1608. Chronologia Catholica .... Magdeburg
(The two editions have the same foliation.)

Calvisius (Seth Kalwitz 1556-1615).
1605. Chronologia .............. Leipzig
1620. Opus Chronologicum ........ Frankfurt on Oder
1650. Opus Chronologicum ........ Frankfurt on Main

We often quote these three editions, which differ greatly. We have also seen a 1685 edition. Frankfurt on Main and Leipzig.

Caussin, C. 1799. Pages 1-12 of 'Histoire' section of Mém. de l'Inst. Nat. des Sci. et Arts (Sci., math. et phys.), 2, Paris, Fructidor an VII. A preliminary account, much slighter than Caussin 1804.
1804. Le Livre de la Grande Table Hakémite. Paris. (French translation of work by Ibn Yūnus.) Not seen. Details of eclipses are given primarily in Newcomb 1878 and Newton 1970.

Cedrenus, G. (11th C). Synopsis Historion (Compendium Historiarum).
Ed. I. Bekker, 2. Bonn 1839.
See also PG 121, 1894.

Chabot. (See Michael the Syrian.)

Chambers, G.F. 1889 History of Astronomy. (4th ed.) London.
1902 The Story of Eclipses. London. (Various editions)
1909 The Story of the Comets. App.II, pp.242-244. London.

CHFMA. 1923. Les Classiques de l'Histoire de France au Moyen Age. Series by Société d'Edition 'Les Belles Lettres', Paris.

Chronicon Scotorum (CS). 1866. Ed. W.M. Hennessy. Rolls Series. London.

Cohen, A.P. and Newton, R.R. Solar Eclipses recorded in China during the Tang Dynasty. Monumenta Serica, 35, 1984.

Colgrave, B. and Mynors, R.A.B. 1969. Bede's Ecclesiastical History of English People. Oxford. (Latin and English)

Collection des mémoires relatifs à l'histoire de France (CMHF). Ed. F. Guizot. Paris. 1823-1835. Many errors. See now CHFMA.

Conde, J.A. Conde has no accent in Spanish, but tends to be spelled Condé in English. We have seen Conde 1840 and Condé 1854; there are further editions and translations beside those mentioned below.

Conde 1840. Historia de la dominacion de los Arabes en España ... One volume, Paris, Baudry. (The Spanish edition appeared originally in three volumes at Madrid in 1820-1 and again in three volumes at Barcelona in 1844.)

Condé 1854. History of the Domination of the Arabs in Spain. Vol.I (of the three-volume translation by Mrs. Foster, London, Bohn, 1854-5).

Dall'Olmo, U. 1978. Meteors, meteor showers and meteorites in the Middle Ages: from European sources, J. Hist. Astr., 9, 123-134.

Dall'Olmo, U. 1979. An Additional list of Aurorae from European sources from AD 450 to 1466. Journ.Geophys.Res., 84, 1525-1535.
1980. Latin Terminology relating to Aurorae, Comets, Meteors and Novae. Journ.Hist.Astron., xi, 10-27.

Delaporte. (See Elias.)

Denys (or Dionysius). Chronique de Denys de Tell Mahré, quatrième partie. Bibliothèque de l'Ecole des Hautes Etudes. Soc. phil. et hist., 112, Paris, 1895.

Dio (Cassius). (AD c.230). Loeb ed., with translation E. Cary. 9 Vols.

Dunthorne, R. 1749. A letter ... concerning the Acceleration of the Moon. Read 1749 June 1. Phil. Trans. 46 (London, 1752), No.492 (for 1749 April-June), pp.162-172.

Einhardi Annales. (See RFA.)

Elias of Nisibis (11th C). Chronographie de Mar Elie Bar Sinaya, Métropolitain de Nisibe, ed. L.J. Delaporte. BEHE 181. Paris, 1910. (Syrian text and Latin translation eds. E.W. Brooks and J-B. Chabot.) In CSCO 1954.

Eusebius (c.AD 325-336). Eusebi Chronicorum libri duo. Ed. A. Schoene. 2 vols., both reprinted 1967 from editions of 1875, 1866 respectively. (See T.D. Barnes 'The Editions of Eusebius' Ecclesiastical History' in Greek, Roman and Byzantine Studies, 21, 1980, 191-201.)

Eusebius-Jerome. Jerome's extension written about 380 is included in Schoene, vol.II. The whole of the Latin (only) has also been published in Eusebii Pamphili Chronici Canones, Latine vertit, adauxit, ad sua tempora produxit S. Eusebius Hieronymus. Ed. J.K. Fotheringham, Oxford 1923.

Fagnan, E. 1901-4. Histoire de l'Afrique et de l'Espagne intitulée Al-Bayano'l-mogrib, traduite et annotée par E. Fagnan. Two vols. Alger. (French translation of work by Ibn Aḍarī.)

Fotheringham, J.K. 1921. Historical Eclipses. Being the Halley lecture delivered May 17, 1921. 32pp. Oxford: Clarendon Press. (cf. MN 81, 1920, 104-126.)

'Fredegar', Chronicle. Some useful editions and translations.

L: 1869, in Bouquet, Recueil 2 (2nd ed.).
L: 1888, SrM 2, ed. B. Krusch.
F: 1823, tr. F. Guizot, CMHF 2.
(Book 5 and continuations, now called Book 4 and continuations.)
L,E: 1960, Book 4 and continuations, ed. and tr. J.M. Wallace-Hadrill. London, Nelson.

Georgius, Monachus (11th C). Byzantine Chronicle (see under AD 891). Bonn ed. 1839. PG 121, 1894.

Gingerich. (See Oppolzer.)

Ginzel, F.K. 1882. Astronomische Untersuchungen über Finsternisse. I. Weiner Akad., Sitzungsberichte (math. naturwiss.), 85 (II), Jahrgang 1882, Vienna 1882, 663-747.
1883. AUF II. Ibid., 88 (II), Jahrgang 1883, Vienna 1884, 629-755.
1884. AUF III. Ibid., 89 (II), Jahrgang 1884, Vienna 1884, 491-559.
1886. Über einige historische, besonders in altspanischen Geschichtsquellen erwähnte Sonnenfinsternisse. Berl. Akad., Sitz., 1886, 963-81.
1887. Über einige von persischen und arabischen Schribstellern erwähnte Sonnen und Mondfinsternisse. Berl. Akad., Sitz., 1887, 709-14.
1899. Spezieller Kanon der Sonnen und Mondfinsternisse für das Ländergebiet der Klassischen Altertumswissenschaften und den Zeitraum von 900 vor Chr. bis 600 nach Chr., Berlin, Mayer and Müller, viii, 271pp., 15 maps.
(Special canon of solar and lunar eclipses for the territories of ancient Classical Science and the period 900 BC to AD 600. Calculations from 10°W - 50°E and 30° - 50°N, with visibility of eclipses at Rome, Athens, Memphis and Babylon. Remarkably reliable tracks.)
1914. Handbuch der mathematischen und technischen Chronologie ... III. Band. Leipzig. (I. Band 1906, II. Band 1911.)
1918. Beiträge zur Kenntnis der historischen Finsternisse Berl. Akad., Abh. (phys. math. Kl.), 1918, No.4 or Abhandl. Preuss. Akad. Wiss.

Glycas, M. (10th C.) Byzantine Chronicler, Bonn ed. 1836. PG 158, 1866. (See under 891.)

Goldstine, H.H. 1973. New and Full Moons 1001 BC to AD 1651. Mem. Amer. Philos. Soc., 94. Philadelphia.

Gray, R. 1965. Eclipse Maps. Jnl. Afr. Hist. 6, 251-262.

Gregory of Tours. History of the Franks. Some useful editions and translations.

L: 1869, in Bouquet, Recueil, 2 (2nd ed.).
L: 1885, SrM 1(1), ed. W. Arndt and B. Krusch.
L: 1951, 2nd ed. of SrM 1(1), ed. B. Krusch and W. Levison.
F: 1823, tr. F. Guizot, CMHF, 1, Books 1-8, 2, Books 9,10.
F: 1963-5, tr. R. Latouche, 2 vols.
E. 1974, tr. L. Thorpe (Harmondsworth, England).

Guizot. (See CMHF).

Halley, E. 1693. Emendationes ac notae in vetustas Albatenii observationes astronomicas ... Phil. Trans. 17 (London, 1694), No.204 (for 1693 October). pp.913-21.

Halma, 1813. Composition Mathématique ... I. (Books I-VI of Syntaxis).
1816. Ditto. Tome II. (Books VII-XIII of Syntaxis).
Greek text and French translation of Ptolemy's Syntaxis (Almagest).
Both volumes reprinted 1927, Paris, Hermann. See also Theon Halma and Heiberg, J.L.

Hammer, C.U., Clausen, H.B. and Dansgaard, W. 1980. Greenland ice-sheet evidence of post-glacial volcanism and its climatic impact. Nature, 288, 230-235.

Harrison, K. 1973. The beginning of the year in England c.500-900 in Anglo-Saxon England, 2, 51-69.
1976. The Framework of Anglo-Saxon History. Cambridge.
1977-78. Epacts in Irish Chronicles. Studia Celtica XII-XIII, 19-32.

Hasegawa, I. 1979. Orbits of Ancient and Medieval Comets. Publ. Astron. Soc. Japan, 31, 257-270.
1980. Catalogue of Ancient and Naked-eye Comets. Vistas in Astronomy, 24, 59-102. (See Table 3.)

Heiberg, J.L. 1898. Ptolemy, Opera. Vol.I, Part I. (Books I-VI of Syntaxis). Leipzig, Teubner.
1903. Ditto. Vol.I, Part II. (Books VII-XIII of Syntaxis). Same Greek text (for German translation, see Manitius). (Vol.II, 1907, contains Opera astronomica minora.)

Ho, P.Y. 1962. Ancient and Medieval Observations of Comets and Novae in Chinese Sources. Vistas in Astronomy, 5, 1962, 127-225.
1966. The Astronomical Chapters of the Chin-Shu. Mouton. Paris.

Hoang, P. 1925. Variétés sinologiques no.56. Catalogue des éclipses de soleil et de lune relatées dans les documents chinois ... Mission Catholique, Shanghai. Ed. L. Gauchet after the death of P. Hoang in 1909. Dates mainly calculated. (= P. Huang.)

Huber, P.J. 1982. Astronomical Dating at Babylon I and Ur III. Occas. Papers on the Near East. Undena Publications, Malibu, Cal., USA.

Hydatius. Eclipses are mentioned in two chronological works. The chief editions are by T. Mommsen in Chr. Min.

(i) Chronicle (AD 379-458):
1861. PL 51, cols.875-90.
1879. PL 74, cols.703-50 (replacing 1861 ed.).
1894. Chr. Min. 2 = AA 11, pp.1-36.

(ii) Fasti (509 BC to AD 468):
1861. PL 51, cols.891-914.
1865. PG 92, cols.1077-98.
1892. Chr. Min. 1 = AA 9, pp.205-47.

Ibn Aḍarī (Ibn al-'dhārī). Al-Bayano-l-Mogrib. Ed. R. Dozy, Leyden, 1848-51. (These are the English transliterations used in Camb. Med. Hist. 3, 1922, 631.) We refer to the French translation, see Fagnan.

Ibn Ḥayyān (c.1070). Al-Muqtabis. There is a Spanish translation by J.E. Guraieb, Cuadernos de Historia de España, XXXI-XXXII, pp.316-321.

Ibn Yūnus (AD 1008). (See Caussin.)

Johnson, S.J. 1874. Eclipses, Past and Future ... Oxford and London.
1896. Historical and Future Eclipses ... London and Oxford.

Jones, C.W. 1943. Bedae Opera de Temporibus. Cambridge, Mass., the Medieval Acad. of America. (Contains, inter alia, the short De Temporibus of 703 and the longer De Temporum Ratione of 725.)

Kelley, D.H. 1976. Deciphering the Maya script. (Austin, Texas and London.) 1983. The Maya Calendar Correlation Problem (in ed. R.M. Leventhal and A.L. Kolata 'Civilization in the Ancient Americas', Univ. of New Mexico Press, pp.157-210.

Kepler, J. 1604. Astronomiae Pars Optica. (Ges. Werke 2, 1939, on which see a note on p.392 of C. Doris Hellman's translation, New York, 1959, of Max Caspar's Kepler.)

Kiang, T. 1971. The Past Orbit of Halley's Comet. Mem. R. Astr. Soc., 76, 27-66.

Kudlek, M. and Mickler, E.H. 1971. Solar and Lunar Eclipses of the Ancient Near East from 3000 BC to 0 (actually AD 59) with Maps. Neukirchen-Vluyn.

Lansberg, P. 1632. Theoricae motuum coelestium. (Opera Omnia, 1663.) Middelburg, Holland.

Latouche, R. 1963-5. (See Gregory of Tours.)

Leo Grammaticus. Byzantine Chronicler. See under AD 891. (Bonn ed. 1842, PG 108, 1861.)

Lycosthenes (Conrad Wolffhart). 1557. Prodigiorum et ostentorum Chronicon., Basle. (Preface dated 1557 Sept.1.)

Malalas, I. 1831. Chronographia c.563. Bonn ed. CSHB (with Latin translation).

Manitius, K. 1912. Handbuch der Astronomie. Band I. (Books I-VI of Syntaxis.)
1913. Ditto. Band II. (Books VII-XIII of Syntaxis.)
German translation of Heiberg's text of Ptolemy's Syntaxis (Almagest), with extensive notes. Both volumes reprinted 1963, Leipzig, Teubner, with introduction and corrections by O. Neugebauer.

Marcellinus Comes. Chronicle (AD 379-518, continuations to 548, etc.).
1861. PL 51, cols.917-948.
1894. Chr. Min. 2 = AA 11, pp.37-108.

Meeus, J., Grosjean, C.C. and Vanderleen, W. 1966. Canon of Solar Eclipses 1898 to 2510. Pergamon, Oxford. 749pp.

Meeus, J. and Mucke, H. 1979. Canon of Lunar Eclipses -2002 to +2526. Wien. 2nd ed., 276pp.
1985. Canon of Solar Eclipses -2002 to +2526. Wien.

MGH. Monumenta Germaniae Historica. Started in 1826 by G.H. Pertz. We refer only to the series SS, SrG, SrM, AA (qq.v.).

Michael the Syrian (12th C). Chronique de Michel le Syrien. Tr. J-B. Chabot. 3 vols. Paris 1901 = Bruxelles 1963.

Migne, J.P. Patrologia Graeca (PG). Paris.
Patrologia Latina (PL). Paris.

Mommsen, T. 1892. Chronica Minora 1 = AA 9.
1894. Chronica Minora 2 = AA 11.
1898. Chronica Minora 3 = AA 13.

Mucke, H. 1972. Bright Comets from -86 to +1950. (In German.) Astronomische Büro, Vienna (2nd ed.).

Muller, P.M. 1975. An Analysis of the Ancient Astronomical Observations with the Implications for Geophysics and Cosmology. (Ph.D. Thesis, School of Physics, Newcastle.) Fiddes Litha Press.

Muller, P.M. and Stephenson, F.R. 1975. The Accelerations of the Earth and Moon from Early Astronomical Observations. pp.459-534 in 'Growth Rhythms and The History of the Earth's Rotation', ed. G.D. Rosenberg and S.K. Runcorn. London etc; Wiley. (Includes references to earlier work by Stephenson.)

Muralt, E. de 1855 repr. 1963. Chronographie Byzantine I, AD 393-1057. Hakkert, Amsterdam. Many errors (cf. Schove 1974).

Muratori, L.A. Rerum Italicarum Scriptores. Milan, 1723-51. (Revision under G. Carducci and V. Fiorini, Città di Castello, S. Lapi, 1900-.)

Nallino, C.A. 1903. al-Battānī sive Albatenii Opus Astronomicum. 1899-1907. Our references are all to Part I, in Publ. R. Oss. di Brera in Milano, N.XL, Part I. Milan, Hoepli, 1903. Arabic text in Part II.

Needham, J. 1959. Science and Civilisation in China. Vol.III. C.U.P.

Neugebauer, O. 1975. A History of Ancient Mathematical Astronomy. Three parts. Berlin, Heidelberg; New York, Springer-Verlag.

Neugebauer, P.V. 1930. Astron. Abhandlungen (Supplementary volume of the AN), 8(2). (Reprints of, and discussions arising from, papers by C. Schoch, 1873-1929.)

Neugebauer, P.V. 1931 presented computations for the earlier eclipses in 'Spezieller Kanon der Sonnenfinsternisse für Vorderasien und Ägypten für die Zeit von 900 v. Chr. bis 4200 v. Chr.' (Astronomische Abhandlungen, Vol.8 No.4.)

Newcomb, S. 1878. Researches on the Motions of the Moon. Part I. (Washington Observations for 1875, Appendix II.)

Newton, R.R. 1969. Secular Accelerations of the Earth and Moon. Science, 166, 825-31.
1970. Ancient Astronomical Observations and the Accelerations of the Earth and Moon. Baltimore and London, Johns Hopkins Press. (cf. Schove 1971.)
1972. Medieval Chronicles and the Rotation of the Earth. Baltimore and London, Johns Hopkins Press.
1976. Ancient Planetary Observations and the Validity of Ephemeris Time. Baltimore and London, Johns Hopkins Press.
1979. The Moon's Acceleration and its Physical Origins. Baltimore and London, Johns Hopkins Press.

Nithard's Histories:

L: 1829, ed. G.H. Pertz, SS 2, 649-72.
L: 1907, ed. E. Müller, in SrG.
L,F: 1926, ed. and tr. P. Lauer, in CHFMA 7. (Reprinted 1964)
F: 1824, tr. F. Guizot (Hist. des Dissensions des Fils de Louis-le-Débonnaire. CMHF, 3).
E: 1970, tr. B.W. Scholz with Barbara Rogers, in Scholz 1970.

O'Connor, F.J. 1952. Solar eclipses visible in Ireland between AD 400 and AD 1000. Proc. Roy. Irish Acad. (A), 55, 61-72.

Oppolzer, T. von. 1887. Canon der Finsternisse. Denkschriften K. Akad. Wiss. (math.-naturwiss. Kl.) 52, Vienna. Tr. of Introduction ed. by O. Gingerich now available as Canon of Eclipses (1962). Dover Pubs., New York. Tracks imprecise. See Ginzel to AD 600, Schroeter from AD 601.

Pauly-Wissowa. (See RE in Abbreviations.)

Pertz, G.H. Founding editor (1826) of MGH. (See MGH above.)

Petavius, D. 1627. Opus de doctrina temporum. Paris.

Pingré, A.G. 1783, 1784. Comètographie ou Traité historique et Théorique des Comètes. Vol.I (to p.265), II (after p.265).

Pingree, D. 1968. The Thousands of Abū Ma'shar. London, Warburg Institute.

Potthast, A. 1895-6. Bibliotheca historica medii aevi. 2nd ed. I. Berlin. New ed. in progress:
Repertorium fontium historiae medii aevi. Rome 1962-70.

Proudfoot, A.S. 1975. The Sources of Theophanes for the Heraclian Dynasty, AD 610-711. Byzantion, 44, 1974-5, 367-439.

Ptolemy. For the editions and translations of the Syntaxis (Almagest) mentioned by us, see Heiberg 1898, 1903 (Greek text), Manitius 1912-3 (German translations), and Halma 1813-6 (Greek text and French translation). (See also under Theon and Toomer.)

RFA. Royal Frankish Annals (741-829; there are original and revised versions).

L: 1826, ed. G.H. Pertz, SS 1, 135-218 (as Annales Laurissenses and Einhardi Annales).
L: 1895, ed. F. Kurze, in SrG (as Annales regni Francorum 741-829).
F: 1824, tr. F. Guizot (as Annales d'Eginhard, in CMHF, 3).
F: 1856, tr. A. Teulet (as Annales des Francs, in Les Oeuvres d'Eginhard).
E: 1970, tr. B.W. Scholz with Barbara Rogers, in Scholz 1970 (the Introduction discusses the construction of the work).
(See also Scholz.)

Rodgers, R.F. 1952. Byzantine Comets. Roy. Astr. Soc. Canada Jnl., 46, 177.

Rome, A. 1952. The calculation of an eclipse of the Sun according to Theon of Alexandria. Proc. Internat. Congress of Mathematicians, Cambridge, Mass., Aug.30-Sept.6, 1950, Providence 1952, 1, pp.209-19. (See also Theon-Rome.)

Samuels, A.E. 1972. Greek and Roman Chronology, etc. Beck'sche Verlag, Munich. 307pp.

Sawyer, J.F.A. 1972. Joshua 10, 12-14 and the solar eclipse of 30 Sept. 1131 BC. Pal. Expl. Qly., 104, 139-146.

Scaliger, J.J. De emendatione temporum. (Titles differ slightly, but these words always occur.) Editions seen:

1583. Paris
1593. Frankfurt
1598. Leiden
1629. Geneva

Schafer, E.H. 1977. Pacing the Void. Univ. of Calif. Press, Berkeley. 352pp.

Schoene. (See Eusebius.)

Scholz, B.W. 1970. Carolingian Chronicles: Royal Frankish Annals and Nithard's Histories. Tr. by B.W. Scholz with Barbara Rogers. Ann Arbor, Univ. of Michigan Press. (See RFA and Nithard.)

Schove, D.J. 1948. Sunspots and Aurorae, 500-250 BC. (Includes some eclipses.) JBAA, 58, 178-190 and 202-204.
1950. Visions and dated Auroral Displays, AD 400-600. (Comet of Columba.) JBAA, 3rd series, 13, 34-49.
1951. Sunspots, Aurorae and Blood Rain: the Spectrum of Time. Isis, 42, 133-138.
1955a. The Earliest British Eclipse Record, AD 400-600. JBAA, 65, 37-43. (AI is primary; AU is later.)
1955b. Halley's Comet, 1 : 1930 BC to AD 1986. The Comet of David and Halley's Comet. JBAA, 65, 285-289, 289-290.
1956a. Halley's Comet and Kamienski's Formula. JBAA, 66, 131-139.

1957. Tree-rings and Medieval Archaeology (with A.W.G. Lowther). Medieval Archaeol., London, I, 78-95. (First millennium date wrong; corrected Schove 1983a, c and d.)
1960. Chronology of Natural Phenomena in East and West. Archives Internat. d'Histoire des Sciences, Paris, 52-53, 263-268 and addition 56-57, 1961, 337.
1961a. The Spectrum of Time (1476-1531). JBAA, 71, 320-322.
1968a. Eclipses, Comets and the Spectrum of Time in Africa. JBAA, 78, 91-98.
1971. Review of Newton (1970). Sky and Telescope, 42, 303-305.
1972. The Leonids: Who saw them first? Sky and Telescope, 43, 156-157.
1974. Chronology and historical geography of famine, plague and other pandemics. Procs. XXIII Internat. Congr. Hist. Medicine, London, pp.1265-72.
1975. Comet Chronology in Numbers, AD 200-1882. JBAA, 85, 401-407.
1976. Maya Chronology and the Spectrum of Time. Nature, 261, 471-473. (Corrected in 1977, 268, 670.)
1977. Maya Chronology and Planetary Conjunctions. JBAA, 77(1), 38-52 and 104. See also 87(5), 438-439 and 522, etc.
1983a. 'Sunspot Cycles', ed. D.J. Schove (Hutchinson Ross, Stroudsburg, USA, distributed by Van Nostrand Reinhold).
1983b. Sunspot, auroral, radiocarbon and climatic fluctuations since 7000 BC. Annales Geophysicae, 1, 4-5, 391-396.
1983c. Tree-rings and Climate. Review of Hughes, M.K. et al. and Baillie, M.G.L. in Internat. Jnl. of Nautical Archaeol. and Underwater Exploration, 12(4), 353-354.
1983d. Recent Progress in Dendrochronology in Bull. Inst. Archaeology, 20, London.
1984a. Maya Eclipses and the Correlation Problem. Estudios de Cultura Maya, XXV, Mexico City. Accepted 1981. Not yet out.
1984b. Maya Correlations, Moon Ages and Astronomical Cycles. Jnl. Hist. Astr., 15(1), No.42, 18-29.

Schove, D.J. and Fairbridge, R.W. 1985. 'Ice-cores, Varves and Tree-rings', Balkema, Rotterdam.

Schroeter, J.F. 1923. Spezieller Kanon der zentralen Sonnen und Mondfinsternisse, welche ... 600 bis 1800 nach Chr. in Europa sichtbar waren, Kristiana (Oslo), in Kommission bei Jacob Dybwad, xxiv, 305pp., maps, tables (USSR incomplete). (The reliable continuation of Ginzel 1899.)

Seyffarth, G. 1878. Corrections to the present theory of the Moon's motions, according to the classic eclipses. Trans. Acad. Sci., St. Louis, 3(4), 401-530. St. Louis, Missouri, 1878.

Sidersky, D. 1944. Nouvelle Etude sur la chronologie de la dynastie Hammurapienne. Revue d'Assyriologie, 48, 146-151.

Smiley, C.H. 1970. 'Solar eclipses and correlation of the Mayan and Christian Calendars'. Paper presented at the 1970 symposium. (Copy available from D.J. Schove.)

Smyth, A.P. 1972. The Earliest Irish Annals. Proc. R. Irish Acad., C no.1, 71, 1-48.

Stephenson, F.R. 1975. Astronomical Verification and Dating of Old Testament Passages referring to Solar Eclipses. Palestine Exploration Quarterly, 107, 107-120.

1982. Historical Eclipses. Scientific American, 247, 170-183.

Stephenson, F.R. and Clark, D.H. 1978. Applications of Early Astronomical Records. Adam Hilger Ltd., Bristol. 114pp.

Stephenson, F.R. and Clark, D.H. 1982. The Historical Supernovae. In M.J. Rees and R.J. Stoneham (eds.), Supernovae: A Survey of Current Research. Dordrecht. 355-370.

Stephenson, F.R. and Houlden, M.A. 1985. (In progress.)

Stephenson, F.R. and Morrison, L.V. 1982. History of the Earth's Rotation since 700 BC. In P. Brosche and J. Sundermann (eds.) Tidal Friction and the Earth's Rotation, II. Springer-Verlag, Berlin. 29-50.

Stothers, R. 1984. Mystery Cloud of AD 536. Nature. (Accepted.)

Stothers, R. and Rampino, M.R. 1983a. Historic Volcanism, European Dry Fogs and Greenland Acid Precipitation, 1500 BC - AD 1500. Science, 222, 411-413.

Stothers, R. and Rampino, M.R. 1983b. Volcanic Eruptions in the Mediterranean before AD 630 from Written and Archaeological Sources. Jnl. Geophys. Res., 88 (B8).

Stratos, A.N. 1968. Byzantium in the Seventh Century, Vol.I, 602-34. Tr. by Marc Ogilvie-Grant. Amsterdam, Hakkert. Vol.II (in progress).

Struyck, Nicolaas (Str.). 1740. Inleiding tot de Algemeene Geographie ... Amsterdam.

Struyck-Ferguson 1773. List of eclipses, based on Struyck 1740, in James Ferguson's Astronomy, 'new edition', pages 175-8. The years and pages for the first three editions are 1756 (171-4), 1757 (169-72), 1764 (175-8). Other editions exist.

Syriac Chronicle, known as that of Zachariah of Mitylene, English translation by F.J. Hamilton and E.W. Brooks, London, 1899. (The eclipses c.547-569 are not in the part due to Zachariah of Mitylene.)

Teeple, J.E. 1930. Maya astronomy. Carnegie Inst. of Washington. 405(2). Washington, DC.

Teulet, A. 1856. (See RFA.)

Theon. (Fl. 4th C.) The Basle 1538 ed. mentioned contains (1) Ptolemy's Syntaxis (Almagest), (2) Theon's Commentary on the same. The dedicatory prefaces are by Simon Grynaeus and Joachim Camerarius respectively. (1) is now entirely superseded by better editions, (2) only partly so.

Theon-Halma 1822. Theonōs Alexandreōs Hypomnēma. Commentaire de Théon d'Alexandrie sur le livre III de l'Almageste de Ptolémée; Tables Manuelles des Astres, traduite pour la première fois du grec en français ... Par M. l'Abbé Halma ... Paris, 1822.

Theon-Rome 1936, 1943. Studi e Testi 72, 106. Rome, Biblioteca Apostolica Vaticana. (54 (1931), 72, 106 are respectively Vols. I, II, III of A. Rome's edition (Greek text and French notes) of "commentaires de Pappus et de Théon d'Alexandrie sur l'Almageste". I, Pappus, commentary on Books 5 and 6; II, Theon, commentary on Books 1 and 2; III, Theon, commentary on Books 3 and 4.)

Theon-Tihon. 1978 ed. Studi e Testi 282, ed. Anne Tihon, le 'Petit Commentaire' de Théon d'Alexandrie aux Tables Faciles de Ptolémée (Histoire du Texte, Edition Critique, Traduction), Città del Vaticano, Biblioteca Apostolica Vaticana, VIII, 381pp.

Theophanes (9th C). 1982. (English translation by H. Turtledove.) The Chronicle of Theophanes, AD 602-813. University of Pennsylvania Press, Philadelphia. 1982. 204pp. Solar eclipses correctly dated.

Toomer, C.J. 1984. Ptolemy's Almagest, trans. English (London).

Tuckerman, B. 1964. Planetary, Lunar and Solar Positions AD 2 to AD 1649 at Five-Day and Ten-Day Intervals. Mem. Amer. Philos. Soc., 59, Philadelphia. 842pp. (Positions for 601 BC to AD 1 are in 56, 1962.)

Turtledove. (See Theophanes.)

Tycho-Curtius 1666 (T-C.) Historia Coelestis of Tycho Brahe (1546-1601), edited by Albertus Curtius (Curtz), under the pseudonym (anagram) of Lucius Barrettus. The eclipses used were inserted by Curtius (see J.L.E. Dreyer, Tycho Brahe, Edinburgh 1890, pp.372-3). The edition used (Ratisbon 1672) seems from Dreyer to differ only in title-page from the original edition (Augsburg 1666). Also ed. Dreyer, Opera Omnia, Copenhagen 1913.

Wallace-Hadrill, J.M. 1960. (See 'Fredegar'.)

Weiss, J. 1914. Elementarereignisse im Gebiete Deutschlands. I. Die Elementarereignisse vom Beginn unserer Zeitrechnung bis zum Jahre 900. 2nd ed. Vienna, Holzhausen.

Weir, J.D. 1982. The Venus Tablets: a fresh approach. Jnl. Hist. Astron., 13, 23-49.

Whitelock, D., Douglas, D.C., and Tucker, S.I. 1961. The Anglo-Saxon Chronicle: A revised translation.

Williams, J. 1871. Observations of Comets ... from the Chinese annals. London. (Many corrections are given in Ho 1962.)

Yeomans, D.K. and Kiang, T. 1981. The long-term motion of comet Halley. MN, 197, 633-646.

Zech, J. 1853. Astronomische Untersuchungen über die wichtigeren Finsternisse, welche von den Schriftstellern des Klassischen Altertums erwähnt werden. Leipzig.

Zhuang, Tian-shan. 1977. Ancient Chinese records of Meteor Showers, Chinese Astronomy, 1, 197-220. (Original in Chinese in Acta Astr. Sinica, 14 (1966), 37-58. Pergamon Press, Oxford, UK.)

## S O U R C E  A N D  N A M E  I N D E X

NUMBERS refer to YEARS not pages
ROMAN FIGURES refer to the INTRODUCTION
a = Asiatic reference appended to the century in question
c = Cometary reference (Appendix B)
m = Maya reference (Appendix C)
A  SOURCES are in CAPITALS
B  PERIOD-LIMITS based on Appendix C

For this selective index some new information has been useful from sources such as the following:

McKitterick, R.  The Frankish Kingdom under the Carolingians, 751-987, Longman, 1983, London
Dunlop, D.M.  Arab Civilization to AD 1500.  Longman, 1971
Morony, M.O.  Iraq after the Muslim conquest.  Princeton U.P., 1984

Sources examined by one or both of us include most of those listed (Vitae and fragmenta excepted) in Weiss, 1914, 89ff or in Muralt, 1963, ix, sources which give further clues to the places where the eyewitness records were made, and to the lost sources hidden in extant chronicles (cf. also the new edition of Potthast). Brief descriptions, useful also for the secondary chronicles omitted in this index are given in Newton 1971 under the countries (e.g. Belgium) of the modern map.

| | | |
|---|---|---|
| EDESSA, Chronicles of | 490/540 etc. | See Joshua, James, etc. |
| Edwin 7 C | | 624 |
| 'EINHARD' (wr. 830) | - 812 | 777-9 (Revised RFA for 741-812) |
| EKKEHARD ed. 1844 | 11-12 C | 760 (Used early source) |
| ELIAS Early 11 C | 480+ | 484, 512, 565c, 601, 604, 693, 745c, 893, 905c, 947c, 970 |
| ENNIN 9 C | 9 C | (See Chinese section) |
| ENNIUS 3 C BC | 400 BC | v (400 BC recte 402 BC) |
| ETHELWEARD 10 C | 9 C/ 10C | 878, 975 |
| EUNAPIUS 3 c - 4 C | 4C/404 | Fragments only (Used by Zosimus e.g. 393?) |
| EUSEBIUS 300/323 | 160/330 | 17, 29 (See 4 C Introd.), 69c recte 660c |
| EUTROPIUS 4 C | | 337 recte 336c |
| EUTCHIUS, Cont. 10 C | 940s | 941c, 947c. See Yahya |
| EXCERPTA VALESIANI | | 520c |
| EVAGRIUS of Antioch 6 C | 580/594 | (Earthquakes) |
| F airbridge, R.W. 1985 | | xxxv, See Schove and Fairbridge |
| FARAJIUS = Abu-l-Faraj 13 C | | = Bar Hebraeus |
| FARFA, Italy | | 807, 967 |
| FASTI VINDOBONENSES (c.576) | Italian Annals | 118, 390c, 393, 418, 418c, 451c, 534 |
| FELIX 4 C | | iv, 303/4 |
| FIRMICUS 4 C | 330s | 334 |
| FLAVIGNY, etc. 9 C + | 780/840+ | 764, 788, 795, 803, 985 |
| Fletcher, A. | | See Preface and xxxiii |
| FLEURY 10 C | 950/970 | 864c, 878, 956 recte 955, 957, 968 |
| FLODOARD | 920/940 | 926, 936 |
| FLORENCE OF WORCESTER | | 995 recte 989 (From lost ASC) |
| FOLCWIN I | 890s | 875c, 878, 891c |
| FOLCWIN II c.962 | 950s-960s | 962 (St Bertin's) |
| FONTANELLE, Frag. 839+ (= St Wandrille) | 830 - 867 | 842c, 864c (Acts of the Abbots ...) |
| Forester, T. 1854 | | 664m 666, 760 |
| Fotheringham J.K. 1908/13 | | xxi, 17, 71, 75/83, 125/26, 186, 240, 292, 334, 346, 364 |
| FREDEGAR and Cont. 7 C | 590/Cont. to 641 | 464 (Hydatius), 590, 603 (cf. BURGUNDIAN Chron.) |
| FULDA, Ann. of c.901 (Ne. Frankfurt-am-Main) | 830/887 and 901 | 837c, 838/9c, 840, 841/2c, 875c, 878, 881/2c |
| GENESIUS | | 594c |
| GEORGE OF PISIDIA | 600/641 | 622 |
| GEORGE THE SYNCELLOS | 280s | |
| GEORGIOS HAMARTOLOS (George the Sinner) (George the Monk I (c.842) | 300/842 | 303/6, 609, 693 ( ˚= Hamartolos) |
| GEORGIUS MONACHUS | | 891, 905c, 908, 912c |
| (George the Monk II) 948 who used Cont. Symeon? | 940s | 891 |
| GESTA (Liege) | | 968 (From Ratpert) |
| Giles, J.A. 1843/9 | | 650, 848, 752, 760 |
| Gingerich,O. 1962 | | x, xvi |
| Ginzburg, J. 1917 | | 646 |
| Ginzel, F.K. 1882/1918 | | xvii to xix, All solar eclipses in Europe AD 5 to AD 600 |
| 1884 | | Also 809 |
| 1886 | | Also 655 |

| | | |
|---|---|---|
| IBN-MISKAWAYH 10 C | 451-983 | (Used Tabit). Eyewitness from 951 |
| IBN-YUNUS 1008 | 820+ | 923-933, 977-1004 |
| INDIAN SOURCES | 940s | 941c |
| INISFALLEN, Ann. | 440s | 445 recte 447, 594, etc. |
| IRISH ANNALS (AU, AI, CS and AT) | 510 to 610 Irish 'Eusebius' in AI etc. 650+ AU, etc | 447, 497, 594, 603, 664, 672, 676c, 688, 691, 719, 734, 753, 763/4, 773, 788, 807, 814, 849, 865, 878, 885, 894, 912, 912c, 921 |
| ISIDORE, Cont. 7 C | 620s-636 & Continuators | 655, 718, 720 |
| ITALIAN ANNALS (Latin) esp. Chron. Venetum See Lupus | 770/990 568/1008 | 534, 787, 939 See also Consular Annals to 468 |
| JAMES of Edessa (& Cont.) c. 692+ | 670/700s Cont. | 512. 604 |
| JAPANESE sources (Nihongi) | 630s | 628a, 676c, etc. |
| Jenkins, R.J.H. 1948 | | 908 |
| Jerome 333/378 | 40/400 | 346, 348, c.398, 400 |
| JOHN of Asia 6 C | 570 | 449, 450, 496 |
| JOHN of Biclaro 6 C | 540s, 565-590 | See App. F |
| JOHN of Ephesus (Syriac) 6 C (Used by Michael) | 530s, 540s | 536 |
| JOHN of Nikiu 7 C | 7 C | 601 |
| Johnson, S.J. 1874/1905 | | 334, 650, 661, 671, 764, 796, 800, 912, 926, 977 |
| JORDANES 6 C | | 563-5. App. F c.469 |
| JOSEPHUS 1 C | 60s | 69c recte 66c |
| JOSHUA BC | | xi |
| JOSHUA THE STYLITE 506 | 494/505 | c.499 (Dust) and c.500c (In Denys) |
| JUVAVENSES (835, 956,975) (Austria) | 800/820 and Cont. | 807, 810, 891, 974 |
| Kelley, D.H. 1976/83 | | 857m, Maya App. |
| Kepler, J. 1604 | | 29, 370, 418, 926 |
| KHUSISTAN Chron. 670s,etc. | 670s | 942c |
| al-KHWARISMI 8 C | 610/750 | (Used by Denys) |
| Kiang, T. 1971 | | 218 |
| Knobel, E.B. 1879 | | 929, 980 |
| Knowles, D. 1969 | | 538, 664 |
| Kroll and Skutsch 1897 | | 334 |
| Krusch, B. 1888 | | 590 |
| Kudlek, M. et al. 1971 | | 59 (Maps to AD 59), xvii Maya App. |
| Kurze, F. 1889 | | 990 |
| Lanebek, J. 1772 | | 447, 692 |
| Lansberg 1632 | | 883, 901 |
| Lappenberg 1839 | | 990 |
| LATIN ANNALS (of Italy) | 530/990 | See also Consular Annals and Italian Annals |
| Latouche, R. 1963/65 | | 485, 560, 567, 582, 590, 929 |
| LAUBACENSES 9-10 C | c.893-926 | 733, 893, 894 (From S.Gall?) |
| LAURESHAMENSES (c.803) RFA (wr. in Alemannia not Lorsch) | 794-803 | 787 (MS of original annals survive) See App. F |
| LAURISSENSES (c.829) RFA (Lorsch) | 788-829 | 764, 807, 809, 810, 812, 817, 818, 820, 824, 828 |
| Lehmann, J. 1977 | (Re 14 C BC) | v |
| LEO theDeacon 990 (Diaconus) and Cont. | 959/976 | 968, 975c, 989c |

| | | |
|---|---|---|
| Leo, the Emperor 886-912 | | 908 |
| Leo, the Grammarian c.912 | | 891, 905c, 908, 912c |
| LEODIENSES (MGH IV) | | 941c |
| LIBER HIST. FRANC. 727 | 8 C | cf. S. Amand, etc. |
| LIBER PONTIFICALIS (Book of Popes) | | 676c, 680, 684c, 716 (See Paul) |
| LIUDPRAND c.969 | 960s | 968 |
| LIVY | BC/AD | iii, xi |
| 'LOISELLIANOS' 814 RFA | (810s) | See RFA |
| LORSCH, So-called of 803 | | See Laureshamenses (RFA) and Laurissenses |
| Louis, Emp. (Vita) | 830s | |
| Loyn, H. | | 934 |
| LUGDUNENSIS (Lyon) c.841 | 840/1 | 840 |
| LUND (and Ann. of Esrom( | (British Isles) | 447, 691 |
| LUPUS of Bari c.1102 | 866-1102 | 961 |
| LUXEUIL | | 810, 812 |
| Lycosthenes (Wolfhart) 1557 | | 810, 814 |
| LYDUS 6 C | | 512 |
| Lynn, W.T. 1891-1911 | | 1st Cent. Intro., 197, 218, 239, 240, 796 |
| MacAirt, S. ed. 1951 | See Irish Annals | e.g. 447 (AT) |
| MacCarthy, B. 1901 | | e.g. 807, 885, 921 (See Irish), 763, 764, 773, 788 |
| MAGDEBURG, Ann. | | 990 |
| MALALAS of Antioch c.563 | 290+, c.516-563 | 295 <u>recte</u> 291, 447, 519c <u>recte</u> 520c, 530c, 556c, 878, 885 |
| Syriac and Slavonic versions | 565/574 lost | (Translation in progress) |
| MAMERTINUS c.290 | 281/291 | Weather, etc. See App. F, 289, 291 |
| Manitius, K. 1912 | | 125/6 |
| MAQDIZI c.966 | | |
| MARCELLINUS, Count c.534, 548 | 389+, 534 & 566 | 390s, 393, 418c, 418, 422c, 442c, 473, 497, 512 |
| MARINUS c.486 | 480s | 484, 486 |
| MARIUS | 560/565c | |
| MASUDI Mid 10 C | c.900/943 | 912c, 939 |
| MATTHEW of Edessa | 952/1136 | xxvii. See Michael |
| Meeus, J. et al. 1969/85 | | xviii, xix, 49, 604 |
| MELLICENSES c.1564 | Byz? | 346, 418 |
| MENANDER (Lost) | Lost 6 C | 563-565 |
| MERCIAN REGISTER, England | 900/920 | 904 (Better than ASC) |
| Message, P.J. | | See Preface |
| METTENSES (Metz) c.905 | 900s | 878 |
| e.g. Ann. Mett. Prior | 803/805 only | |
| Michael, C | | Preface |
| MICHAEL the Syrian 12 C | | 512, 603, 604, 626/9, 644, 684c, 693, 812, 839c, 891 |
| MISCELLA HISTORIA (Several) | 6 C, 8 C 9 C | 563-5, 745c, 760, 760c, 812, 813 (See Muralt x amd xiv). App. F, 743 |
| MOISSAC (Aquitaine) 10 C | | |
| Monceaux 1901 | | 212 |
| MONTPELLIER MS | 440s | 447 |

| | | |
|---|---|---|
| Petrie, F. 1848 | | 760 |
| PHILOSTORGIUS 395-425 | 390-425 | 390c, 400c, 418 and 418c (Partly lost) |
| PHLEGON 2 C | 20/120 | 29 |
| PHOTIUS 9 C | | Used Philostrogius |
| PICTISH CHRON. 10 C | 880/970 | 885, 966 |
| Pingre, A.G. 1783 | | xxviii |
| Pingree, D. 1968 | | 610, 622, 630, 634, 644, 655, 660, 666 |
| PISIDIA, Geo. of 7 C | 610/630 | 622 |
| Pithou, P. 1588 | 445 to 455 | See Prosper |
| PLINY I 77 | 40/77 | 46 (54c and 55c i.e. <u>two</u>), 59, 71, 76c (<u>recte</u> 77?) |
| Plummer, C. 1892 | | ix, 796, 878, 904 |
| PLUTARCH c.90/100 | | ii, 75/83 |
| Poole, E. | | 878 |
| PRISCUS 5 C | -c. 471 | |
| PROCLUS 5 C | 160s | 164, 474 |
| PROCOPIUS 554 | 315, 530/ c.554 | 536, 540c <u>recte</u> 539c |
| PROSPER d.463 (Continued Jerome) | 410/455 | 393, 421 <u>recte</u> 418, 418c App. F 422 |
| Proudfoot, A.S. 1975 | | 610, 617, 626/9, 644 |
| PRUDENTIUS 835-861 (Royal Annalist at Troyes) | 850s | 837/9c, 838, 861 |
| PTOLEMY 125/136 | 120s/136 | 125/136 |
| QUEDLINBURGENSES, Ann. c. 1025 | 990s | 789/730c (from Bede), 984c, 990 (= Hildensheim) |
| RATISBON, Cont. | 880s | 998 |
| RATPERT (c.883) | c.883 | |
| RAVENNA, Ann. of | e.. 110/580 | e.g. 534 See Latin and Italian Annals |
| Rawlings, G.C.I. | | 693 |
| REGINO d.906 and Cont. (c.967) at Trier | c.873/906 | 878, 905c <u>recte</u> 941c |
| Rhys (and Evans) 1890 | | 690, 807 |
| Riccioli 1653 | | 306, 316-20, 337, 360, 364, 410, 421 <u>recte</u> 418, 891, 901, 926 |
| RICHER of Reims | (888+), 960 to 995 | 926 (from Flodoard) |
| Riley, H.T. | | 755, 796 |
| Rodgers, R.F. 1952 | | Byz. Comets |
| ROGER OF WENDOVER | 8 C - 9 C Additions to ASC | 752, c.760, 848, 879 See Addenda |
| Rome, A. 1952 | | 364 |
| Rosenhead, L. | | Preface |
| ROYAL FRANKISH ANNALS | 788, 796 | Various 817c |
| RFA | 808-829 | See Lorsch, et c. |
| Sabine, E. 1953 | | 360 |
| S. AMANDI, Ann. of Trier, etc. | 715+ | Brief Annals |
| S. BERTIN, Ann. of (Aachen, etc) | 835-882, etc. | 837/8c, 838, 861, 891 RFA Western Continuation |
| S. COLOMB. SENON | | 809 |
| S. DENIS, Chron. | 9 C | (cf. RFA) 812 |

| | | |
|---|---|---|
| THIETMAR (c.1018) (Saxony) | c.970s to 1078 | 990 (From Quellinburg?) |
| Thompson, J.E.S. 1950 | | vii, 837m, 846m |
| Thorpe, B. ed. 1848 | | 760, 904 |
| TREVIRENSIS 10 C (Cont. Regino at Trier) | 950-967 | |
| Trillmich, W. 1874 | | 990 |
| Tuckerman, B. 1964 | | |
| Turtledove, H. tr. 1982 | | |
| Tycho-Curtius 1666 | | 291, 360, 364, 410, 421 recte 418, 450/3, 650, 661, 693, 881, 891, 901, 926 |
| ULSTER, Ann. | 664+ | See Irish Annals 591/4, etc. |
| Usener, H. 1914 | | 617 |
| Van den Bergh, G. 1955 | | xviii, 617 |
| VEDASTANI, Ann. c.900 | 877-900 | 878, 891c |
| VELLEIUS 1 C | | 1st C Addendum and App. F |
| VICTOR, Aurelius 4 C | | 46, 316/320 |
| VICTOR, Tonnennsis c.567 | 560s | 319, 467c |
| VITELLIUS 1 C | | 68/9 |
| Virji, K.J. 1955 | | 819a |
| Wade-Evans, A.W. 1938 | | 624 |
| Wainwright, F.T. 1945 | | 904 |
| WALES (sources) | 990s | 447, 807. See Brut |
| Wallace-Hadrill, J.M. 1960 | | 590 |
| WEINGARTENSES c.936 | 930s/936 | 878 (cf. Augienses) |
| Weir, J.D. 1982 | | xxvii |
| Weiss, J. 1914 | | |
| WEISSEMBURGENSES c. 846 | 840s | 729/730c (Used Bede), 840 |
| WESSEX (Charter) | 870s | 878 |
| Westland, C.J. 1943 | | 45 |
| White, W.H. 1946 | | 75/83 |
| Whitelock, D. 1955 | | 848 |
| Whitelock, D. 1961 | | 538, 661, 878 |
| Whittaker, C.R. 1969/70 | | 186, 238 |
| WIDUKIND c.973 | 930/980 | 934, 937 recte 941c |
| Williams, J. 1860/1871 | | 810 and Chinese comets |
| Wood, I. | | 581 |
| Wright, F.A. 1930 | | 969 |
| XANTENSES, Ann. c.873 | c.840/870 | 832, 835, 837c, 840, 841/2c, 865/6, 869 recte 868c |
| XIPHILINUS 11 C | (Lost sources) | 68/9, 218 (cf. Dio) |
| YAHYA, cont. of Eutychius | 960s | 941c, 947c |
| Yeomans, D.K. and Kiang, T. 1981 | | xxxii, xxxiii |
| 'ZACHARIAH' 569 | | 512, 536/7, 601 |
| Zech 1853 | | 364 |
| Zhuang, T. 1977 | | xxxiv |
| ZONARAS to 1118? | 1-10 C | xxvi, 68/9, 787, 908 |
| ZOSIMUS 395/410 | 270/410 | 393 (Eunapius?) See App. F |

## II PLACE INDEX

(See heading to SOURCE AND NAME INDEX for explanation)